Biomedical Materials for Multi-functional Applications

Series Editors

Arnab Chanda, Centre for Biomedical Engineering, Indian Institute of Technology Delhi, New Delhi, India

Sarabjeet Sidhu, Department of Mechanical Engineering, Sardar Beant Singh State University, Gurdaspur, India

The series 'Biomedical Materials for Multi-functional Applications' is intended for the audience with an interest in biomaterials for various applications, including bio-implants, bio-instrumentation, biomechanics, and tissue engineering. The book titles under this series will bring different areas of biomedical engineering, bioengineering, materials science, mechanical engineering, and biosciences under one umbrella of biomedical materials. The books in this series will provide deep insights into the state-of-art of biomedical materials and deal with diverse and interdisciplinary aspects of biomaterials, ranging from basic science to host responses, as well as real-time applications. Some of the topics include but are not limited to: functional materials, smart materials, biomimetic materials, polymers, composites, ceramics, metals, alloys, and nanomaterials. This series will provide the most recent advances in biomedical materials, including design, synthesis, processing, and characterization. This series will accept both authored and edited volumes, including monographs, textbooks, reference works, and handbooks. This series aims to publish new advances in biomedical materials that have been made in the fabrication and development of biomaterials for different applications and the advancement of principles, theories, and designs. Each book title in the series will give a comprehensive overview of pioneering topics, and the readers will gain new knowledge and insights about the topic. As a collection, the series on 'Biomedical Materials for Multi-functional Applications' is anticipated to benefit a wide audience in academia, the research community, and industry.

Anurag Dixit · Anil Kumar · Dayanidhi K. Pathak
Editors

Biodegradable Metallic Materials

Design, Development and Characterization

Springer

Editors
Anurag Dixit
G. B. Pant DSEU Campus
New Delhi, Delhi, India

Anil Kumar
G. B. Pant DSEU Campus
New Delhi, Delhi, India

Dayanidhi K. Pathak
G. B. Pant DSEU Campus
New Delhi, Delhi, India

ISSN 2731-9695 ISSN 2731-9709 (electronic)
Biomedical Materials for Multi-functional Applications
ISBN 978-981-95-2400-6 ISBN 978-981-95-2401-3 (eBook)
https://doi.org/10.1007/978-981-95-2401-3

This Springer imprint is published by the registered company Springer Nature Singapore Pte Ltd.
The registered company address is: 152 Beach Road, #21-01/04 Gateway East, Singapore 189721, Singapore

Preface

The proposed book entitled *Biodegradable Metallic Materials: Design, Development and Characterization* provides a comprehensive overview in the evolving research field of biomaterials covers biodegradable metals for biomedical applications. The book basically delves into the composition, design strategies, fabrication techniques, clinical translation, and regulatory considerations respectively for iron, magnesium and zinc based biodegradable materials for biomedical applications. This book will be the first book focusing on biodegradable metal-based biodegradable materials and their production methodologies to provide contemporary comfort and customized treatment of patients. In general, this book is intended for those working in the domains of materials science, biomedical engineering, and medicine who participates in the creation and application of medical implants. By shedding light on new materials and production processes in medical technology, it seeks to greatly advance ongoing efforts. The primary audience for the proposed book may be Mechanical and Biomedical Engineers, Academicians, Healthcare Researchers, and professionals having an interest in combined area of mechanical and biomedical engineering.

Chapter 1 discusses and compares in depth mechanical characteristics, degradation behavior, biocompatibility, and bioactivity behavior of alloys based on magnesium, zinc, and iron as well as their biomedical applications. The review's overall findings highlight the increasing importance of biodegradable metallic implants and demonstrate how they can be used to meet a range of clinical requirements.

Chapter 2 discusses compositional approaches like the addition of manganese, palladium, and silver, and structural design approaches like porosity optimization, grain refinement, and surface modification. The intention behind these approaches is to enhance corrosion rates, improve tissue integration, and offer biocompatibility.

Chapter 3 underscores the significance of the various fabrication techniques including powder metallurgy, additive manufacturing, casting, and electrochemical deposition, each offering control over porosity, microstructure, and overall mechanical performance. Emphasis is placed on the relationship between microstructural features and in vivo performance, underlining the need for precise material design.

Chapter 4 highlights both completed work and possible future directions in the development of biodegradable iron-based materials. It discusses iron metabolism and biocompatibility, clarifies corrosion mechanisms, critically evaluates degradation rates attained through various methodologies, and summarizes this achievement.

Chapter 5 explores the use of advanced computational tools, including CALPHAD, MD, DFT and ML for composition optimization and properties prediction of Mg alloys. The methodologies discussed herein highlight their potential to streamline alloy development processes, reduce time and costs, and enhance the precision and sustainability of material innovation.

Chapter 6 chapter examines Mg-based biodegradable implants and recent advancements in materials research, production, and characterization. Advancements in research and innovation in biomedical engineering have the potential to transform medical implants and enhance patient outcomes.

Chapter 7 includes the clinical trials, benefits, and difficulties of magnesium based biomaterials for bio applications—particularly tissue regeneration and repair. This study also indicates that to enhance complex tissue repair, future studies will concentrate on surface modification, alloy composition optimization, and the investigation of cutting-edge technologies like 3D printing. MBMs have a great deal of promise for tissue engineering and regenerative medicine, which calls for more research and interdisciplinary cooperation to optimize their therapeutic benefits.

Chapter 8 provides a comprehensive overview of recent developments related to the composition of different Zn-based biodegradable alloys. It discusses the effect of designed composition on the biocompatibility, degradation rate and mechanical performance of the developed material. Additionally, it discusses both conventional and advanced manufacturing techniques to fabricate desired Zn alloys based implants and parts.

Chapter 9 includes other possible materials for biodegradable implants and why Zn is preferable over them through comparing their properties. Furthermore, areas where Zn is used as an implant are thoroughly explained. Various fabrication methods such as additive manufacturing, powder metallurgy and casting are discussed in reference to optimize the structural integrity and degradation behavior of Zn based implants.

Chapter 10 aims to provide a comprehensive overview of the current state of research, highlighting the key advances, challenges, and future directions in the field of Zinc-Based Biodegradable Materials for Biomedical Applications.

Chapter 11 discusses recent advances, futuristic opportunities and challenges for biodegradable metallic materials. The recent developments in biodegradable implants in the biomedical field are examined in this chapter, concentrating mostly on orthopedic and cardiovascular applications. The mechanical characteristics, corrosion mechanism, and degrading behavior of these materials, both in vitro and in vivo, are thoroughly studied.

The goal of this book, which is devoted to groundbreaking research in the application of biodegradable metals in bioengineering, is to highlight the present trends and the potential of biodegradable metallic materials for various applications. The most common biodegradable metals discussed in this book are iron, magnesium, and zinc, which have led to a lot of studies in creating new alloys for degradable biological applications. The chapters in this book, written by eminent experts from around the world, will highlight recent developments and challenges in biodegradable metallic materials. This book is expected to be a fundamental textbook for research in biodegradable metallic materials, fabrication of implants, tissue engineering, and biomedical engineering. A viewpoint on the application of biodegradable metals for biomedical purposes in the biomedical engineering age closes the book.

New Delhi, India
April 2025

Anurag Dixit
anurag.dixit@dseu.ac.in

Anil Kumar
anil.kumar1@dseu.ac.in

Dayanidhi K. Pathak
dayanidhikrishana.pathak@dseu.ac.in

Contents

Editors and Contributors

About the Editors

Dr. Anurag Dixit is working as an Assistant Professor in Department of Mechanical and Automation Engineering at G. B. Pant DSEU Campus, New Delhi (Formerly: G. B. Pant Government Engineering College, New Delhi). He has about fifteen years of teaching and research experience. He has research interest in the area of Biomaterials, Finite Element Analysis of Textile Composites, Large Deformation of Metals and Composites at low, medium, and high rates of Loading. His researches find application in design for safety in defense applications, aerospace, mobility, marine industries, and in the design of metal forming processes. He has published more than 30 Science Citation Indexed (SCI) Journal research articles in various international journals of repute. He has been granted a research project in the capacity of Co-principal Investigator, funded by DRDO, Government of India, of Rs. 36 Lacs. He is also serving as the editor for the Springer book series *Biomedical Materials for Multifunctional Applications.*

Dr. Anil Kumar is working as an Assistant Professor in Department of Mechanical and Automation Engineering at Delhi Skill and Entrepreneurship University, New Delhi. He has over 16 years of teaching experience. Prior to joining DSEU, he has worked with Hindustan Everest Tools Limited as Production Officer. His area of research is Production and Industrial Engineering and Composite Materials. He has published about 20 research papers and chapters in various SCI/Scopus Indexed Journal/books/conferences. He was also the university topper in his MBA degree program. Dr. Kumar has completed various short term courses and workshops in different training institutes of India in the field of engineering education. He is also a lifetime Member of the Robotics Society of India.

Dr. Dayanidhi K. Pathak is working as Assistant Professor in Department of Mechanical and Automation Engineering at G. B. Pant DSEU Campus, New Delhi (Formerly: G. B. Pant Government Engineering College, New Delhi). His research interests include Additive Manufacturing, Biomaterials, Biodegradable Metals, Degradation of Biomaterials, Next Generation Bioimplants, Conventional and Unconventional Machining Processes, and Powder Metallurgy. His research finds application in the biomedical field, viz. metal based implants, scaffolds, and surgical devices. He has more than sixteen years of teaching experience. He has published a number of articles in Science Citation Indexed (SCI) journals. He has been the recipient of the Gold Medal during M.Tech. He is also the editor for the Springer book series *Biomedical Materials for Multifunctional Applications*. He is also a member of the Robotics Society of India and SAE India.

Contributors

Shitanshu Arya Department of Mechanical Engineering, Indian Institute of Technology Delhi, Hauz Khas, New Delhi, India

Priyabrata Das Department of Mechanical Engineering, Indian Institute of Technology Delhi, New Delhi, India

Anurag Dixit Department of Mechanical and Automation Engineering, G. B. Pant DSEU Campus, New Delhi, India

Shubhra Dixit Amity University, Noida, Uttar Pradesh, India

Pooja Dwivedi Department of Applied Science, Bharati Vidyapeeth's College of Engineering, New Delhi, India

Vipin Goyal Department of Mechanical Engineering, Indian Institute of Technology Indore, Indore, India

Surbhi Gupta Amity University, Noida, Uttar Pradesh, India

Anil Kumar Department of Mechanical and Automation Engineering, G. B. Pant DSEU Campus, New Delhi, India

Arun Kumar Department of Mechanical Engineering, Indian Institute of Technology Delhi, Hauz Khas, New Delhi, India

Mukesh Kumar Department of Mechanical and Automation Engineering, G. B. Pant DSEU Campus, New Delhi, India

Ajay Lamba G.B. Pant DSEU Campus, New Delhi, Delhi, India

Sachin Maheshwari Department of Mechanical Engineering, Netaji Subhas University of Technology, New Delhi, India

R. K. Mishra Shri Mata Vaishno Devi University, Jammu, India

Pulak Mohan Pandey Department of Mechanical Engineering, Indian Institute of Technology Delhi, New Delhi, India

Dayanidhi K. Pathak Department of Mechanical and Automation Engineering, G. B. Pant DSEU Campus, New Delhi, India

Ajay Sharma Amity University, Noida, Uttar Pradesh, India

Arshad Noor Siddiquee Department of Mechanical Engineering, Jamia Millia Islamia, New Delhi, India

Bhajneet Singh G.B. Pant DSEU Campus, New Delhi, India

Jasvinder Singh Department of Production and Industrial Engineering, Punjab Engineering College (Deemed to Be University), Chandigarh, India

Abhishek Tevatia Department of Mechanical Engineering, Netaji Subhas University of Technology, Dwarka, New Delhi, India

Siddharth Tevatia Department of Periodontology, I.T.S Dental College, Greater Noida, Uttar Pradesh, India

Gaurav Tripathi Department of Mechanical Engineering, Indian Institute of Technology Delhi, Hauz Khas, New Delhi, India

Girish Verma Department of Mechanical Engineering, Indian Institute of Technology Indore, Indore, India

Chapter 1
Introduction to Biodegradable Metallic Materials in Biomedical Applications

Anurag Dixit and Bhajneet Singh

1 Introduction

1.1 Historical Background and Evolution

Since the 1920s, when stainless steel was first developed, metal implants have seen significant advancements and applications in therapy. The establishment of ASTM Committee F04 on Medical and Surgical Materials and Devices in 1962 has greatly supported research, application, and the standardization of these implants. Today's medical implants are crafted from a variety of corrosion-resistant metals, including cobalt-chromium alloys, titanium and its alloys, and 316L stainless steels. New manufacturing techniques such as amorphization and nano-processing have even removed nickel from newer metallic biomaterials. Additionally, the concept of biodegradable metals—designed for temporary implants that gradually break down—has emerged as an alternative to inert materials. The modern use of metals in implants dates back over a century to when Lane introduced the metal plate for bone fracture repair [1]. Metals were initially chosen for their load-bearing capabilities in the treatment of diseases, injuries, and organ failures, despite early challenges with corrosion and inadequate strength [2, 3]. The advent of corrosion-resistant 18-8 stainless steel in the 1920s addressed the corrosion problem and spurred further development and clinical use of metal implants. In order to standardize metals for medical use, the American Society for Testing and Materials (ASTM) formed Committee F04 in 1962, which has since led major advancements in devices, tests, materials, and medical equipment.

Today, implants are made from hundreds of different metals, typically categorized into stainless-steel alloys, cobalt-chromium alloys, titanium and its alloys, and precious alloys. Figure 1 illustrates several examples of metal-based medical implants. Although the structural support and inert nature of metals have been

A. Dixit (✉) · B. Singh
G.B. Pant DSEU Campus, New Delhi, India
e-mail: anurag.dixit@dseu.ac.in

A. Dixit et al. (eds.), *Biodegradable Metallic Materials*, Biomedical Materials for Multi-functional Applications, https://doi.org/10.1007/978-981-95-2401-3_1

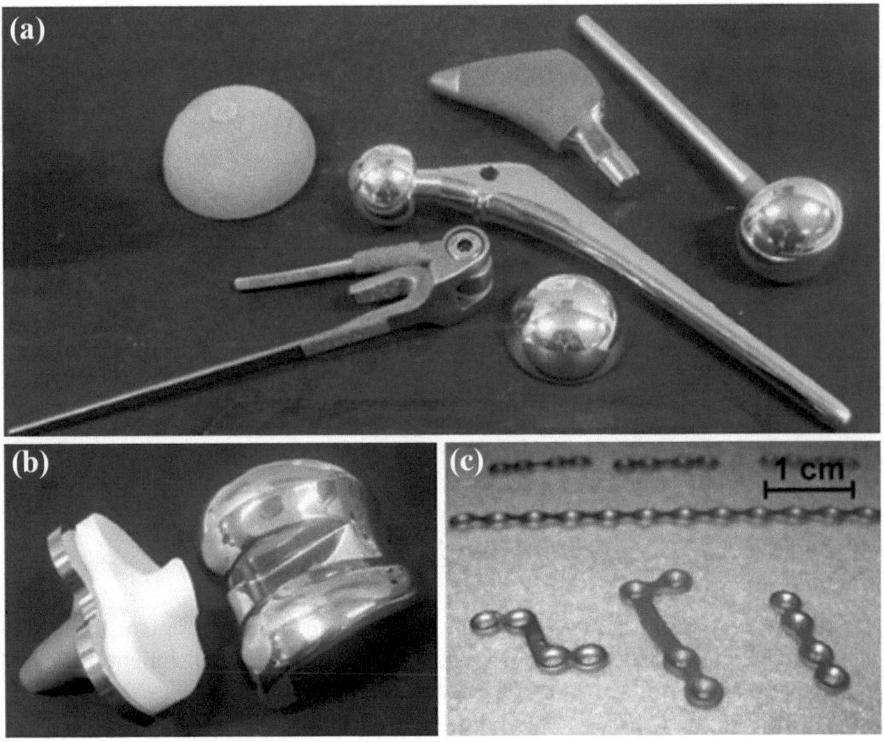

Fig. 1 Example of metallic implants designed for various applications: **a** hip and elbow, **b** knee, and **c** craniofacial. Courtesy of MediTeg, Universiti Teknologi Malaysia [6]

crucial, current trends favor implants that offer additional bioactive properties, such as promoting bone growth and ensuring blood compatibility [4, 5].

1.2 Importance of Biodegradable Metallic Materials

Biodegradable metallic materials are particularly important in applications such as bone repair, cardiovascular stents, and pediatric surgeries, where tissue continues to grow after implantation. These materials can be implanted and gradually degrade over time, eliminating the need for a second surgery once the tissue has healed. Additionally, they may aid tissue regeneration by releasing beneficial ions as they break down [6, 7].

1.3 Advantages of Biodegradable Metals

- **Reduced surgical interventions**: Once the tissue heals, a second surgery to remove the implant is unnecessary, which is especially beneficial for children whose bodies are still developing.
- **Enhanced biocompatibility**: Metals such as zinc, iron, and magnesium can be engineered to degrade at controlled rates, minimizing adverse tissue reactions while naturally being biocompatible. In addition, some, like magnesium, may release ions that support new bone formation and repair.
- **Customizable degradation rates**: By adjusting the alloy composition, researchers can design biodegradable metals to break down at rates that align with the healing time of the targeted tissue.
- **Wide-ranging medical applications**: These metals are used across various fields, including dental implants, cardiovascular stents, and orthopedic procedures for repairing bone fractures.

2 Properties of Biodegradable Metallic Materials

Materials used in biomedical applications include metals, metal composites, polymers, and ceramics; however, metals are most commonly chosen due to their strong structural properties and inert nature [8, 9]. Today, metals are widely employed in various fields, such as orthopedic implants and dentistry. Significant research efforts have been dedicated to designing and developing biodegradable metal implants, particularly for orthopedic and cardiovascular applications, largely because metals offer superior mechanical properties compared to biodegradable polymers. This research has mainly concentrated on metals that naturally exhibit excellent biocompatibility, such as magnesium, iron, and zinc.

2.1 Magnesium-Based Alloys

Throughout the twentieth century, significant efforts have been made to exploit the unique benefits of magnesium-based materials for medical uses [10–12]. Magnesium and its alloys, in particular, have garnered substantial interest in bone healing research [13]. Devices such as screws, pins, and plates made from magnesium alloys have been designed to provide crucial support during bone repair surgeries [14]. These devices not only offer initial mechanical stability but also gradually degrade, reducing their load-bearing capacity over time [15, 16]. This controlled degradation helps shift stress more naturally to the surrounding bone, thereby promoting healing and remodeling [17, 18]. In addition, elements like zinc (Zn), manganese (Mn), calcium (Ca), and zirconium (Zr), known for their non-toxicity, are often alloyed with magnesium. Beyond serving as bone fixation devices, magnesium and its alloys

Table 1 Summary of the physical and mechanical properties of magnesium alloys in comparison with human bone [16]

	Density (g/cm^3)	Elastic modulus (GPa)	Tensile strength (MPa)	Fracture toughness ($MPa^{1/2}$)	Total elongation (%)
Cortical bone	1.8–2.1	3–20	35–283	3–6	1.07–2.10
Cancellous bone	1.0–1.4	NA	1.5–38	NA	NA
Magnesium alloys	1.74–2.0	41–45	150–400	15–40	2–20

Note Different values are due to differences in ethnicity, age, testing conditions, etc. *NA* not applicable

are being investigated as scaffold materials in bone tissue engineering, where cells are seeded onto an artificial three-dimensional scaffold [19–21].

These scaffolds provide a supportive structure that encourages cell adhesion and growth, essential for new tissue formation [22]. For optimal bone regeneration, the scaffold's pore sizes should ideally mimic those found in natural bone—typically ranging from tens to hundreds of microns [23, 24]. Moreover, a high porosity with interconnected pores is necessary to facilitate nutrient transport and the removal of metabolic waste [25].

2.1.1 Mechanical Properties of Mg Alloys

A major advantage of magnesium alloys is their mechanical resemblance to natural bone, as highlighted in Table 1 [26]. With an elastic modulus of around 45 GPa, magnesium-based materials compare favorably to cortical bone, which has a modulus between 7 and 20 GPa. This similarity helps minimize stress shielding effects [27]. Due to their lower mechanical strength, magnesium alloys have an edge over biopolymers like poly-ether-ether ketone (PEEK), poly(l-lactic acid) (PLLA), and polyglycolic acid (PGA) [28–30]. Magnesium alloys also exhibit excellent strength, distinct degradation behaviors, and high impact resistance [31]. Furthermore, their density is close to that of bone (approximately 1.8–2.1 g/cm^3), with specific alloys like WE43 ranging from 1.74 to 1.84 g/cm^3 [27].

2.1.2 Degradation Behavior of Mg Alloy

Extensive reviews and studies have examined the degradation behavior of magnesium-based materials (MBMs) [32–35]. The corrosion of MBMs occurs when magnesium reacts with water, leading to the formation of hydrogen gas (H_2) and magnesium hydroxide ($Mg(OH)_2$). The corrosion process can be broadly summarized by the following reactions [33, 36]:

Anodic reaction:

$$2Mg \rightarrow 2Mg^{+} + 2e^{-}$$

Cathodic reaction:

$$2H_2O + 2e^{-} \rightarrow H_2 + 2OH^{-}$$

Chemical reaction:

$$2Mg^{2+} + 2H_2O \rightarrow 2Mg^{2+} + 2OH^{-} + H_2$$

Overall reaction:

$$Mg + 2H_2O \rightarrow Mg^{2+} + 2OH^{-} + H_2$$

Product formation:

$$Mg^{2+} + 2OH^{-} \rightarrow Mg(OH)_2$$

Product dissolution:

$$Mg(OH)_2 \rightarrow Mg^{2+} + 2OH^{-}$$

Multiple factors contribute to the complex degradation process of magnesium, including the initiation of degradation, the degradation rate, the formation and composition of corrosion products, the elimination of these products under shear stress accompanied by flow, as well as localized pitting and hydrogen evolution. This dynamic process interacts continuously with the surrounding tissue, triggering a foreign body reaction that involves macrophages and the formation of foreign body giant cells [37]. Furthermore, the local physiological environment, including fluid flow and mechanical loading, influences the breakdown of magnesium; therefore, it is critical to comprehend the important factors at play. These reactions release charged molecules that can affect molecular and cellular reactions in addition to producing solid $Mg(OH)_2$ and H_2 gas.

2.1.3 Biocompatibility and Bioactivity Behavior of Mg Alloys

Magnesium-based materials (MBMs) are valued for their degradability, making them excellent candidates for therapeutic devices in cardiovascular and orthopedic applications. The capability of MBMs to interact with body tissues without seriously harming them is known as biocompatibility. It is theoretically possible for human tissue to both tolerate and benefit from proper interactions with MBM implants. However, it has been observed that the in vivo behavior of MBMs can differ from in vitro results. When magnesium reacts with water—whether in an organic tissue environment or in cell cultures—it produces magnesium hydroxide ($Mg(OH)_2$) and hydrogen gas (H_2).

Both laboratory (in vitro) and animal/human (in vivo) studies have been conducted to investigate the biological responses to magnesium-based materials [38–42]. Typically, MBM implantation leads to a significant deposition of calcium phosphate (Ca–P), forming a mineral layer between the implant and the tissue. Additionally, gas pockets containing varying concentrations of O_2, H_2, CO_2, and/or N_2 can form, and the local pH of bodily fluids tends to increase [43, 44]. While gas pockets are quickly released in vitro, in vivo they are often trapped by the surrounding tissue.

Within nanoseconds of contact with tissue, the surface of magnesium-based materials (MBMs) quickly binds to the extracellular matrix through protein integrins—heterodimeric receptors found on cell membranes—a behavior also seen with many nondegradable biomaterials. These integrins play a key role in intracellular signaling, thereby influencing a wide range of cellular processes [45, 46]. Additionally, the adsorption of blood serum proteins onto the biomaterial surface suggests that MBMs are susceptible to the Vroman effect, particularly in cardiovascular applications [47].

Unlike non-biodegradable materials, magnesium reacts with the aqueous environment during protein adhesion, leading to the formation of magnesium hydroxide ($Mg(OH)_2$) and the release of hydrogen gas (H_2), which in turn increases the Mg^{2+} concentration. This physiologically active magnesium is known to catalyze over 300 enzymes, including those essential for ATP production and for synthesizing DNA and RNA [48]. To further enhance biocompatibility, biocompatible elements such as rare earth metals and niobium are often alloyed with MBMs and other permanent biomaterials like titanium-based alloys [49, 50]. However, both in vitro and in vivo cytotoxicity assessments are necessary to fully understand the molecular mechanisms and biological responses to these additional components.

Magnesium alloys stand out as excellent implant materials because of their favorable biocompatibility and controlled degradation characteristics. However, a major obstacle is the disparity in corrosion behavior between in vitro and in vivo investigations, which makes it more difficult to anticipate biological reactions from laboratory research alone. Moreover, the rapid corrosion of MBMs and the byproducts they generate pose additional challenges. Addressing these issues requires innovative methods to thoroughly investigate the cellular and molecular reactions—such as tissue growth, immune responses, signaling pathways, and cell adhesion—that are critical for assessing the long-term clinical safety of MBM implants [51].

2.2 *Zinc-Based Alloys*

In the past decade, much research has focused on pure and alloyed forms of iron and magnesium as potential biodegradable metals for medical applications. However, these systems have exhibited significant limitations in clinical settings. Zinc and its alloys have recently been suggested as viable substitutes, expanding the range of metals that decompose. The primary corrosion product of zinc is Zn^{2+}, which is tightly regulated within physiological systems and is essential for many fundamental cellular functions [52–56]. Zinc ions released from an implant may promote healthy bone formation while preventing the proliferation of harmful smooth muscle cells and reducing arterial restenosis.

Zinc, much like iron and magnesium, is a vital trace element in the human body, playing a critical role as a component of over 300 enzymes and numerous other proteins. Adequate zinc levels are necessary for cell growth, division, function, and the optimal metabolism of proteins and nucleic acids. Thus, zinc ions released during the degradation process can be integrated into normal metabolic processes without causing adverse effects [57].

- **Electrode Potential**: Zinc has an electrode potential of – 0.762 V, which is intermediate between that of iron (– 0.444 V) and magnesium (– 2.372 V). This indicates its notable chemical activity. Consequently, pure zinc forms moderately stable passive layers due to its corrosion products, resulting in a degradation rate that is faster than that of iron and its alloys, yet slower than the rapid degradation observed with magnesium alloys.
- **Processing Advantages**: Zinc and its alloys have low melting temperatures, low chemical reactivity, and excellent machinability, which simplify their casting and processing. For example, melting zinc alloys in air is more straightforward compared to magnesium-based alloys.

2.2.1 Mechanical Properties of Zn Alloys

For implantable biomaterials such as stents and bone plates, it is essential that their mechanical properties are compatible with the tissues they replace [58]. The mechanical properties of bone and vascular tissues are contrasted with those of zinc and contemporary zinc-based alloys in Table 2 [59]. When compared to cortical bone, it is clear that pure zinc's mechanical qualities are inadequate for use in bone and artery devices. This issue can be addressed by incorporating alloying elements and refining the grain size through thermomechanical processing.

The ultimate tensile strength and elongation of zinc alloys can vary significantly, ranging from 87 to 399 MPa and 0.9% to about 170%, respectively. Even minor alloying additions can substantially enhance these properties. For example, adding only 0.15% Mg enhances the elongation of pure zinc from 0.32 to 22% and raises its ultimate tensile strength from 18 to 250 MPa. Moreover, the strength and ductility of zinc alloys are improved by processing methods including hot rolling and hot

Table 2 Katarivas Levy et al. [59] mechanical properties of bone and arterial tissues, compared with current biomedical zinc alloys

	Mechanical properties			
Tissue/alloy	Yield strength (YS) (MPa)	Tensile strength (UTS) (MPa)	Elongation (%)	Hardness (HV)
Cortical bone	104.9–114.3	35–283	5–23	–
Cancellous bone	–	1.5–38	–	–
Arterial wall	–	0.5–1.72	–	–
Zn[c]	10	18	0.32	38
Zn*,[c]	35	60	3.5	–
Zn**,[c,d]	30–110	50–140	5.8–36	39
Zn–Mg				
Zn–0.15Mg*	114	250	22	52
Zn–0.5Mg*	159	297	13	65
Zn–1Mg/ZnMg1	180	340	6	75–86
Zn–1Mg*,[f]	175	250	12	–
Zn–1.2Mg	116	130	1.4	93
Zn–1.2Mg*	220	362	21	96
Zn–1.5Mg	112	150	1.3	155
Zn–3Mg/ZnMg^3	–	104	2.3	201
Zn–3Mg*	291	399	1	117
Zn–3Mg***	–	88	8.8	175
Zn–1.5Mg–0.1Ca	173	241	1.72	150
Zn–1Mg–0.5Ca	–	150	1.34	116
Zn–1Mg–1Ca[c]	80	130	1	90
Zn–1Mg–1Ca*,[c]	205	250	5.2	–
Zn–1Mg–1Ca**,[c]	138	197	8.5	105
Zn–1Mg–0.1Sr	109	133	1.4	94
Zn–1Mg–0.5Sr	129	144	1.1	109
Zn–1.5Mg–0.1Sr	130	209	2.0	145
Zn–1Mg–1Sr[c]	85	135	1.2	85
Zn–1Mg–1Sr*,[c]	200	250	7.3	–
Zn–1Mg–1Sr**,[c]	140	200	9.7	90
Zn–1Mg–0.1Mn	114	132	98	1.11
Zn–1.5Mg–0.1Mn	115	122	149	0.77
Zn–Ca				
Zn–1Ca	119	165	2	73
Zn–1Ca–1Sr[c]	83	140	1.1	90
Zn–1Ca–1Sr*,[c]	210	260	6.8	–

(continued)

Table 2 (continued)

	Mechanical properties			
Tissue/alloy	Yield strength (YS) (MPa)	Tensile strength (UTS) (MPa)	Elongation (%)	Hardness (HV)
Zn–1Ca–1Sr**,c	145	203	8.6	85
Zn–Sr				
Zn–1Sr	120	171	2	61
Zn–Al				
Zn–0.5Al*	119	203	33	59
Zn–1Al*	134	223	24	73
Zn–1Al**,d	190	220	24	–
Zn–3Al**,d	200	240	30	–
Zn–5Al**,d	240	300	16	–
ZnAl4Cu1	171	210	1	80
ZA0.1Mg	–	87	1.6	79
ZA0.3Mg	–	93	1.7	89
ZA0.5Mg/ Zn–0.5Al–0.5Mg	–	92–102	1.73–2.1	94
Zn–0.5Al–0.5Mg–0.1Bi	–	102	2.4	102
Zn–0.5Al–0.5Mg–0.3Bi	–	108	2.7	109
Zn–0.5Al–0.5Mg–0.5Bi	–	98	1.97	99
ZA4–1	75	180	~ 112	50
ZA4–3	110	200	~ 130	55
ZA6–1	175	275	~ 170	65
Zn–Cu				
Zn–1Cu*	149	186	21	–
Zn–2Cu*	199	240	46	–
Zn–3Cu*	213	257	47	–
Zn–4Cu*	227–250	270	51	–
Zn–3Cu–0.1Mg*,b	340	355	5	–
Zn–3Cu–0.5Mg*,b	390	400	2	–
Zn–3Cu–1Mg*,b	427	441	0.9	–
Zn–Ag				
Zn–2.5Ag*,a	174	200	35	–
Zn–5Ag*,a	236	250	36	–
Zn–7Ag*,a	258	287	32	–
Zn–Li				
Zn–2Li**,e	240	360	14.2	98

(continued)

Table 2 (continued)

	Mechanical properties			
Tissue/alloy	Yield strength (YS) (MPa)	Tensile strength (UTS) (MPa)	Elongation (%)	Hardness (HV)
Zn–4Li**,e	420	440	13.7	115
Zn–6Li**,e	470	560	2.2	136
Zn–1Li	238	274	17	97

*Hotextrusion; **Hotrolling; ***Homogenization

extrusion [60]. For instance, Zn–1 $Mg^{-1}Ca$ as-cast has a yield strength of 80 MPa, a UTS of 130 MPa, and a 1% elongation. Zn-based ternary alloy samples, on the other hand, exhibit improvements to 205 MPa (YS), 250 MPa (UTS), and 5.2% elongation following hot extrusion, and 138 MPa (YS), 197 MPa (UTS), and 8.5% elongation following hot rolling [61]. Thus, conventional metallurgical methods prove effective in achieving the necessary strength criteria for zinc alloys.

2.2.2 Degradation Behavior of Zn Alloy

In 1961, Iranian farmers who were fed a zinc-deficient diet consisting of milk, potatoes, and unrefined flat bread were found to have anemia, hypogonadism, and dwarfism. This finding underscored zinc's critical role in human health and spurred growing interest in its biochemical and therapeutic properties [62]. Zinc is now acknowledged as one of the most prevalent vital minerals in the human body. All organs, tissues, fluids, and secretions contain zinc; however, the majority of it is found in skeletal muscle and bone (86%), followed by the skin (6%), liver (5%), brain (1.5%), and other tissues [63].

For biodegradable metals in a neutral physiological environment, Zheng et al. [64] proposed a model involving both cathodic and anodic processes. In metal corrosion, byproducts such as hydroxides, oxides, and hydrogen gas are formed [65]. Figure 2 illustrates this process for zinc.

- **Anodic Reaction**: Zinc is oxidized to Zn^{2+} cations upon contact with bodily fluids:

$$2Zn \rightarrow 2Zn^{2+} + 4e^-$$

- **Cathodic Reaction**: The electrons produced are consumed by the reduction of dissolved oxygen:

$$O_2 + 2H_2O + 4e^- \rightarrow 4OH^-$$

- **$Zn(OH)_2$ Formation**: Zn^{2+} reacts with OH^- to form $Zn(OH)_2$:

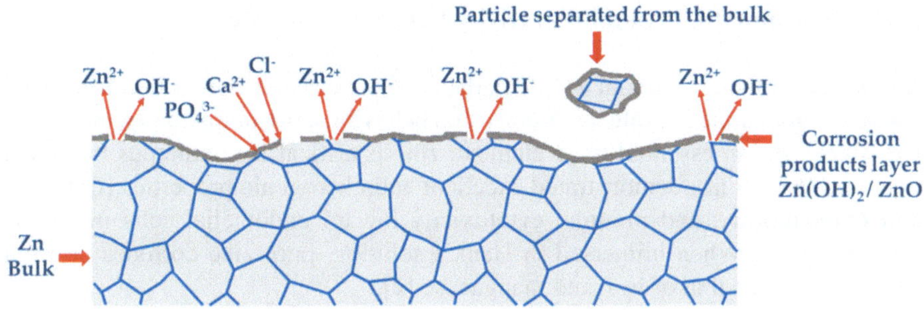

Fig. 2 Model of the zinc degradation process in physiological fluids [59]

$$2Zn^{2+} + 4OH^- \rightarrow 2Zn(OH)_2$$

- **ZnO Formation:** $Zn(OH)_2$ then converts into ZnO with the release of water:

$$Zn(OH)_2 \rightarrow ZnO + H_2O$$

These reactions yield $Zn(OH)_2$ and ZnO as corrosion products on the metal surface without generating hydrogen gas—a stark contrast to the problematic hydrogen gas evolution seen in magnesium corrosion [66].

It is also important to note that the physiological environment is quite harsh due to high chloride ion levels. Chloride ions can convert the surface into soluble chloride salts, disturbing the balance between dissolution and the formation of the protective corrosion layer, as shown in the following reactions:

- $6Zn(OH)_2 + Zn^{2+} + 2Cl^- \rightarrow 6Zn(OH)_2 \cdot ZnCl_2$
- $4ZnO + 4H_2O + Zn^{2+} + 2Cl^- \rightarrow 4Zn(OH)_2 \cdot ZnCl_2$.

Bowen et al. [66] found that, in the early stages (1.5 and 3 months), zinc wires (99.99% pure) implanted in rat aortas developed only thin coatings of zinc oxide. As the corrosion process advanced to 4.5 and 6 months, the corrosion layer thickened and evolved into three distinct phases: an outer calcium phosphate layer and inner layers composed of zinc oxide and zinc carbonate. The calcium phosphate formed on the external surface without contributing significantly to bulk formation, suggesting that it did not markedly affect zinc biocorrosion. The inner compact layer featured both zinc oxide and zinc carbonate, with ZnO appearing in areas separated by the carbonate phase. This complex, multilayered corrosion structure hints that the corrosion products of zinc in bodily fluids may be somewhat analogous to those of magnesium.

2.2.3 Biocompatibility and Bioactivity Behavior Zinc Alloys

Zinc is a promising material for degradable, bioresorbable stents because of the extensive study done to evaluate its corrosion behavior, cytocompatibility, and bioactive effects. As an essential trace element, zinc is critical for numerous biological processes [67]. It has demonstrated excellent antiatherogenic properties [68], good hemocompatibility, and minimal cytotoxicity toward endothelial cells in various in vivo studies. When immersed in Hank's solution, pure zinc corrodes at a rate that falls between that of iron and manganese [2].

Torne et al. [53] found that a passivation coating made up of inorganic materials and biomolecules developed on the zinc surface of Zn-based implants when they were submerged in plasma and whole blood contrary to phosphate-buffered saline (PBS) or Ringer's solution. Electromechanical measurements further helped elucidate the mechanisms behind the development of this protective barrier in whole blood. Importantly, the concentration of protein in human blood serum (approximately 60–80 g L^{-1}) has a significant impact on implant corrosion behavior, meaning that the presence of proteins can alter the degradation patterns of biodegradable zinc alloys [69, 70].

High levels of zinc ions released during breakdown, however, have been shown in tests to be cytotoxic and may impair fibroblast activity. For example, Su et al. [68] reported that pure zinc hindered the proliferation and differentiation of MC3T3-E1 cells, with similar effects observed in MG-63 cells. Although pure zinc has been shown to have some genotoxic effects, alloying zinc with other metals—such as in Zn–3Mg—can reduce this risk, as suggested by Kubasek et al. [71]. Moreover, Yin et al. [72] found that the Zn-0.8Mg alloy did not significantly harm U-2OS cells, and additional alloys like Zn-0.05Mg and Zn-0.5Mg have even been observed to promote cell proliferation.

Figure 3 provides an overview of in vivo studies for cardiovascular applications. For example, Bowen et al. [66] implanted pure Zn stents into rabbits' abdominal aortas and found that the stents had high mechanical integrity and did not exhibit any notable inflammation, thrombosis, platelet aggregation, or intimal hyperplasia. According to Zhou et al. [73], Zn–0.8Cu alloy stents placed in coronary arteries for two years did not cause inflammation or thrombus formation. In a different investigation, Hehrlein et al. [74] placed Zn–3Ag alloy stents in young pigs' iliofemoral arteries and saw good healing without vascular occlusion or thrombosis. In rats with Zn–xMg alloys implanted in their abdominal aortas, Jin et al. [75] also observed very little intimal activation and inflammation. Furthermore, rat aorta treatments with Zn–0.1Li alloys produced comparable positive results.

2.3 *Iron Alloys*

Iron has recently gained attention as a biodegradable metal, especially with the use of additive manufacturing techniques that enable the creation of customized porous

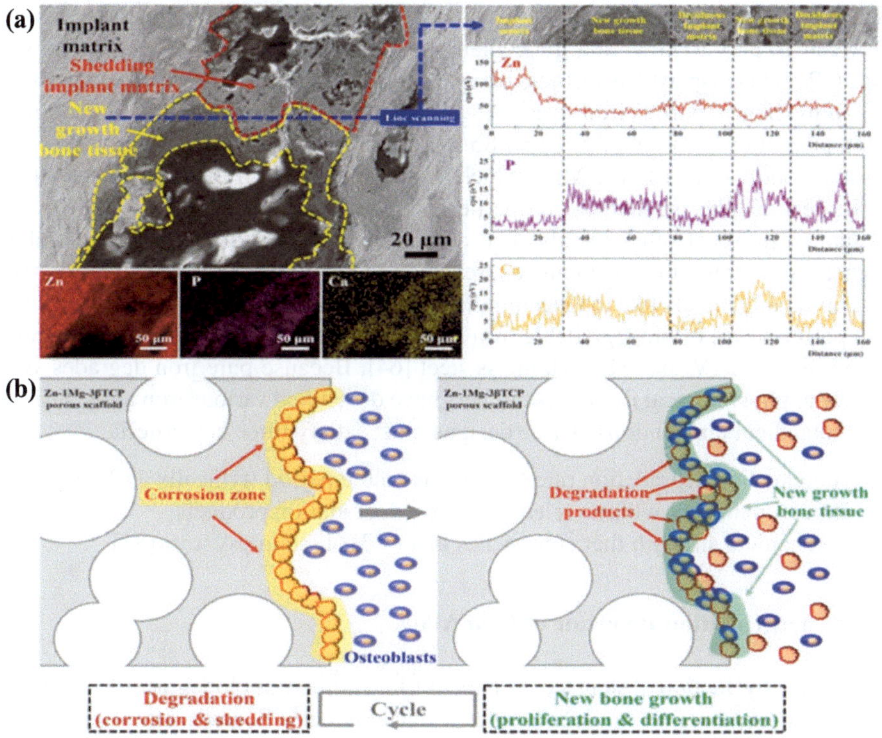

Fig. 3 Mechanism of in vivo bone formation using porous Zn–Mg-TCP scaffolds. Panel **a** presents a diagram outlining the process of osteogenesis and the associated degradation products, while panel **b** displays the microstructure of these degradation products alongside the development of new bone tissue, with corresponding EDS analysis [76]

structures. These structures not only enhance the corrosion rate but also mimic the hierarchical porosity of natural bone. Although pure iron tends to degrade too slowly, the addition of alloying elements can adjust its degradation rate while preserving its desirable mechanical properties [77–79]. Iron alloys, recognized as biodegradable materials, offer high strength, ductility, and formability, making them well-suited for applications such as stents, which require fabrication into thin, complex geometries while retaining adequate mechanical integrity.

2.3.1 Mechanical Properties of Biodegradable Porous Iron

Iron-based biodegradable materials are considered potential alternatives to permanent metallic implants, particularly in non-load-bearing applications, due to their mechanical properties resembling those of human bone. For example, cortical (or compact) bone has a low porosity of 3–5%, whereas cancellous (or trabecular) bone is highly porous (50–90%) and rich in marrow and blood vessels. Compact iron typically has a Young's modulus of about 210 GPa, while cancellous bone ranges between 10 and 20 GPa, and trabecular bone is even lower, from 3×10^{-4} to 3×10^{-3} GPa. This disparity in stiffness between compact iron and bone can lead to stress shielding; hence, porous iron structures are engineered to lower the elastic modulus. With porosities between 45.6% and 86.9%, these structures achieve a compressive elastic modulus of 218–845 MPa, which is closer to that of trabecular bone [80].

In general, the mechanical properties of biodegradable iron and its alloys are similar to those of other metals such as 316L stainless steel. Iron often exhibits higher yield stress, ultimate tensile strength, and fracture strain compared to other biodegradable metals like magnesium. For load-bearing bone implants, the mechanical performance of biodegradable iron should be comparable to materials like pure titanium, Ti6Al4V, and 316L stainless steel [64]. Because pure iron degrades very slowly in physiological fluids, researchers have developed various iron alloy designs for biomedical applications. These designs must satisfy three key criteria:

1. A controlled degradation rate over a reasonable period (typically 1–2 years)
2. Biocompatibility with no release of toxic degradation products
3. Mechanical strength that approaches that of 316L stainless steel [81]

2.3.2 Degradation Behavior of Iron Alloy

Since the introduction of stent technology in 1987, numerous biodegradable materials have been explored to produce implants that gradually degrade after performing their function, eventually being absorbed and eliminated by the body. The very slow degradation rate of pure iron in physiological fluids has led to the development of several iron alloy formulations tailored for biomedical use. The design of these Fe-based alloys for biodegradable stents focuses on achieving:

1. Mechanical strength comparable to 316L stainless steel
2. Biocompatibility that avoids releasing toxic byproducts
3. A controlled degradation rate over a practical time frame (1–2 years) [81].

Furthermore, research has shown that Fe–Mn alloys produced via powder metallurgy degrade faster in vitro than those made by casting due to increased porosity. Similarly, electroformed iron degrades more rapidly than casting-fabricated Armco iron, as its finer microstructure and higher density of grain boundaries make it more susceptible to corrosion [82]. Figure 4 schematically illustrates the degradation mechanism of Fe-based alloys, where liberated metal ions react with hydroxyl ions to form ferrous hydroxide ($Fe(OH)_2$) or hydrous ferrous oxide ($FeOnH_2O$) [83]. This degradation process is applicable to any alloy that contains iron.

The effects of different alloying elements, specifically Mn, Co, Al, W, Sn, B, C, and S, on the biodegradability and biocompatibility of pure iron were recently investigated by Liu et al. [84]. According to their findings, Co, W, C, and S are especially well-suited for incorporation into iron-based biomaterials because they improve

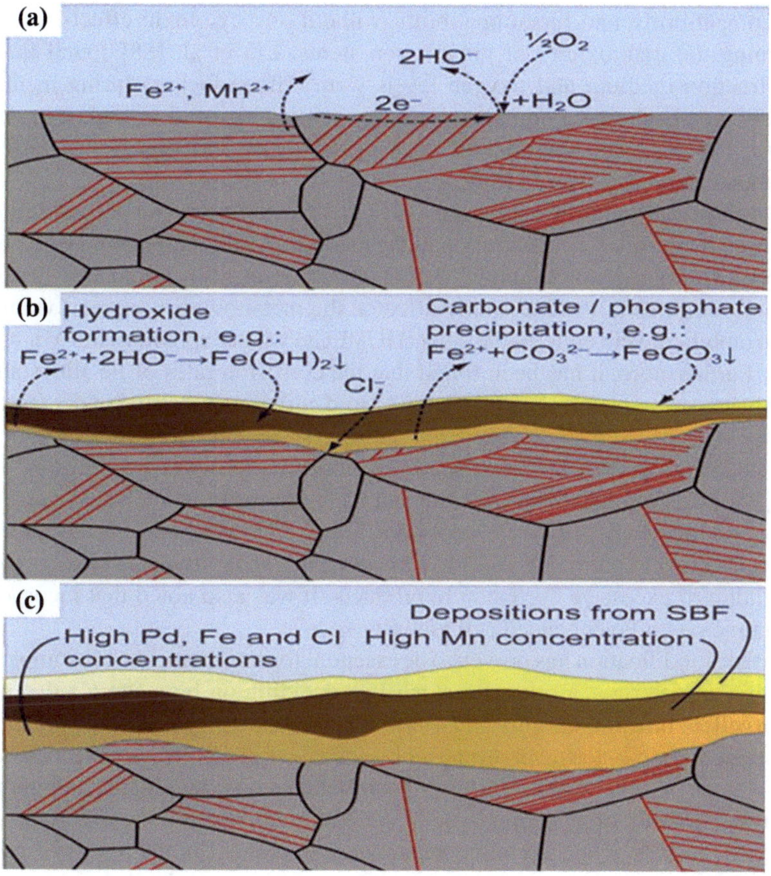

Fig. 4 The degradation process and the formation of corrosion products for TWIP-1Pd alloys [83]

mechanical characteristics, provide advantageous corrosion rates, and provide superior biocompatibility. To accelerate iron degradation for degradable stents, an electroforming technique was used to produce pure iron [85]. The resulting fine grains and microstructural defects made the electroformed iron more prone to corrosion, leading to a higher release of Fe^{2+} ions. Notably, the corrosion of electroformed iron was uniform, with no localized degradation observed.

2.3.3 Biocompatibility and Bioactivity Behavior Fe Alloys

Numerous in vitro studies have demonstrated the excellent biocompatibility of iron (Fe). For instance, pure electroformed iron used in cardiovascular stents did not inhibit the metabolic activity or proliferation of rat smooth muscle cells (SMCs) [86]. Similarly, Chen et al. [87] reported that industrial-grade pure Fe exhibited good hemocompatibility and biocompatibility without any cytotoxic effects. In studies examining the cytotoxicity of nitride iron stents, Lin et al. [88] found that both the extraction medium and oxygen levels were critical factors during incubation, with L929 cells showing no cytotoxicity even when exposed to high iron concentrations. However, the presence of alloying elements was observed to affect the biocompatibility of Fe alloys [89].

Research on alloying effects has revealed that the inhibitory impact is significantly influenced by the alloy concentration in the medium. Multicomponent Fe alloys have also been examined; for example, Schinhammer et al. [90] found that the extract medium's composition significantly affected the metabolic activity and viability of human umbilical vein endothelial cells (HUVECs) when exposed to Fe–Mn–C (-Pd) alloys. Furthermore, it has been found that the corrosion rates of Fe alloys and cell viability are strongly correlated [89], with rapid corrosion resulting in decreased cell viability.

In studies on porous Fe-based scaffolds under static in vitro conditions, Orinak et al. [91] observed notable morphological changes, increased cell survival, and the scaffolds' blood compatibility. Fibroblasts, however, perished after a day, indicating that cytotoxicity experiments must to be carried out under dynamic settings to avoid the buildup of excessive corrosion byproducts. It was also noted that incorporating bioceramics with Fe enhanced cell viability.

Surface modification has proven to be essential for improving the biocompatibility of Fe-based materials, a critical factor when deciding on biomaterials that require both excellent biocompatibility and a slow corrosion rate. By inserting pure Fe stents into the descending aortas of white rabbits, Peuster et al. [92] carried out the first in vivo tests of iron stents. No thromboembolic events, significant inflammation, neointimal growth, or systemic toxicity were observed during the follow-up period (Fig. 5). However, when the same stents were implanted in the aortas of pigs and observed for over a year, there was no evidence of iron overload or organ toxicity, and the corrosion products did not cause any local adverse effects. In a subsequent short-term study, Fe stents implanted in young pigs' coronary arteries for 28 days demonstrated safety and effectiveness, with histomorphometric analyses showing no fibrin deposition, severe inflammation, thrombosis, or particle embolization.

The bioactivity of Fe-bioceramic composites in Indonesian sheep leg bones was assessed by Ulum et al. [93]. After 35 days of implantation, the inflammation induced by the stents subsided, and minimal tissue reaction was observed during healing. The microstructure of iron-based composites and pure iron's sintered and polished surfaces (as seen in Fig. 5a) revealed bioceramic phases with irregular morphologies within the Fe matrix and extensive sintering. Additionally, Rabeeh et al. [94] demonstrated in vitro that dispersing gold particles on the surface of iron accelerated its degradation, as illustrated in Fig. 5b. In general, the rate of degradation of Fe-based materials is quite modest.

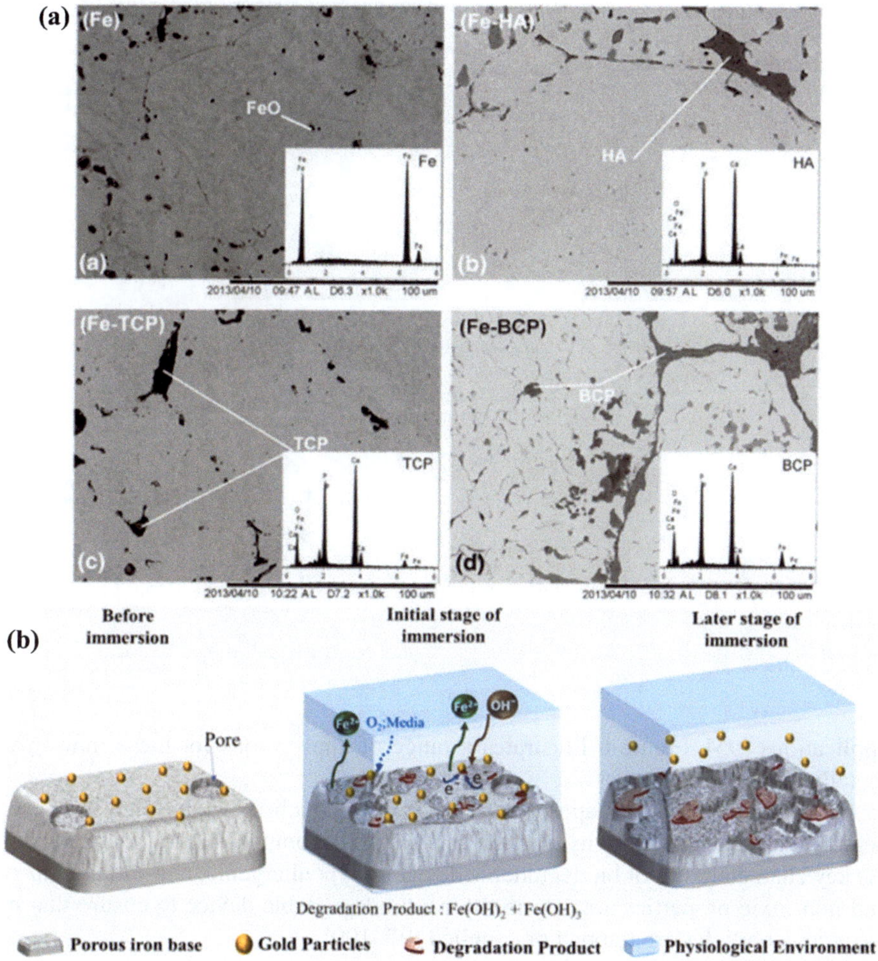

Fig. 5 **a** SEM images and EDS spectra for (**a**) pure Fe, (**b**) Fe–HA, (**c**) Fe–TCP, and (**d**) Fe–BCP [93]. **b** In vivo immersion test along with a schematic representation of the degradation progression for gold-sputtered Fe [94]

3 Biomedical Applications of Biodegradable Metallic Materials

Biodegradable metallic materials—such as those based on magnesium (Mg), zinc (Zn), and iron (Fe) alloys—are revolutionizing biomedical applications by providing temporary structural support that safely degrades within the body. Their ability to dissolve naturally over time eliminates the need for additional surgeries to remove implants and aligns well with the body's intrinsic healing processes. This makes them highly suitable for use in orthopedic, cardiovascular, and tissue engineering

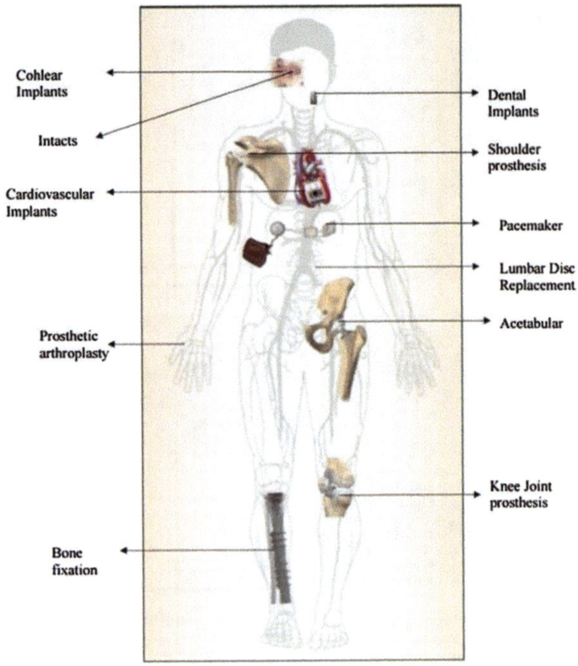

Fig. 6 Applications of metallic biomaterials used as implants in different regions of the human body [96]

applications [95]. Figure 6 illustrates a range of applications for these innovative metallic biomaterials.

High biocompatibility, appropriate mechanical qualities, controlled corrosion behavior, antibacterial activity, and the capacity to decompose naturally are some of the key characteristics of biodegradable devices. Hypoallergenic, non-inflammatory, and non-toxic properties are essential for a biodegradable device to ensure that no hazardous particles are trapped or expelled [97–100].

3.1 Orthopedic Applications

Biodegradable metallic materials are essential in orthopedics as they provide temporary structural support necessary for bone healing and regeneration. They are widely used in fracture fixation devices such as screws, plates, pins, intramedullary nails, and wires, as illustrated in Fig. 7. For an orthopedic implant to be effective, it must offer an optimal combination of mechanical strength, biocompatibility, and controlled degradation, thereby promoting proper bone healing while minimizing the risk of negative side effects.

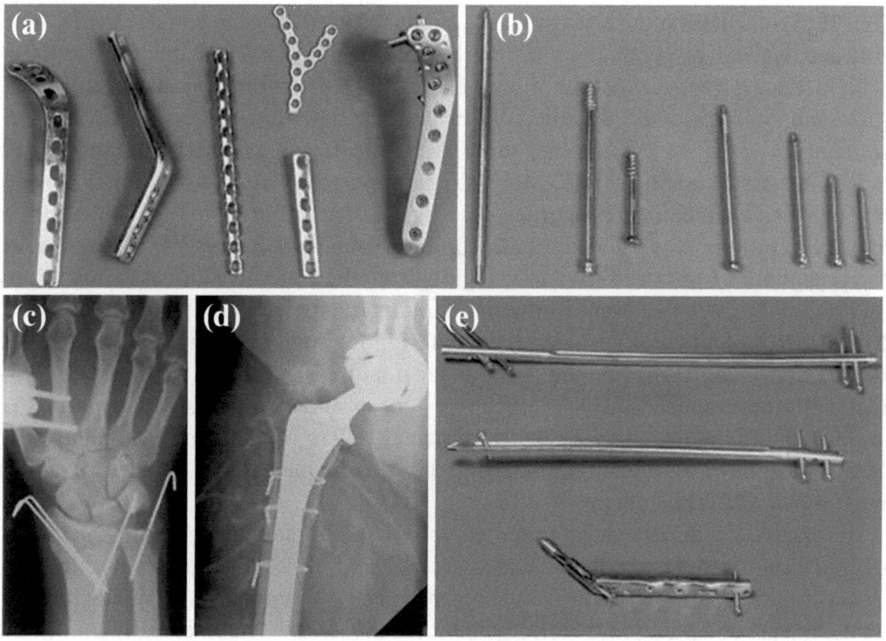

Fig. 7 **a** Plates, **b** screws, **c** pins, **d** wires, **e** intramedullary nails [101]

3.1.1 Magnesium-Based Alloys

Magnesium-based alloys are extensively utilized in orthopedics because they combine biodegradability, biocompatibility, and mechanical properties that closely resemble natural bone. Alloys like WE43 and AZ31 are commonly used in devices for fracture fixation—such as screws, plates, and pins—and are also incorporated into porous scaffolds to support bone regeneration [102]. These materials encourage osteogenesis and help minimize stress shielding; however, their rapid degradation in physiological conditions (pH 7.4–7.6, high chloride content) can cause a premature loss of mechanical strength and produce hydrogen gas, potentially disrupting tissue healing. To mitigate these issues, alloying elements like zinc (Zn), calcium (Ca), and rare earth metals (e.g., neodymium, Nd) are added to enhance both corrosion resistance and mechanical stability [103]. Additionally, applying surface coatings—such as hydroxyapatite (HA) or biodegradable polymers like poly(lactic-co-glycolic acid) (PLGA)—further improves performance by regulating degradation rates and reducing hydrogen evolution [104, 105].

Recent advancements in Mg-based implants for orthopedics have focused on surface modifications and innovative designs aimed at enhancing corrosion resistance, bioactivity, and mechanical properties. For example, Santos et al. demonstrated that ZnO coatings applied via High Power Impulse Magnetron Sputtering

(HiPIMS) on ZK60 and ZK60-Mm alloys reduced corrosion current density by fourfold and twofold, respectively. This improvement in corrosion resistance was coupled with increased hydroxyapatite formation—crucial for osseointegration—suggesting that such coatings could significantly extend the functional lifespan of Mg-based implants in vivo [106]. In another study, wollastonite nanoceramics reinforced with a 10% magnetite nanoparticle (MNP) coating increased apatite formation by 29% and enhanced mechanical properties, making the material suitable for applications in sports and orthopedics where durability is vital [107]. Furthermore, He et al. reviewed the potential of Mg-based implants, highlighting their biomechanical compatibility and ability to promote bone formation and angiogenesis. They proposed a hybrid Mg-PEEK (polyether ether ketone) cage design for spinal fusion that leverages the bioactivity of Mg alongside the mechanical stability of PEEK to meet both biological and structural needs [108]. Collectively, these studies emphasize the promise of surface modifications—like ZnO coatings and MNP reinforcement—and innovative hybrid designs in significantly improving the performance of Mg-based implants for bone repair and replacement.

Further research has concentrated on optimizing implant designs and elucidating degradation mechanisms. For instance, titanium wire-based geometries with three-dimensional pore networks have been proposed (as shown in Fig. 8). Finite element analysis and computational fluid dynamics indicate that these designs mimic the properties of trabecular bone, making them ideal for treating bone defects [109]. Additionally, investigations into the degradation behavior of Mg alloys in the presence of organic components, such as albumin and glucose, revealed that chicken egg albumin (CEA) provides superior corrosion protection compared to bovine serum albumin (BSA) [110]. The studies also showed that glucose competes with BSA, thereby affecting both corrosion rates and stress corrosion cracking (SCC). These findings underscore the importance of accounting for organic interactions and fluid dynamics in implant design to ensure biocompatibility and long-term performance. Together, these advancements pave the way for safer and more effective biodegradable Mg-based implants in orthopedic applications.

3.1.2 Zinc-Based Alloys

Zinc-based alloys have attracted considerable attention as biodegradable materials for orthopedic implants because of their balanced degradation rates, excellent biocompatibility, and non-toxicity. Unlike magnesium- or iron-based biodegradable alloys, zinc alloys offer unique advantages, including adjustable mechanical strength, moderate corrosion behavior, good cytocompatibility, dose-dependent osteoinductive properties, and inherent bacteriostatic activity [111]. Recent research has addressed the limitations of pure zinc—such as its relatively low tensile strength (100–150 MPa) and poor elongation (0.3–2%)—by incorporating alloying elements like silver (Ag), manganese (Mn), aluminum (Al), and copper (Cu) [112]. For example, Xiao et al. demonstrated that Zn–Ag alloys, especially Zn–2Ag and Zn–4Ag, exhibit enhanced mechanical strength, improved antibacterial performance,

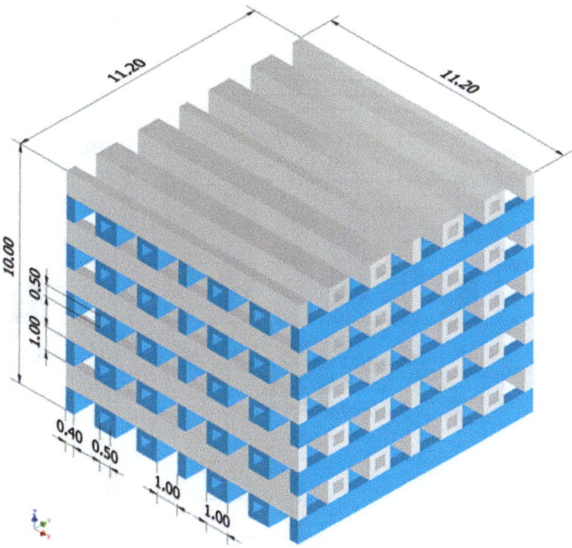

Fig. 8 Architecture of the developed implant [109]

and increased osteogenic activity, making them promising for bone repair [113]. The benefits of alloying are illustrated in Fig. 9. Similarly, Jia et al. reported that Zn–Mn alloys, particularly Zn–0.8Mn, deliver exceptional ductility (with an elongation of 83.96%) while promoting significant bone regeneration in vivo, all while maintaining favorable biosafety [114]. These advancements highlight the potential of alloying to fine-tune the mechanical and biological properties of zinc-based materials for load-bearing applications.

In addition to alloying, researchers have explored surface modifications and composite fabrication to further enhance the performance of Zn-based materials. For example, applying polymethyl methacrylate (PMMA) coatings on Zn–Mg alloys has been shown to reduce their corrosion rates, thereby proving their suitability for orthopedic implants [115]. Moreover, adding trace amounts of Gd can refine the microstructure of zinc, promoting more uniform degradation and improved osseointegration [116]. Studies have also revealed that Zn–Al–Cu alloys, such as Zn–2Al–8Cu, offer enhanced mechanical strength—with a compressive yield strength of 245.17 MPa—and improved corrosion resistance, making them viable candidates for medical implants [117]. These innovations, along with in vivo evidence of osteogenic potential and biocompatibility, position Zn-based alloys as promising alternatives to traditional biodegradable materials like magnesium (Mg) and iron (Fe).

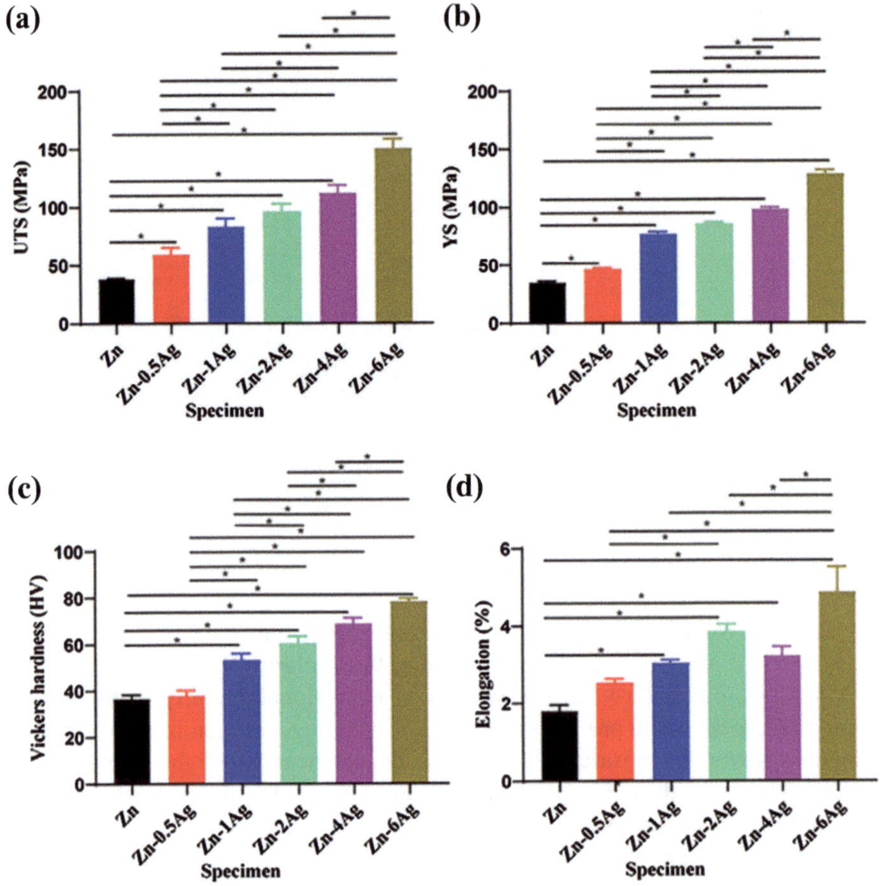

Fig. 9 Mechanical properties of Zn–Ag alloys are depicted, including **a** ultimate tensile strength, **b** yield strength, **c** Vickers hardness, and **d** elongation rate [113]

3.1.3 Iron-Based Alloys

Iron-based alloys are highly valued in orthopedic applications because of their excellent biocompatibility, degradability, and superior mechanical properties—particularly their higher strength compared to Mg and Zn—which makes them ideal for load-bearing implants [118, 119]. Recognized as suitable biodegradable materials for osteosynthesis, these alloys are capable of temporarily replacing bone under high load conditions while maintaining a balance between high strength and moderate corrosion rates. Open-cell iron foams, fabricated via a replication method using foamed polyurethane combined with powder metallurgy, have shown promising results. Research comparing pure carbonyl iron samples with those incorporating

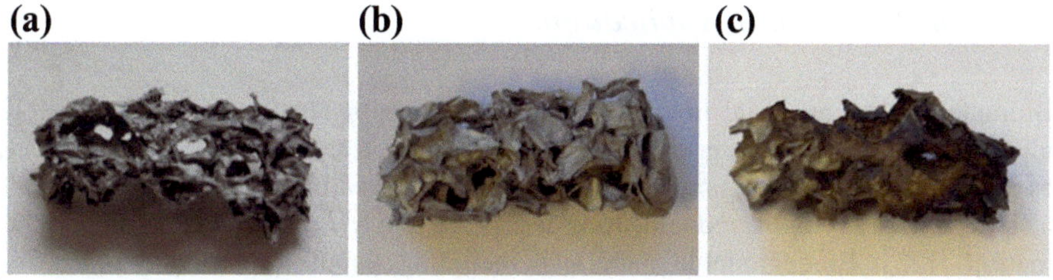

Fig. 10 Carbonyl iron-based cellular materials produced via powder metallurgy, with sample dimensions of approximately 0.9–1 cm in diameter and 3.5–4 cm in length: **a** pure Fe, **b** Fe with carbon nanotubes (Fe-CNTs), and **c** Fe with magnesium (Fe–Mg) [120]

carbon nanotubes (CNTs) and magnesium additives (as illustrated in Fig. 10) demonstrated that the porous structure of the foam increases biodegradation rates. Potentiodynamic polarization tests performed in Hank's solution and physiological saline at 37 °C indicated that the corrosion rates rank as follows: Fe–Mg > Fe > Fe–CNTs, with the Fe–Mg alloy degrading the fastest. Additionally, cytotoxicity tests revealed a modest but measurable proliferation of osteoblastic cells on these iron-based materials, indicating their acceptable biocompatibility [120].

Laser powder bed fusion (LPBF) has enabled the fabrication of Fe–Mn biodegradable alloys, further expanding their potential for orthopedic use. For instance, an LPBF-produced multiphase Fe–18Mn alloy (comprising γ-austenite, ε-martensite, and α-ferrite) was annealed at 650 °C to achieve a uniform microstructure. This alloy demonstrated outstanding mechanical properties, including a tensile strength of 1 GPa, a fracture elongation of 16%, and a very high work-hardening rate of 8500 MPa—surpassing most Fe–Mn steels and Fe-based high-entropy alloys. Its exceptional work-hardening ability is attributed to various deformation mechanisms (such as stacking faults, martensitic transformations from γ to ε to α', and nano-deformation twins) and the interactions among these features and the transformed martensite. These findings highlight the promise of additively manufactured Fe–Mn alloys as durable, stable bone substitutes that combine high strength, controlled degradation, and superior work-hardening performance [121].

Additional progress has been made in Fe–Mn–Zn alloys by examining how varying the zinc (Zn) content influences their microstructure and mechanical properties. In nanocrystalline alloys with compositions (70–x) Fe–30Mn–xZn (with $x = 0$, 3, 6, and 9 wt%), produced via mechanical alloying and hot pressing, the Fe–30Mn–9Zn variant achieved the highest relative density (94.21%), the highest bending strength (1277.61 MPa), and the lowest corrosion rate (0.044 mm/year) due to its low porosity. Furthermore, increasing the Zn content notably improved the alloy's antibacterial activity against *Staphylococcus aureus* and *Escherichia coli*, making it a promising candidate for infection-resistant orthopedic implants. In contrast, the Fe–30Mn alloy (without additional Zn) exhibited the highest hardness (548.66 HV) and wear resistance as a result of solid solution hardening [122].

3.2 Cardiovascular Applications

Biomaterials are essential in cardiovascular treatments, such as stents that restore blood flow in occluded arteries. As shown in Fig. 11, a stent is guided to the artery, expanded to push plaque, and left in place to support healing while degrading harmlessly over 6–12 months, reducing long-term risks like restenosis and chronic inflammation [123]. Biodegradable materials, including magnesium (Mg)-based, zinc (Zn)-based, and iron (Fe)-based alloys, offer a promising alternative to traditional corrosion-resistant stents. These materials provide temporary mechanical support during vessel healing and then safely degrade, eliminating the need for permanent implants and addressing complications like late-stage thrombosis and in-stent restenosis [119, 124–126].

3.2.1 Magnesium-Based Stents

Magnesium-based alloys have demonstrated significant potential in stent applications due to their biocompatibility and ability to biodegrade. For instance, clinical trials like the PROGRESS-AMS study have shown that magnesium scaffolds are completely absorbed within two months, with no signs of thrombosis or calcification. However, the downside of their rapid degradation is that the mechanical support they provide diminishes too quickly, potentially leading to early recoil and restenosis before the vessel has fully healed. Research efforts are focused on decelerating the degradation rate while preserving mechanical strength and biocompatibility. Although preclinical studies have indicated that magnesium alloys trigger minimal inflammation and exhibit no toxicity, further improvements are necessary to ensure they provide sustained support during the healing process [119].

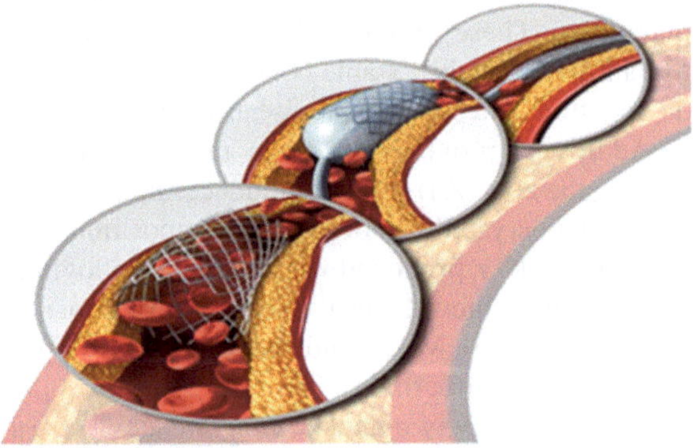

Fig. 11 Stent implantation into a coronary vessel [123]

3.2.2 Zinc-Based Stents

Zinc-based alloys have recently gained attention as promising materials for biodegradable stents because of their anti-inflammatory and anti-atherogenic properties. Zinc helps stabilize the membranes of endothelial cells, which prevents apoptosis and reduces the risk of restenosis, as shown in animal studies where zinc supplementation was associated with smaller atherosclerotic lesions. Additionally, zinc scaffolds degrade at a rate that is well-suited for vascular healing, maintaining approximately 70% of their cross-sectional area after four months. However, a notable drawback is their low radial strength—around 120 MPa—which is below the approximately 300 MPa needed for adequate vascular support. Ongoing research is aimed at improving the mechanical properties of zinc alloys while retaining their beneficial biological effects [119].

3.2.3 Iron-Based Stents

Iron-based alloys are emerging as another viable option for biodegradable stents, offering superior ductility and strength compared to magnesium-based alloys. Pure iron stents, such as the NOR-I scaffold, have been found to be free of thrombosis and other adverse effects in animal models. However, their degradation is excessively slow, with stent struts remaining visible even 18 months after implantation. To address this issue, iron-manganese alloys (e.g., Fe-35Mn) have been developed, striking a balance between the sluggish degradation of pure iron and the rapid breakdown of magnesium alloys. These iron-manganese alloys possess excellent mechanical properties, are MRI-compatible, and promote better endothelial cell attachment, making them promising candidates for vascular applications. Nonetheless, further research is required to fine-tune their degradation rates and ensure long-term biocompatibility [119].

In a recent study, iron-based stents were found to degrade more quickly in vitro and in a pig model using a metal-polymer composite method. As illustrated in Fig. 12, PLA-coated iron sheets exhibited significantly higher corrosion rates in Hank's solution compared to bare iron sheets, indicating that the PLA coating effectively speeds up iron corrosion. The study also highlighted the critical role of serum proteins, especially albumin, in mitigating reactive oxygen species and inflammatory responses during iron corrosion. The composite stent was successfully implanted in a patient during clinical trials, with follow-up data confirming its safety, efficacy, and appropriate biodegradability, thus demonstrating its potential for cardiovascular applications [127].

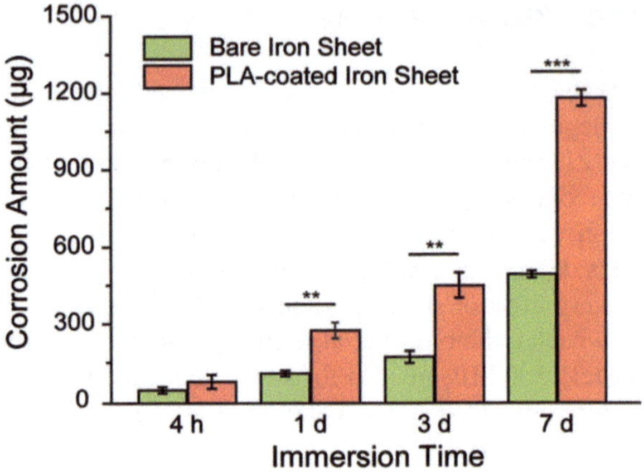

Fig. 12 The corrosion amounts of the PLA-coated iron sheet and bare iron sheet (10 mm × 10 mm × 0.1 mm) after being immersed in Hank's solution at 37 °C for 4 h, 1 day, 3 days, and 7 days [127]

3.3 *Tissue Engineering*

Tissue engineering has become a groundbreaking strategy for overcoming significant challenges in regenerative medicine, such as extensive bone defects, limited availability of donor tissue, immune rejection, and the risk of pathogen transmission. A core element of this approach is the use of three-dimensional (3-D) biocompatible scaffolds that serve as frameworks for cell proliferation, differentiation, and tissue development while gradually degrading to allow for natural tissue integration [128]. Recent advancements have involved incorporating bioactive elements—such as zinc (Zn), magnesium (Mg), and iron (Fe) ions—into these scaffolds to boost their functionality. Naturally present in bone, these ions play crucial roles in bone metabolism (as depicted in Fig. 13) and have been shown to improve properties like mechanical strength, degradation rates, and osteogenesis [129]. In addition, biodegradable metals like Mg-, Zn-, and Fe-based alloys are increasingly recognized for their ability to provide both structural support and bioactive benefits in various medical applications [130]. Together, these innovations create a synergistic approach to tissue engineering, opening new avenues for breakthroughs in bone repair, cardiovascular treatments, and beyond.

3.3.1 Magnesium-Based Scaffolds

Magnesium and its alloys are particularly effective in hard tissue engineering because their mechanical properties closely mimic natural bone, and their inherent ionic content plays critical roles in physiological functions. Their ability to biodegrade in body fluids makes them excellent candidates for creating biocompatible, bioactive, and biodegradable scaffolds suited for load-bearing applications [131]. Researchers have extensively studied alloy systems such as Mg–Ca, Mg–Zn, and Mg–Sr to understand their microstructures, mechanical performance, and degradation behaviors

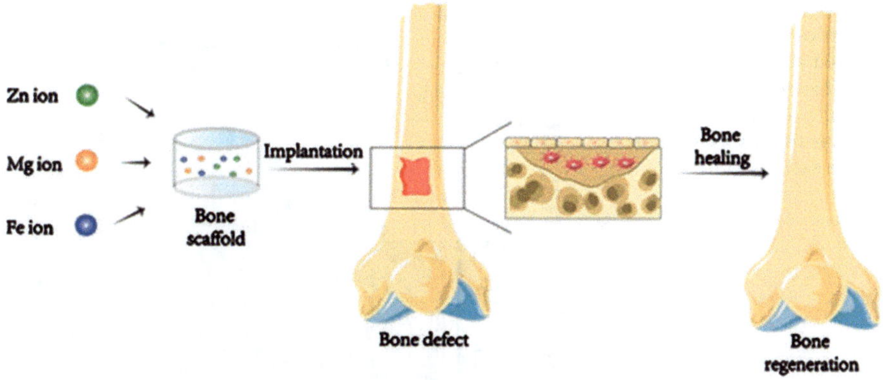

Fig. 13 Mg, Zn, and Fe ions and their role in bone regeneration [129]

[130]. Furthermore, the degradation rates of Mg-based alloys can be finely tuned by altering alloying elements, modifying microstructures, or applying surface treatments like calcium phosphate or polymer coatings [132, 133]. Recently, porous magnesium and its alloys have emerged as promising metallic scaffolds for bone tissue engineering, with ongoing refinements in fabrication techniques expected to further enhance their performance as hard tissue substitutes [131].

Because infections pose a significant risk in treating bone injuries, Mg-based composites with antibacterial properties have been developed as biodegradable biomaterials. For example, researchers have fabricated porous Mg–Ca–Mn–Zn–xAg (MCMZ–xAg) scaffolds with varying silver contents (0, 0.5 wt%, 1 wt%, and 2 wt%) using the space holder technique. The findings indicated that scaffolds with a low silver concentration (0.5 wt%) offered superior corrosion resistance, satisfactory mechanical properties, and excellent antibacterial activity against *E. coli* and *S. aureus*. However, higher silver levels increased cytotoxicity, underscoring the importance of optimizing silver content. The 0.5 wt% Ag scaffolds also demonstrated interconnected porosity and enhanced cell adhesion, making them promising candidates for bone repair and substitution [134]. Currently, Mg-based implants are utilized in various applications, including microclips for laryngeal microsurgery, orthopedic systems, and cardiovascular devices, illustrating their versatility in tissue engineering.

3.3.2 Zinc-Based Scaffolds

Zinc-based alloys are gaining popularity for bone tissue engineering due to their moderate degradation rates—positioned between those of magnesium and iron—which provide a balanced compromise between degradation and mechanical support. Recent studies have shown that porous Zn-based scaffolds offer good biocompatibility and more appropriate degradation rates compared to Mg-based or Fe-based

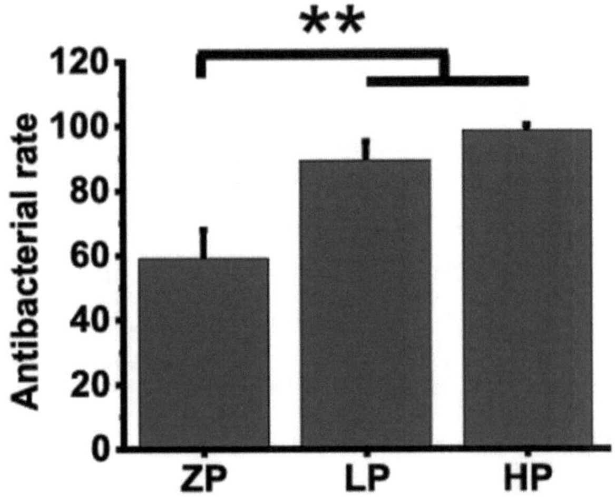

Fig. 14 Antibacterial properties of scaffolds cultured with *E. coli* for 24 h in Lysogeny Broth (LB) medium; zero porosity (ZP), low porosity (LP), and high porosity (HP) [136]

scaffolds. However, there remains a gap between the current performance of porous Zn-based scaffolds and the ideal requirements for bone tissue engineering [135].

Using additive manufacturing and casting techniques, Cockerill et al. produced Zn scaffolds with two distinct pore sizes (900 μm and 2 mm), achieving fully interconnected pore structures with precise topological control. Their results demonstrated that as pore size and overall porosity increased, mechanical strength decreased and corrosion rates accelerated. Nevertheless, ex vivo pretreatment improved cell adhesion and growth, and in vitro testing confirmed the scaffolds' good biocompatibility. Additionally, higher porosity was associated with significant antibacterial effects, as shown in Fig. 14, further supporting the use of Zn-based scaffolds for bone tissue engineering [136]. Despite these promising outcomes, further work is needed to optimize the biocompatibility, mechanical properties, and degradation behavior of Zn-based scaffolds for clinical applications.

3.3.3 Iron-Based Scaffolds

Iron-based alloys are highly regarded for their robust mechanical strength and slow degradation, which make them ideal for applications that require long-lasting structural support. As degradable metallic biomaterials, Fe-based alloys introduce a new class of bioactive implants that serve temporary roles. They are engineered to provide support during tissue healing and then gradually break down, potentially replacing the traditional corrosion-resistant metals used in orthopedic, cardiovascular, and pediatric implants [118]. However, pure iron degrades very slowly (about 0.16 mm/year in osteogenic conditions) and its ferromagnetic properties can be problematic for permanent implants. To overcome these challenges, researchers are developing innovative coatings—such as calcium phosphate-based ceramics and polymer layers—to boost the surface bioactivity of Fe-based alloys, which have historically been less effective at stimulating bone formation compared to bioinert anti-corrosion steels [137].

4 Conclusion and Future Scope

Biodegradable metallic materials mark a major advancement in biomedical engineering by providing solutions that not only work effectively but also harmonize with the body's natural healing processes. Continued research and development in this field hold the promise of revolutionizing medical implants, offering safer, more efficient, and patient-friendly options across a wide range of applications. This article discusses the biomaterials currently in use, pointing out potential shortcomings and offering solutions. These issues that present implants confront can be resolved by new or modified materials, which can provide biomaterials that reduce the risk of medical complications while also potentially providing patients with more realistic, aesthetically acceptable results. The biodegradability of bioresorbable materials cannot be confirmed by the available human in vivo research. This prevents the materials from being valued for marketing purposes. One of the numerous difficulties is that during the tissue remodeling process, the mechanical characteristics of magnesium-based alloys deteriorate more quickly than anticipated. To ascertain the mechanical and degradation dynamics of materials like magnesium- and iron-based alloys, more research is required.

The contact between the implant and host tissue should guarantee the biosafety of the deteriorating substance. The selection of the necessary bioresorbable composites and ceramic implants depends on a number of additional factors, including age, physical condition, infection risk, and fracture type. Nowadays, biomedical and pharmaceutical applications effectively employ biodegradable/bioresorbable polymers.

Biomaterials have become a cornerstone of modern medicine, driving innovations in diagnostics, therapeutics, and regenerative medicine. As the field evolves, the future of biomaterials in biomedical applications is vast and multidisciplinary, with a focus on innovation, sustainability, and patient-centric solutions. Key areas for future research include:

- **Advanced Alloy Design and Multifunctional Materials**: The creation of new alloys and multifunctional materials is crucial for meeting the diverse demands of biomedical applications. Future work should explore multi-principal element alloys (MPEAs) and high-entropy alloys (HEAs) to fine-tune mechanical properties, degradation rates, and biocompatibility.
- **Smart and Responsive Biomaterials**: The next generation of biomaterials will be intelligent and responsive, adapting to physiological changes. Incorporating shape-memory alloys and self-healing materials can enhance devices like stents, while integrating biosensors for real-time monitoring of parameters (such as glucose or inflammation markers) represents a promising direction.

- **Additive Manufacturing and 3D Bioprinting**: Additive manufacturing (AM) and 3D bioprinting are transforming the fabrication of biomaterials, allowing for the creation of complex, patient-specific structures. Future research should focus on improving techniques like laser powder bed fusion (LPBF) and fused deposition modeling (FDM) to produce implants with superior mechanical and biological properties.
- **Nanotechnology and Nanomaterials**: Nanotechnology offers exciting opportunities to enhance biomaterials. Nanomaterials with high surface areas and unique properties can improve applications in drug delivery, imaging, and tissue engineering.
- **Surface Modification and Coatings**: Surface engineering is critical for boosting the performance of biomaterials. Techniques such as plasma electrolytic oxidation (PEO), chemical vapor deposition (CVD), and electrochemical coatings can improve corrosion resistance, biocompatibility, and bioactivity. Future research should aim to develop multilayer coatings that combine mechanical strength with additional therapeutic functions, like drug delivery or antibacterial properties.
- **AI-Driven Design**: Artificial intelligence can accelerate the screening of material properties and aid in the development of personalized biomaterials tailored to the specific needs of individual patients.

References

1. Lane WA (1895) Some remarks on the treatment of fractures. BMJ 1(1790):861
2. Lambotte A (1909) Technique et indication des prothèses dans le traitement des fractures. Presse Med 17:321
3. Sherman WO (1912) Vanadium steel bone plates and screws. Surg Gynecol Obstet 14:629–634
4. Niinomi M (ed) (2019) Metals for biomedical devices. Woodhead Publishing
5. Nagel JH (2000) The biomedical engineering handbook
6. Hermawan H, Hermawan H (2012) Introduction to metallic biomaterials. In: Biodegradable metals: from concept to applications. pp 1–11
7. Yang K, Ren Y (2010) Nickel-free austenitic stainless steels for medical applications. Sci Technol Adv Mater 11(1):014105
8. Amirtharaj Mosas KK, Chandrasekar AR, Dasan A, Pakseresht A, Galusek D (2022) Recent advancements in materials and coatings for biomedical implants. Gels 8(5):323
9. Vaiani L, Boccaccio A, Uva AE, Palumbo G, Piccininni A, Guglielmi P et al (2023) Ceramic materials for biomedical applications: an overview on properties and fabrication processes. J Funct Biomater 14(3):146
10. Allavikutty R, Gupta P, Santra TS, Rengaswamy J (2021) Additive manufacturing of Mg alloys for biomedical applications: current status and challenges. Curr Opin Biomed Eng 18:100276
11. Zhang T, Wang W, Liu J, Wang L, Tang Y, Wang K (2022) A review on magnesium alloys for biomedical applications. Front Bioeng Biotechnol 10:953344
12. Dutta S, Gupta S, Roy M (2020) Recent developments in magnesium metal–matrix composites for biomedical applications: a review. ACS Biomater Sci Eng 6(9):4748–4773
13. Yang Y, He C, Dianyu E, Yang W, Qi F, Xie D et al (2020) Mg bone implant: features, developments and perspectives. Mater Des 185:108259

14. Laires MJ, Monteiro CP, Bicho M (2004) Role of cellular magnesium in health and human disease. Front Biosci 9(262):76
15. Dong J, Lin T, Shao H, Wang H, Wang X, Song K, Li Q (2022) Advances in degradation behavior of biomedical magnesium alloys: a review. J Alloy Compd 908:164600
16. Lu Y, Deshmukh S, Jones I, Chiu YL (2021) Biodegradable magnesium alloys for orthopaedic applications. Biomater Transl 2(3):214
17. Amukarimi S, Mozafari M (2021) Biodegradable magnesium-based biomaterials: an overview of challenges and opportunities. MedComm 2(2):123–144
18. Agarwal S, Curtin J, Duffy B, Jaiswal S (2016) Biodegradable magnesium alloys for orthopaedic applications: a review on corrosion, biocompatibility and surface modifications. Mater Sci Eng, C 68:948–963
19. Mohamed A, El-Aziz AM, Breitinger HG (2019) Study of the degradation behavior and the biocompatibility of Mg–0.8 Ca alloy for orthopedic implant applications. J Magnes Alloy 7(2):249–257
20. Ahmadi M, Tabary SB, Rahmatabadi D, Ebrahimi MS, Abrinia K, Hashemi R (2022) Review of selective laser melting of magnesium alloys: advantages, microstructure and mechanical characterizations, defects, challenges, and applications. J Mater Res Technol 19:1537–1562
21. Barbosa TP, Naves MM, Menezes HHM, Pinto PHC, de Mello JDB, Costa HL (2017) Topography and surface energy of dental implants: a methodological approach. J Braz Soc Mech Sci Eng 39:1895–1907
22. Augustin J, Feichtner F, Waselau AC, Julmi S, Klose C, Wriggers P et al (2022) Effect of pore size on tissue ingrowth and osteoconductivity in biodegradable Mg alloy scaffolds. J Appl Biomater Funct Mater 20:22808000221078170
23. Becerra LHC, Rodríguez MALH, Solís HE, Arroyo RL, Castro AT (2020) Bio-inspired biomaterial Mg–Zn–Ca: a review of the main mechanical and biological properties of Mg-based alloys. Biomed Phys Eng Express 6(4):042001
24. Teoh SH (2000) Fatigue of biomaterials: a review. Int J Fatigue 22(10):825–837
25. Al-Maharma AY, Patil SP, Markert B (2020) Effects of porosity on the mechanical properties of additively manufactured components: a critical review. Mater Res Express 7(12):122001
26. Dehghanghadikolaei A, Ibrahim H, Amerinatanzi A, Elahinia M (2019) Biodegradable magnesium alloys. In: Metals for biomedical devices. Woodhead Publishing, pp 265–289
27. Antoniac I, Miculescu M, Mănescu V, Stere A, Quan PH, Păltânea G et al (2022) Magnesium-based alloys used in orthopedic surgery. Materials 15(3):1148
28. Zhang J, Yang S, Yang X, Xi Z, Zhao L, Cen L et al (2018) Novel fabricating process for porous polyglycolic acid scaffolds by melt-foaming using supercritical carbon dioxide. ACS Biomater Sci Eng 4(2):694–706
29. Shuai C, Li Y, Wang G, Yang W, Peng S, Feng P (2019) Surface modification of nanodiamond: toward the dispersion of reinforced phase in poly-L-lactic acid scaffolds. Int J Biol Macromol 126:1116–1124
30. Panayotov IV, Orti V, Cuisinier F, Yachouh J (2016) Polyetheretherketone (PEEK) for medical applications. J Mater Sci—Mater Med 27:1–11
31. Song J, Chen J, Xiong X, Peng X, Chen D, Pan F (2022) Research advances of magnesium and magnesium alloys worldwide in 2021. J Magnes Alloys 10(4):863–898
32. Neil WC, Forsyth M, Howlett PC, Hutchinson CR, Hinton BRW (2009) Corrosion of magnesium alloy ZE41–the role of microstructural features. Corros Sci 51(2):387–394
33. Song GL, Atrens A (1999) Corrosion mechanisms of magnesium alloys. Adv Eng Mater 1(1):11–33
34. Smith CE, Xu Z, Waterman J, Sankar J (2013) Cytocompatibility assessment of MgZnCa alloys. Emerg Mater Res 2(5):283–290
35. Atrens A, Liu M, Abidin NIZ (2011) Corrosion mechanism applicable to biodegradable magnesium implants. Mater Sci Eng, B 176(20):1609–1636
36. Witte F, Hort N, Vogt C, Cohen S, Kainer KU, Willumeit R, Feyerabend F (2008) Degradable biomaterials based on magnesium corrosion. Curr Opin Solid State Mater Sci 12(5–6):63–72

37. Anderson JM, Rodriguez A, Chang DT (2008, April) Foreign body reaction to biomaterials. In: Seminars in immunology, vol 20, no 2. Academic Press, pp 86–100
38. Kuhlmann J, Bartsch I, Willbold E, Schuchardt S, Holz O, Hort N et al (2013) Fast escape of hydrogen from gas cavities around corroding magnesium implants. Acta Biomater 9(10):8714–8721
39. Mcbride ED (1938) Absorbable metal in bone surgery: a further report on the use of magnesium alloys. J Am Med Assoc 111(27):2464–2467
40. Zhang S, Li J, Song Y, Zhao C, Zhang X, Xie C et al (2009) In vitro degradation, hemolysis and MC3T3-E1 cell adhesion of biodegradable Mg–Zn alloy. Mater Sci Eng, C 29(6):1907–1912
41. Wang J, Liu L, Wu Y, Maitz MF, Wang Z, Koo Y et al (2017) Ex vivo blood vessel bioreactor for analysis of the biodegradation of magnesium stent models with and without vessel wall integration. Acta Biomater 50:546–555
42. Liu L, Koo Y, Collins B, Xu Z, Sankar J, Yun Y (2017) Biodegradability and platelets adhesion assessment of magnesium-based alloys using a microfluidic system. PLoS ONE 12(8):e0182914
43. CP M (1942) Chemical gas gangrene from metallic magnesium. Indust Med 11:71–78
44. Rodewald UC, Chevalier B, Poettgen R (2007) Rare earth–transition metal–magnesium compounds—an overview. J Solid State Chem 180(5):1720–1736
45. Schwartz MA, Schaller MD, Ginsberg MH (1995) Integrins: emerging paradigms of signal transduction. Annu Rev Cell Dev Biol 11:549–599
46. Grzesik WJ (1997) Integrins and bone–cell adhesion and beyond. Arch Immunol Ther Exp 45(4):271–275
47. Vroman L, Adams AL, Fischer GC, Munoz PC (1980) Interaction of high molecular weight kininogen, factor XII, and fibrinogen in plasma at interfaces
48. Coté CJ (ed) (1993) A practice of anesthesia for infants and children. Saunders
49. Ehtemam-Haghighi S, Liu Y, Cao G, Zhang LC (2016) Phase transition, microstructural evolution and mechanical properties of Ti–Nb–Fe alloys induced by Fe addition. Mater Des 97:279–286
50. Ehtemam-Haghighi S, Prashanth KG, Attar H, Chaubey AK, Cao GH, Zhang LC (2016) Evaluation of mechanical and wear properties of TixNb7Fe alloys designed for biomedical applications. Mater Des 111:592–599
51. Seitz JM, Durisin M, Goldman J, Drelich JW (2015) Recent advances in biodegradable metals for medical sutures: a critical review. Adv Healthcare Mater 4(13):1915–1936
52. Wang C, Yang HT, Li X, Zheng YF (2016) In vitro evaluation of the feasibility of commercial Zn alloys as biodegradable metals. J Mater Sci Technol 32(9):909–918
53. Törne K, Larsson M, Norlin A, Weissenrieder J (2016) Degradation of zinc in saline solutions, plasma, and whole blood. J Biomed Mater Res B Appl Biomater 104(6):1141–1151
54. Plum LM, Rink L, Haase H (2010) The essential toxin: impact of zinc on human health. Int J Environ Res Public Health 7(4):1342–1365
55. Liu X, Sun J, Qiu K, Yang Y, Pu Z, Li L, Zheng Y (2016) Effects of alloying elements (Ca and Sr) on microstructure, mechanical property and in vitro corrosion behavior of biodegradable Zn–1.5 Mg alloy. J Alloy Compd 664:444–452
56. Ji C, Ma A, Jiang J, Song D, Liu H, Guo S (2024) Research status and future prospects of biodegradable Zn-Mg alloys. J Alloys Compd 174669
57. Niu J, Tang Z, Huang H, Pei J, Zhang H, Yuan G, Ding W (2016) Research on a Zn–Cu alloy as a biodegradable material for potential vascular stents application. Mater Sci Eng, C 69:407–413
58. Gu XN, Zheng YF (2010) A review on magnesium alloys as biodegradable materials. Front Mater Sci Chin 4:111–115
59. Katarivas Levy G, Goldman J, Aghion E (2017) The prospects of zinc as a structural material for biodegradable implants—a review paper. Metals 7(10):402
60. Gong H, Wang K, Strich R, Zhou JG (2015) In vitro biodegradation behavior, mechanical properties, and cytotoxicity of biodegradable Zn–Mg alloy. J Biomed Mater Res B Appl Biomater 103(8):1632–1640

61. Li H, Yang H, Zheng Y, Zhou F, Qiu K, Wang X (2015) Design and characterizations of novel biodegradable ternary Zn-based alloys with IIA nutrient alloying elements Mg, Ca and Sr. Mater Des 83:95–102
62. Roohani N, Hurrell R, Kelishadi R, Schulin R (2013) Zinc and its importance for human health: an integrative review. J Res Med Sci 18(2):144
63. Jackson MJ (1989) Physiology of zinc: general aspects. Zinc in human biology. Springer, London, pp 1–14
64. Zheng YF, Gu XN, Witte FJMS (2014) Biodegradable metals. Mater Sci Eng R Rep 77:1–34
65. Kaur G, Kaur G (2017) Biodegradable metals as bioactive materials. In: Bioactive glasses: potential biomaterials for future therapy. pp 77–98
66. Bowen PK, Drelich J, Goldman J (2013) Zinc exhibits ideal physiological corrosion behavior for bioabsorbable stents
67. Li P, Zhang W, Dai J, Xepapadeas AB, Schweizer E, Alexander D et al (2019) Investigation of zinc-copper alloys as potential materials for craniomaxillofacial osteosynthesis implants. Mater Sci Eng, C 103:109826
68. Su Y, Wang K, Gao J, Yang Y, Qin YX, Zheng Y, Zhu D (2019) Enhanced cytocompatibility and antibacterial property of zinc phosphate coating on biodegradable zinc materials. Acta Biomater 98:174–185
69. Kraus T, Moszner F, Fischerauer S, Fiedler M, Martinelli E, Eichler J et al (2014) Biodegradable Fe-based alloys for use in osteosynthesis: outcome of an in vivo study after 52 weeks. Acta Biomater 10(7):3346–3353
70. Sam S, Joseph B, Thomas S (2023) Exploring the antimicrobial features of biomaterials for biomedical applications. Results Eng 17:100979
71. Kubásek J, Vojtěch D, Jablonská E, Pospíšilová I, Lipov J, Ruml T (2016) Structure, mechanical characteristics and in vitro degradation, cytotoxicity, genotoxicity and mutagenicity of novel biodegradable Zn–Mg alloys. Mater Sci Eng, C 58:24–35
72. Yin YX, Zhou C, Shi YP, Shi ZZ, Lu TH, Hao Y et al (2019) Hemocompatibility of biodegradable Zn-0.8 wt%(cu, Mn, Li) alloys. Mater Sci Eng C 104:109896
73. Zhou C, Li HF, Yin YX, Shi ZZ, Li T, Feng XY et al (2019) Long-term in vivo study of biodegradable Zn–Cu stent: a 2-year implantation evaluation in porcine coronary artery. Acta Biomater 97:657–670
74. Hehrlein C, Schorch B, Kress N, Arab A, von Zur Mühlen C, Bode C, Epting T, Haberstroh J, Mey L, Schwarzbach H, Kinscherf R, Stachniss V, Schiestel S, Kovacs A, Fischer H, Nennig E. Zn-alloy provides a novel platform for mechanically stable bioresorbable vascular stents. PLoS One. 2019 Jan 2;14(1):e0209111. https://doi.org/10.1371/journal.pone.0209111. PMID: 30601854; PMCID: PMC6314592
75. Chen Z, Jin M, He H, Dong J, Li J, Nie J et al (2023) Mesenchymal stem cells and macrophages and their interactions in tendon-bone healing. J Orthop Transl 39:63–73
76. Yao R, Wang H, Shan R, Liu L, Zhao Y, Sun Y et al (2023) Biodegradable porous Zn-1Mg-3βTCP scaffold for bone defect repair: in vitro and in vivo evaluation. J Mater Sci Technol 162:189–202
77. Chou DT, Wells D, Hong D, Lee B, Kuhn H, Kumta PN (2013) Novel processing of iron–manganese alloy-based biomaterials by inkjet 3-D printing. Acta Biomater 9(10):8593–8603
78. Hermawan H, Purnama A, Dube D, Couet J, Mantovani D (2010) Fe–Mn alloys for metallic biodegradable stents: degradation and cell viability studies. Acta Biomater 6(5):1852–1860
79. Hong D, Chou DT, Velikokhatnyi OI, Roy A, Lee B, Swink I et al (2016) Binder-jetting 3D printing and alloy development of new biodegradable Fe-Mn-Ca/Mg alloys. Acta Biomater 45:375–386
80. Vaz MF, Canhão H, Fonseca JE (2011) Bone: a composite natural material. In: Advances in composite materials—analysis of natural and man-made materials
81. Hermawan H, Moravej M, Dubé D, Fiset M, Mantovani D (2007) Degradation behaviour of metallic biomaterials for degradable stents. Adv Mater Res 15:113–118
82. Sharma P, Pandey PM (2019) Corrosion rate modelling of biodegradable porous iron scaffold considering the effect of porosity and pore morphology. Mater Sci Eng, C 103:109776

83. Francis A, Yang Y, Virtanen S, Boccaccini AR (2015) Iron and iron-based alloys for temporary cardiovascular applications. J Mater Sci—Mater Med 26(3):1–16. https://doi.org/10.1007/S10856-015-5473-8/METRICS
84. Liu B, Zheng YF (2011) Effects of alloying elements (Mn Co, Al, W, Sn, B, C and S) on biodegradability and in vitro biocompatibility of pure iron. Acta Biomater 7(3):1407–1420
85. Moravej M, Purnama A, Fiset M, Couet J, Mantovani D (2010) Electroformed pure iron as a new biomaterial for degradable stents: in vitro degradation and preliminary cell viability studies. Acta Biomater 6(5):1843–1851
86. Jurgeleit T, Quandt E, Zamponi C (2015) Magnetron sputtering a new fabrication method of iron based biodegradable implant materials. Adv Mater Sci Eng 2015(1):294686
87. Chen CZ, Shi XH, Zhang PC, Bai B, Leng YX, Huang N (2008) The microstructure and properties of commercial pure iron modified by plasma nitriding. Solid State Ionics 179(21–26):971–974
88. Lin W, Qin L, Qi H, Zhang D, Zhang G, Gao R et al (2017) Long-term in vivo corrosion behavior, biocompatibility and bioresorption mechanism of a bioresorbable nitrided iron scaffold. Acta Biomater 54:454–468
89. Xu X, Li M, Liu Q, Jia Z, Shi Y, Cheng Y et al (2014) Facile immobilization of heparin on bioabsorbable iron via mussel adhesive protein (MAPs). Prog Nat Sci: Mater Int 24(5):458–465
90. Schinhammer M, Gerber I, Hänzi AC, Uggowitzer PJ (2013) On the cytocompatibility of biodegradable Fe-based alloys. Mater Sci Eng, C 33(2):782–789
91. Oriňaková R, Gorejová R, Macko J, Oriňak A, Kupková M, Hrubovčáková M et al (2019) Evaluation of in vitro biocompatibility of open cell iron structures with PEG coating. Appl Surf Sci 475:515–518
92. Peuster M, Hesse C, Schloo T, Fink C, Beerbaum P, von Schnakenburg C (2006) Long-term biocompatibility of a corrodible peripheral iron stent in the porcine descending aorta. Biomaterials 27(28):4955–4962
93. Ulum MF, Arafat A, Noviana D, Yusop AH, Nasution AK, Kadir MA, Hermawan H (2014) In vitro and in vivo degradation evaluation of novel iron-bioceramic composites for bone implant applications. Mater Sci Eng, C 36:336–344
94. V.P., M. R., & Hanas, T. (2022). Enhancing biointerfacial properties of porous pure iron by gold sputtering for degradable implant applications. Mater Today Commun 31:103492
95. Prakasam M, Locs J, Salma-Ancane K, Loca D, Largeteau A, Berzina-Cimdina L (2017) Biodegradable materials and metallic implants—a review. J Funct Biomater 8(4):44. https://doi.org/10.3390/JFB8040044
96. Oladele IO, Onuh LN, Agbeboh NI, Alewi DD, Lephuthing SS (2023) The relationship and functional links between human age, growth, and biomedical implants: a review on the application of bulk and nanomaterials. Nano Select 4(7):419–441
97. Aggarwal D, Kumar V, Sharma S (2022) Drug-loaded biomaterials for orthopedic applications: a review. J Control Release 344:113–133. https://doi.org/10.1016/J.JCONREL.2022.02.029
98. Azadani MN, Zahedi A, Oluwole, Bowoto K, Bankole, Oladapo I (2022) A review of current challenges and prospects of magnesium and its alloy for bone implant applications. Prog Biomater 11(1):1–26. https://doi.org/10.1007/s40204-022-00182-x
99. Ding W (2016) Opportunities and challenges for the biodegradable magnesium alloys as next-generation biomaterials. Regenerative Biomater 3(2):79–86. https://doi.org/10.1093/RB/RBW003
100. Venezuela J, Dargusch MS (2019) The influence of alloying and fabrication techniques on the mechanical properties, biodegradability and biocompatibility of zinc: a comprehensive review. Acta Biomater 87:1–40. https://doi.org/10.1016/j.actbio.2019.01.035
101. Jin W, Chu PK (2019) Orthopedic implants. Encycl Biomed Eng 1(3):425–439
102. Bairagi D, Mandal S (2022) A comprehensive review on biocompatible Mg-based alloys as temporary orthopaedic implants: current status, challenges, and future prospects. J Magnes Alloys 10(3):627–669

103. Radha R, Sreekanth D (2017) Insight of magnesium alloys and composites for orthopedic implant applications—a review. J Magnes Alloys 5(3):286–312. https://doi.org/10.1016/J.JMA.2017.08.003
104. Mahmud S, Rahman M, Kamruzzaman M, Khatun H, Ali MO, Haque MM (2023) Recent developments in hydroxyapatite coating on magnesium alloys for clinical applications. Results Eng 17:101002. https://doi.org/10.1016/J.RINENG.2023.101002
105. Zhu Y, Liu W, Ngai T (2022) Polymer coatings on magnesium-based implants for orthopedic applications. J Polym Sci 60(1):32–51. https://doi.org/10.1002/POL.20210578
106. Santos MJA, Júnior WO, Ribeiro JKL, Maciel NC, Bastos IN, Pereira JN, da Costa Souza PN, Pinto HC, da Silva EP (2024) Zinc oxide coating impact on corrosion of ZK60 magnesium alloys in simulated body fluid. J Alloy Compd 1002:175339. https://doi.org/10.1016/J.JALLCOM.2024.175339
107. Kong J, Kolooshani A, Kolahdouz A, Nejad MG, Toghraie D (2024) Fabrication and characterization of magnesium implants coated with magnetic nanoparticles-wollastonite-hydroxyapatite for medical and sports injury applications: finite element analysis. Ceram Int 50(3):5755–5765. https://doi.org/10.1016/J.CERAMINT.2023.11.364
108. He X, Li Y, Zou D, Zu H, Li W, Zheng Y (2024) An wittein orthopaedics and a prospect of its application in spine fusion. Bioactive Mater 39:456–478. https://doi.org/10.1016/J.BIOACTMAT.2024.04.026
109. Manescu V, Paltanea G, Antoniac A, Gruionu LG, Robu A, Vasilescu M, Laptoiu SA, Bita AI, Popa GM, Cocosila AL, Silviu V, Porumb A (2024) Mechanical and computational fluid dynamic models for magnesium-based implants. Materials 17(4):830. https://doi.org/10.3390/MA17040830
110. Imani A (2024) Corrosion behaviour of a biodegradable rare-earth magnesium alloy in modified simulated body fluid. University of British Columbia
111. Chen K, Gu X, Zheng Y (2024) Feasibility, challenges and future prospects of biodegradable zinc alloys as orthopedic internal fixation implants. Smart Mater Manuf 2:100042. https://doi.org/10.1016/J.SMMF.2023.100042
112. Kong L, Heydari Z, Lami GH, Saberi A, Baltatu MS, Vizureanu P (2023) A comprehensive review of the current research status of biodegradable zinc alloys and composites for biomedical applications. Materials 16(13):4797. https://doi.org/10.3390/MA16134797
113. Xiao X, Wang B, Liu E, Liu H, Liu L, Xu W, Ge S, Shao J (2023) Investigation of zinc-silver alloys as biodegradable metals for orthopedic applications. J Market Res 26:6287–6303. https://doi.org/10.1016/J.JMRT.2023.09.025
114. Jia B, Yang H, Han Y, Zhang Z, Qu X, Zhuang Y, Wu Q, Zheng Y, Dai K (2020) In vitro and in vivo studies of Zn-Mn biodegradable metals designed for orthopedic applications. Acta Biomater 108:358–372. https://doi.org/10.1016/J.ACTBIO.2020.03.009
115. Diaa AA, El-Mahallawy N, Shoeib M, Mouillard F, Ferté T, Masson P, Carradò A (2024) Biodegradable PMMA coated Zn–Mg alloy with bimodal grain structure for orthopedic applications—a promising alternative. Bioactive Mater 39:479–491. https://doi.org/10.1016/J.BIOACTMAT.2024.05.031
116. Yang H, Jia B, Qu X, Dai K, Zheng Y (2022) Modified biodegradation behavior induced beneficial microenvironments for bone regeneration by low addition of gadolinium in zinc. Adv Healthcare Mater 11(21):2201184. https://doi.org/10.1002/ADHM.202201184
117. Anand N, Manglam MK, Kar M (2024) Investigation of biodegradable zinc-based alloy with varying compositions of aluminum and copper for orthopedic applications. J Mater Eng Performance 1–10.https://doi.org/10.1007/S11665-024-10171-9/METRICS
118. Gorejová R, Haverová L, Oriňaková R, Oriňak A, Oriňak M (2019) Recent advancements in Fe-based biodegradable materials for bone repair. J Mater Sci 54(3):1913–1947. https://doi.org/10.1007/S10853-018-3011-Z/METRICS
119. Heiden M, Walker E, Stanciu L (2015) Magnesium, iron and zinc alloys, the trifecta of bioresorbable orthopaedic and vascular implantation—a review. J Biotechnol Biomater 5(2):1
120. Oriňáková R, Oriňák A, Bučková LM, Giretová M, Medvecký L, Labbanczová E, Kupková M, Hrubovčáková M, Koval K (2013) Iron based degradable foam structures for potential

orthopedic applications. Int J Electrochem Sci 8(12):12451–12465. https://doi.org/10.1016/S1452-3981(23)13279-2
121. Liu P, Huang Q, Shan Q, Jiao Z, Wang Q, Ma Y, Zhou R, Baker I, Wu H (2024) Achieving exceptional work-hardening capability of additively-manufactured multiphase Fe–Mn alloys via multiple deformation mechanisms. Int J Plast 173:103871. https://doi.org/10.1016/J.IJPLAS.2023.103871
122. Dag IE, Kilic F, Panigrahi M, Avar B (2025) Microstructure, mechanical properties, and antibacterial performance of novel Fe–Mn–Zn nanocrystalline alloys produced by mechanical alloying. Adv Eng Mater 2500049.https://doi.org/10.1002/ADEM.202500049
123. Mostaed E, Sikora-Jasinska M, Drelich JW, Vedani M (2018) Zinc-based alloys for degradable vascular stent applications. Acta Biomater 71:1. https://doi.org/10.1016/J.ACTBIO.2018.03.005
124. Udrişte AS, Burduşel AC, Niculescu AG, Rădulescu M, Grumezescu AM (2024) Coatings for cardiovascular stents—an up-to-date review. Int J Mol Sci 25(2):1078
125. Zhang ZQ, Yang YX, Li JA, Zeng RC, Guan SK (2021) Advances in coatings on magnesium alloys for cardiovascular stents—a review. Bioactive Mater 6(12):4729–4757
126. Niu J, Huang H, Pei J, Jin Z, Guan S, Yuan G (2021) Research and development strategy for biodegradable magnesium-based vascular stents: a review. Biomater Transl 2(3):236
127. Zhang H, Zhang W, Qiu H, Zhang G, Li X, Qi H, Guo J, Qian J, Shi X, Gao X, Shi D, Zhang D, Gao R, Ding J (2022) A biodegradable metal-polymer composite stent safe and effective on physiological and serum-containing biomimetic conditions. Adv Healthcare Mater 11(22). https://doi.org/10.1002/adhm.202201740
128. Hutmacher DW, Goh JCH, Teoh SH (2001) An introduction to biodegradable materials for tissue engineering applications. Ann Acad Med Singapore. https://annals.edu.sg/an-introduction-to-biodegradable-materials-for-tissue-engineering-applications-2/
129. Chen Z, Zhang W, Wang M, Backman LJ, Chen J (2022) Effects of zinc, magnesium, and iron ions on bone tissue engineering. ACS Biomater Sci Eng 8(6):2321–2335. https://doi.org/10.1021/ACSBIOMATERIALS.2C00368
130. Modrák M, Trebuňová M, Balogová AF, Hudák R, Živčák J (2023) Biodegradable materials for tissue engineering: development, classification and current applications. J Funct Biomater 14(3):159. https://doi.org/10.3390/JFB14030159
131. Yazdimamaghani M, Razavi M, Vashaee D, Moharamzadeh K, Boccaccini AR, Tayebi L (2017) Porous magnesium-based scaffolds for tissue engineering. Mater Sci Eng, C 71:1253–1266. https://doi.org/10.1016/J.MSEC.2016.11.027
132. Uddin MS, Hall C, Murphy P (2015) Surface treatments for controlling corrosion rate of biodegradable Mg and Mg-based alloy implants. Sci Technol Adv Mater 16(5):24. https://doi.org/10.1088/1468-6996/16/5/053501
133. Dorozhkin SV (2014) Calcium orthophosphate coatings on magnesium and its biodegradable alloys. Acta Biomater 10(7):2919–2934. https://doi.org/10.1016/J.ACTBIO.2014.02.026
134. Bakhsheshi-Rad HR, Dayaghi E, Ismail AF, Aziz M, Akhavan-Farid A, Chen X (2019) Synthesis and in-vitro characterization of biodegradable porous magnesium-based scaffolds containing silver for bone tissue engineering. Trans Nonferrous Metals Soc China 29(5):984–996. https://doi.org/10.1016/S1003-6326(19)65007-7
135. Zhao L, Yuan P, Zhang M, Wang X, Qi Y, Wang T, Cao B, Cui C (2023) Preparation and properties of porous Zn-based scaffolds as biodegradable implants: a review. J Mater Sci 58(20):8275–8316. https://doi.org/10.1007/S10853-023-08561-W
136. Cockerill I, Su Y, Sinha S, Qin YX, Zheng Y, Young ML, Zhu D (2020) Porous zinc scaffolds for bone tissue engineering applications: a novel additive manufacturing and casting approach. Mater Sci Eng, C 110:110738. https://doi.org/10.1016/J.MSEC.2020.110738
137. Schinhammer M, Hänzi AC, Löffler JF, Uggowitzer PJ (2010) Design strategy for biodegradable Fe-based alloys for medical applications. Acta Biomater 6(5):1705–1713. https://doi.org/10.1016/J.ACTBIO.2009.07.039

Chapter 2
Composition and Design Strategies for Iron-Based Biodegradable Alloys

Shubhra Dixit, Dayanidhi K. Pathak, Anurag Dixit, and R. K. Mishra

1 Introduction

A notable achievement in the realm of biomaterials is the creation of biodegradable materials for medical purposes, which provides an innovative substitute for temporary implantable devices. Because of their advantageous mechanical characteristics and biocompatibility, iron-based biodegradable materials, in particular, have become attractive options for use in tissue engineering, orthopedics, and cardiovascular applications. These materials' main goal is to offer mechanical support while the body heals and then safely degrades within it. This eliminates the need for a second surgery to remove the implant and lowers the possibility of long-term issues that are frequently linked to permanent implants [1, 2].

There are numerous important benefits of using iron (Fe) as a base metal in biodegradable systems. It is a vital trace element involved in many physiological functions, such as oxygen transport and cellular metabolism, and it has a high tensile strength and exceptional ductility. Pure iron does, however, have a major drawback: its naturally slow rate of degradation in physiological settings is far lower than the rate needed for therapeutic applications [3]. Therefore, researchers have used a variety of composition and design tactics to speed up iron's degradation while maintaining or even improving its mechanical and biological performance in order to make it appropriate for temporary biomedical implants.

S. Dixit
Amity University, Noida, Uttar Pradesh, India

D. K. Pathak (✉) · A. Dixit
G.B Pant DSEU Campus, New Delhi, India
e-mail: dayanidhikrishana.pathak@dseu.ac.in

R. K. Mishra
Shri Mata Vaishno Devi University, Jammu, India

A. Dixit et al. (eds.), *Biodegradable Metallic Materials*, Biomedical Materials for Multi-functional Applications, https://doi.org/10.1007/978-981-95-2401-3_2

Alloying is one of the most extensively studied methods for addressing the gradual deterioration of pure iron. Iron's microstructure, biological characteristics, and corrosion behavior can all be systematically modified by adding alloying elements. The following elements have been studied for this purpose: manganese (Mn), carbon (C), palladium (Pd), silver (Ag), calcium (Ca), zinc (Zn), and sulfur (S). For instance, adding manganese at 20–35 weight percent increases corrosion rates while simultaneously decreasing the alloy's magnetic susceptibility, which is advantageous for MRI compatibility [4].

Other elements, such as Pd and Ag, have two functions: they speed up corrosion and have antibacterial qualities, which makes them ideal for implants in areas where infections are common [5]. Furthermore, scientists have created Fe-based composites by mixing iron with bioactive ceramics (like hydroxyapatite) [6], and calcium silicate was found to have enhanced bioactivity and regulated degradation as compared to pure iron [7]. Iron's electrochemical potential can also be changed by adding rare earth elements or tiny amounts of noble metals, which promotes quicker and more consistent corrosion.

Studies on these alloys using thermodynamics and kinetics have shown that the degradation mechanism is greatly influenced by the presence of second-phase particles and micro-galvanic coupling. For example, carbide precipitates in Fe–Mn–C alloys can serve as cathodic sites that improve the anodic dissolution of the iron matrix locally [8]. These realizations are essential for optimizing the behavior of materials in physiological settings.

In addition to compositional modification, the performance of iron-based biodegradable materials is greatly influenced by microstructural and geometric design. One of the most successful design changes is porosity. Furthermore, to increase the surface area available for corrosion reactions, regulated porosity promotes tissue ingrowth and vascularization, both of which are necessary for the implant to successfully integrate with the host tissue [9].

Techniques including gas foaming, powder metallurgy, freeze casting, and 3D printing (additive manufacturing) can be used to create porous iron structures. The manufacture of patient-specific implants with optimal mechanical performance and degradation profiles is made possible by the additional advantage of fine architectural control that 3D printing provides [10]. Furthermore, gradient porosity designs made possible by sophisticated printing processes can mimic the inherent variety in mechanical characteristics present in soft tissues and bone.

Surface modification is another essential design approach. It has been demonstrated that coating iron-based materials with corrosion accelerators (such as biodegradable polymers loaded with nitric oxide donors or enzymes) or bioactive layers (such as collagen or calcium phosphate) can modify degradation rates while also improving cell adhesion, proliferation, and differentiation [11]. Furthermore, the surface can be nanostructured by anodization or laser ablation to encourage favorable interactions between biological tissues and the implant surface.

Corrosion behavior is also greatly impacted by grain refinement and crystallographic texture control, which are accomplished by thermomechanical processing or extreme plastic deformation. More grain boundaries from finer grains serve as anodic sites, speeding up the pace of degradation. Despite their complexity, these techniques provide a great deal of control over the final characteristics of the material [12].

The therapeutic effectiveness of iron-based biodegradable materials depends on their biocompatibility. Iron ions released during degradation are part of normal cellular metabolism, but local ion toxicity, oxidative stress, or tissue pH alteration may occur due to high ion concentrations. Therefore, degradation rate vs. ion release is a challenge [13]. In vivo studies have confirmed that inflammation against iron-based implants is typically mild and transient, though long-term deposition of degradation products—especially in the liver and spleen—is a problem to be tackled in upcoming research [2].

Iron-based stents used in cardiovascular applications must resorb within 6–12 months while retaining their radial strength as the arteries heal. In orthopedic environments, degradability of implants like screws or pins needs to occur over a longer duration (12–24 months) to facilitate bone remodeling. These user-specific demands highlight the necessity for tailored material solutions. In brief, iron-based biodegradable materials show great promise as next-generation temporary implants. Success in developing them relies on an integrated approach encompassing engineered composition, microstructural design, surface treatment, control of porosity and geometry, and refinement to a perfecting point of near-zero corrosion, along with meeting long-term safety standards. That challenges exist—a fine-tuning in corrosion rate being one—is unimportant: improvement in material and biomedical engineering increasingly brings them toward clinical realization.

The current chapter is intended to discuss different composition and design approaches for the production of iron-based biodegradable materials for biomedical applications like tissue regeneration, drug delivery, and smart sensing capabilities with modified mechanical behavior and enhanced degradation rates.

2 Fe-Based Alloys and Composites for Biomedical Applications

Iron has been used extensively as an engineering material because of its many benefits, including its remarkable mechanical qualities, cost-effectiveness, and machinability [14]. Because of these qualities, iron-based compounds are a great option for creating biodegradable materials for use in medicine. According to the literature, Fe-based materials are mainly used in three areas: scaffolds in bone tissue engineering [15], intravascular stents, and bone implant materials for surgical procedures [6]. In bone procedures, the use of nonbiodegradable metal implants, like screws and plates [16], has been linked to negative consequences, including toxicity [17]. Problems, including pain or abnormalities at the implantation site, may arise if such implants are removed. Because they have mechanical characteristics comparable to stainless steel and satisfy the requirements for bone healing, Fe-based metallic implants overcome these drawbacks [18]. Scaffolds that create ideal conditions for tissue restoration are another essential use for Fe-based materials. Porous scaffolds made of iron are mostly used to repair hard tissues like bones because of their exceptional mechanical integrity and biodegradability. The drawbacks of nonbiodegradable polymer scaffolds are addressed by scaffolds composed of iron-based materials [13]. Changing the composition of metals through a process known as alloying is a popular way to modify their corrosion behavior. To improve a property, alloying entails purposefully adding alloying elements, often known as contaminants. In order to improve the corrosion rate of iron, Schinhammer et al. [19] suggested two alloy design options. The first tactic is to introduce a less noble element at a concentration that is within the element's solubility limit in iron to increase the alloy's susceptibility to corrosion. This tactic works best when Fe is combined with the less noble Mn to form the active Fe–Mn alloy. Noble elements that can produce finely dispersed phases that can act cathodically toward the Fe matrix, such as palladium (Pd) or gold (Au), are added as part of the second tactic. The development of these micro-galvanic corrosion sites may be a useful strategy to accelerate the Fe matrix's overall rate of corrosion.

2.1 Structural Changes of Fe-Based Biomaterials

The rate of degradation of Fe-based materials is comparatively slow [20]. The biodegradation of Fe-based materials at the local implantation site is depicted in Fig. 1. Despite continuous corrosion, the Fe implant is encapsulated by fibrous connective tissues at the tissue level, which slows down biodegradation. In biological applications, increasing the rate of Fe degradation has been difficult [21]. To enhance the corrosion behavior of Fe, a variety of approaches have been used, including alloying, surface modification, adding another phase, and other processing techniques. Alloying is thought to be the most promising technique among these for improving Fe's resistance to corrosion [22]. The effects of alloying iron with other elements can change based on the alloying element's mechanical characteristics. The most appropriate components for creating biodegradable Fe alloys were found to be Mn, Al, C, W, B, Co, and S, among others examined by Lui and Zheng [23].

"Adding a second phase" is another efficient way to increase iron's resistance to corrosion. According to Cheng and Zheng [24], the corrosion potential was reduced by the addition of a second phase. Due to their solubility in the body, bioceramics such as TCP, HA, and BCP are frequently employed as a second phase. It has been observed that surface modification and bioceramics can improve the corrosion behavior and biocompatibility of iron implants. The development of an oxide layer on the surface of pure iron greatly improves corrosion resistance in Simulated Body Fluid (SBF) [25]. Additionally, the formation of a nitride layer in an aqueous solution of NaCl improves the mechanical stability and corrosion resistance of Fe stents [26]. Nitriding increased the corrosion density of pure Fe stents but was not very successful in improving corrosion behavior [27]. Low degradation products were produced by

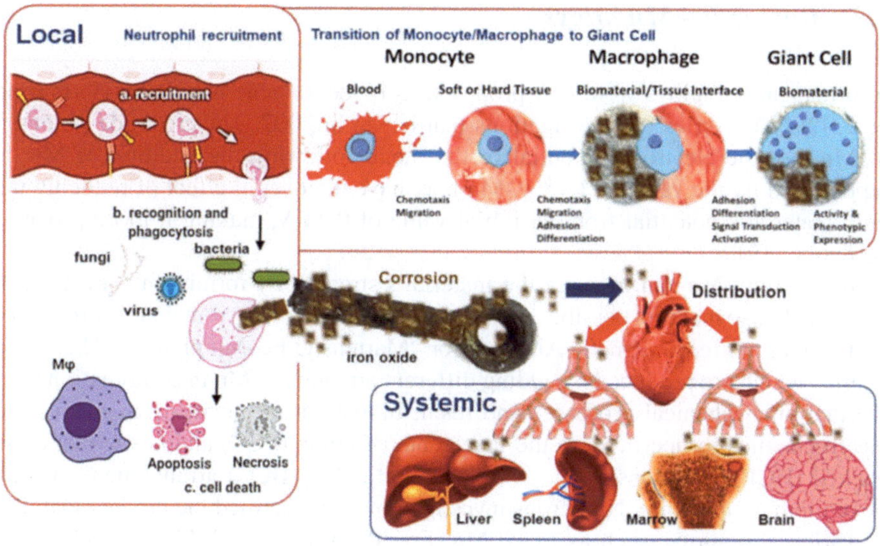

Fig. 1 Ion and Fe-oxide dispersion and active biodegradation of iron-based implants. Reused with permission from Elsevier [26]

nitrided iron-based coronary stents, which showed a gradual rate of degradation with nearly total corrosion occurring 13 months after implantation [2].

Fe stents may lose their mechanical integrity as a result of corrosion. On the surface of pure iron, nitride or oxide layers are formed to stop this. Microhardness, tensile strength, stiffness, and radial strength are all improved by this alteration [27]. Spark Plasma Sintering (SPS) is one of the most sophisticated sintering techniques. Fe materials made with SPS have exceptional corrosion resistance, whereas metallic materials made with this process have very poor mechanical qualities [28]. Additionally, rolling and electrodeposition improve Fe's mechanical qualities. Bioceramics have remarkable biocompatibility. One of the best biodegradable materials is the Fe-bioceramics combination [6]. Another benefit of powder metallurgy is that it permits the combination of metal and non-metal in any ratio. Pure iron's metallic qualities are greatly enhanced by mechanical treatments, particularly those that cause plastic deformation. These treatments can also slow down the corrosion process. It is strongly advised to create porous materials in order to accelerate corrosion. Because of its mechanical qualities, Fe can be produced as a metallic foam without losing its strength or form [9]. The forthcoming section describes various compositions of Fe-based materials for the application in bioengineering.

2.2 Binary Fe–Mn alloys

Iron-manganese alloys were first proposed as biodegradable metals by Hermawan et al. [29] in a series of investigations that began in 2007. The electrode potential of Mn, not the electrode potential of the metal following the addition of Mn, is represented by the value of 1.18 V. Mn may have the overall effect of reducing the bulk material's potential from an initial value of 0.44 V, making it more prone to corrosion [19].

In addition, Mn can increase the material's strength by forming a solid Fe–Mn solution. It also makes it possible to eliminate the material's ferromagnetic properties by forming antiferromagnetic Austenite or "Martensite Fe–Mn phases." These two phases, which are produced by adding different amounts of Mn in combination with thermal or mechanical processes and result in distinct mechanical, magnetic, and electrochemical properties, are the face-centered cubic (FCC) and hexagonal close packed (HCP) variations of the body-centered cubic (BCC) arrangement of pure Fe at room temperature. Numerous investigations have verified the biodegradability of various Fe alloys, such as Fe–20Mn, Fe–30Mn, and Fe–35Mn, in addition to Hermawan's study. There have also been reports of biodegradable phosphate Fe–30Mn (Fe/P–30Mn). Most of these research demonstrate that alloys containing Mn corrode more quickly than pure iron [4]. The fact that most of the Fe–Mn alloys under study have Mn concentrations of at least 20 weight percent is not surprising. This is due to the fact that at this Mn composition, the antiferromagnetic behavior of Fe–Mn is guaranteed. This magnetic characteristic is crucial because it makes the implant compatible with MRIs. Research suggests that the corrosion behavior of the Fe–Mn alloy is influenced by the Mn concentration. According to Hermawan et al. [30], cold-rolled Fe–Mn corroded more quickly when the Mn concentration rose steadily from 20 weight percent to 35 weight percent Mn. Furthermore, Dargusch et al. [31] found that in sintered porous Fe–Mn alloys (such as Fe–20Mn, Fe–30Mn, and Fe–35Mn), the rate of corrosion increased steadily as the Mn content increased. The corrosion behavior of sintered Fe–20Mn, Fe–30Mn, and Fe–35Mn was also examined by Kupkova et al. [32]. They found that raising the Mn concentration from 20 weight percent to 30 weight percent improved the alloy's corrosion rate, but raising it from 30 to 35 weight percent did not significantly speed up degradation.

2.3 Tertiary Fe–Mn–X alloys

2.3.1 Addition of C

Since the resulting alloy composition matches that of the twinning-induced plasticity (TWIP) steels, adding carbon (C) to Fe–Mn is a common method. Advanced high-strength steels with exceptional strength and ductility are known as TWIP steels. In addition to high concentrations of Mn (18–30%) and trace levels of other impurities like silicon (Si), aluminum (Al), and nitrogen (N), typical TWIP steels also contain Fe and C. The mechanical properties and formability of a biodegradable TWIP steel would then make it appealing. Reports on biodegradable TWIP alloys, including ternary and quaternary combinations, are included in this area [4]. The corrosion of Fe–Mn–C in various physiological fluids as well as in CO_2-rich and CO_2-lean fluids [33] was also confirmed. Regretfully, no inference regarding the impact of C on the rate of Fe–Mn deterioration can be drawn from the majority of the aforementioned research since they compared the TWIP alloys to Fe.

However, Xu et al. [34] verified that adding 1 weight percent C to Fe–30Mn increased the rate of corrosion and improved the antiferromagnetic characteristics. The alloy was made by hot forging after vacuum induction casting.

2.3.2 Addition of Pd, Ag, Cu

According to et al. [19], adding Pd to Fe–Mn can produce cathodic precipitates that accelerate the alloy's rate of corrosion. In a 48-hour immersion test in SBF, they verified that the addition of Pd (1 weight percent) increased the corrosion rate of Fe–10Mn by at least three times. The presence of the Pd-rich precipitates in the matrix was guaranteed by their distribution along dislocation lines, where they were shown to be generating accelerated corrosion. One interesting possibility is to include silver (Ag) in Fe–Mn alloys. Since Ag is immiscible in Fe at ambient temperature, a multi-phase system with distinct Ag particles embedded in the Fe matrix is anticipated instead of a true solid solution. This microstructure is perfect since it is anticipated that the significant potential difference between Ag and Fe will produce a strong galvanic pair, which would subsequently result in increased corrosion. Although there may be some toxicity concerns, particularly with the Ag nanoparticles, employing Ag has the added benefit of being a recognized antibacterial that may help avoid implant infections [4]. Mandal et al. [35] looked at how the characteristics of Fe–Mn were affected by the addition of Cu, another element possessing antibacterial qualities. They found that the corrosion rate of Fe–25Mn–10Cu is almost six times greater than that of Fe–35Mn. Furthermore, the Cu-bearing alloys have antibacterial qualities.

3 3D Printing of Fe-Based Biomaterials for Biomedical Applications

Researchers are working very hard to combine additive manufacturing (AM) technology with porous structures because of their many intriguing qualities. This will allow them to use AM technologies to 3D print porous materials. The primary obstacle is the inability of AM technologies to fabricate porous features at the submicron and nanoscale. Most AM techniques can achieve the resolution needed to print macroscale porous structures, as seen in Fig. 2. AM has the ability to process biodegradable iron, enabling the creation of intricate geometric patterns and a variety of surface finishes [21]. Researchers developed an integrated and multidisciplinary approach for 3D printing porous structures by seamlessly integrating AM technologies with traditional porous material manufacturing procedures to overcome the resolution constraint. To produce the micro- and nanoscale porous networks, the feedstock materials are first 3D printed utilizing AM technologies, and then classic templating or solvent exchange techniques are used. The 3D production of multi-scale porous structures that enable control over pore size, shape, organization, and morphology is also made possible by this combination. Rabeeh et al. [36] provided a detailed overview of the developments in the processing and manufacturing of Fe-based biomaterials. Compressed gas is used in this high-pressure powder processing method to move metal powder across a wide temperature range. High-velocity collisions from the powder cause the particles to deform, stick to the substrate, and create a coating. The corrosion rate can be improved by the microstructure's ability to reach grain sizes in the micron to submicron range. The upcoming section discusses the various approaches with regard to the development of Fe-based biomaterials to improve the corrosion behavior to make them a suitable candidate for biomedical applications.

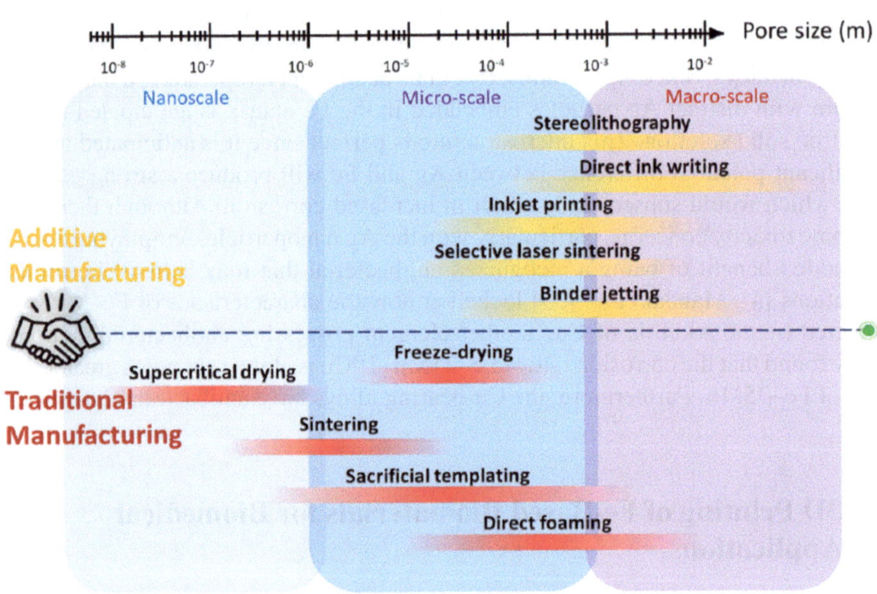

Fig. 2 The complementary nature of traditional manufacturing techniques with additive manufacturing in the creation of three-dimensional porous structures. Reused with permission from Elsevier [37]

3.1 Development of Fe-Based Porous Structure

The new type of biomaterials must primarily have characteristics that are similar to those of human bone. It has been demonstrated that porous biomaterials can successfully meet these needs [38]. Additionally, through their porous network, porous materials enable biological anchoring for the surrounding bone tissue to adhere and grow [25]. In general, foams are another name for three-dimensional porous structures [20]. The foams might be made of composite, metal, ceramic, or polymer materials. However, by appropriately altering the foam structures' design, these materials' mechanical properties could be enhanced [15]. In this context, iron foam has been prepared using a variety of fabrication techniques. It may be concluded that the development of an iron foam with a Topologically Ordered Porous Structure (TOPS) is necessary. Sharma and Pandey [15] proposed a new manufacturing process that combines pressureless microwave sintering and 3D printing to create topologically organized open-cell porous iron foam (TOPIF). Figure 3a represents the methodology that amalgamates the AM process with microwave sintering. Several unit cell shapes of iron foams were manufactured, including pyramid, truncated octahedron, and cubic (Fig. 3b). They have found the dimensional variation ranged from 5 to 22%, and porosities ranged from 45.6 to 86.9% of the TOPIF structures. Figure 3c shows the compressive stress-strain curve of the TOPIF samples. Three regions made up the majority of the obtained stress-strain curves: the densification zone, the plateau region, and the linear elastic region. First, an elastic response was seen, with the stress increasing linearly with the applied strain. Because the cell structures in the TOPIF sample failed, it was discovered that the stress in the second region fluctuated between the mean values. The third area is the widespread rise in stress brought on by the cell structures' densification. It was possible to see that the plateau regions created are not smooth after looking at the stress-strain curves. Because of the brittle fracture of the unit cell structures in the manufactured TOPIF samples, there is a significant difference in stress. The stress-strain curves derived from flexural testing are displayed in Fig. 3d. It was found that all of the TOPIF samples made using the developed approach experienced significant plastic deformation prior to failure. Compared to C-TOPIF and TO-TOPIF samples, P-TOPIF samples showed a greater amount of hardening before failure. They Sharma and Pandey [15] have also conducted a Gibson-Ashby model-based analysis, and the results showed good agreement with the experimental findings. According to a comparison analysis, the manufactured TOPIF samples had the perfect qualities needed for a human cancellous bone augmentation surgery.

A functionally graded porous iron investigation was described by Li et al. [39]. Selective laser melting (SLM), a powder bed-based process from AM, was used to create iron scaffolds. S0.2 = 84.8 $\pm$ 0.1, Dense-in = 70.6 $\pm$ 0.4, Dense-out = 71.0 $\pm$ 0.2, and S0.4 = 58.4 $\pm$ 2.0 were the scaffolds' porosities as determined by μCT. The abbreviations are as follows: "S0.2" denotes a uniform strut thickness of 0.2 mm, "Dense in" denotes a strut thickness of 0.2–0.4 mm from the center to the perimeter, "Dense out" denotes a thickness of 0.4–0.2 mm in reverse, and "S0.4" denotes a consistently thick strut. The topological design of the scaffolds under study

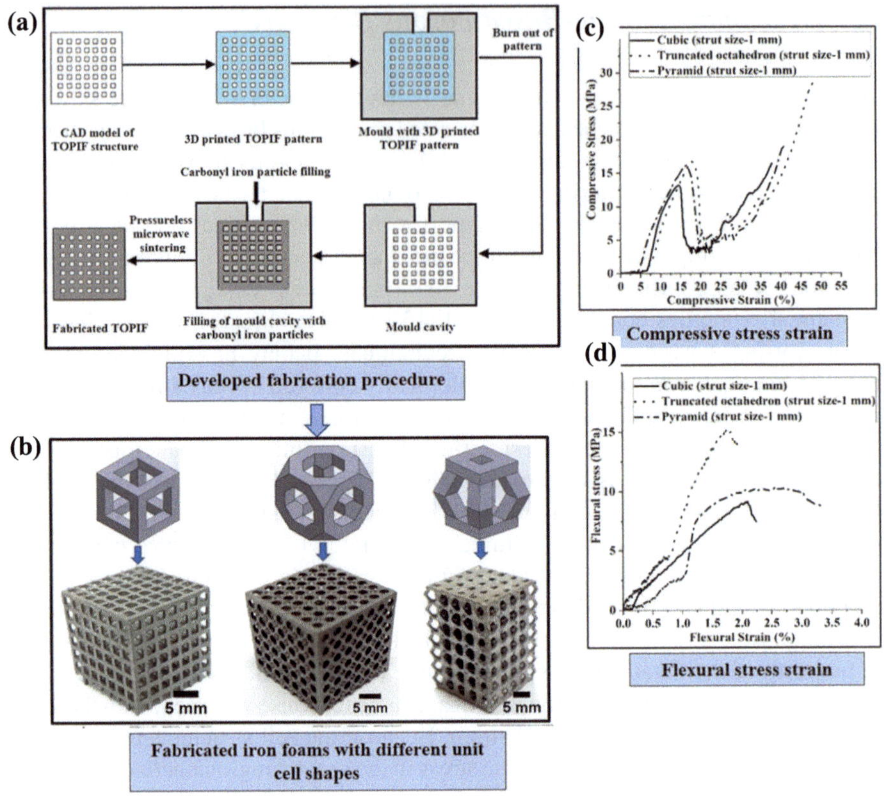

Fig. 3 **a** The developed methodology, **b** fabricated iron foams of different unit cells, **c** compressive stress-strain curves, **d** flexural stress-strain curves. Reused after permission from Elsevier [15]

is depicted in Fig. 4a, b [39]. When four distinct scaffold designs were evaluated, iron showed a smooth compressive behavior devoid of variations or unexpected failures. The mechanical characteristics before and after degradation after submersion in bodily fluid simulations are depicted in Fig. 4 [29]. The mechanical qualities are reliant on topology, according to the results. Topologically ordered porous metallic structures can be created using AM; these configurations have a completely interconnected porous structure that replicates the mechanical characteristics of the bone.

Dipesh et al. [9, 20] used pressureless microwave sintering and a new micro-extrusion-based 3D printer to study the topological structures (OPTS). Additionally, by using different unit cells, such as hexagonal and cubic, with variable strut thicknesses, the compressive mechanical performance of the manufactured Fe-based OPTS was assessed. The generated Fe-based OPTS with various unit cells were morphologically characterized using optical microscopy pictures. Pressureless sintering of loose CIPs and post-processing of the generated OPTS samples produced

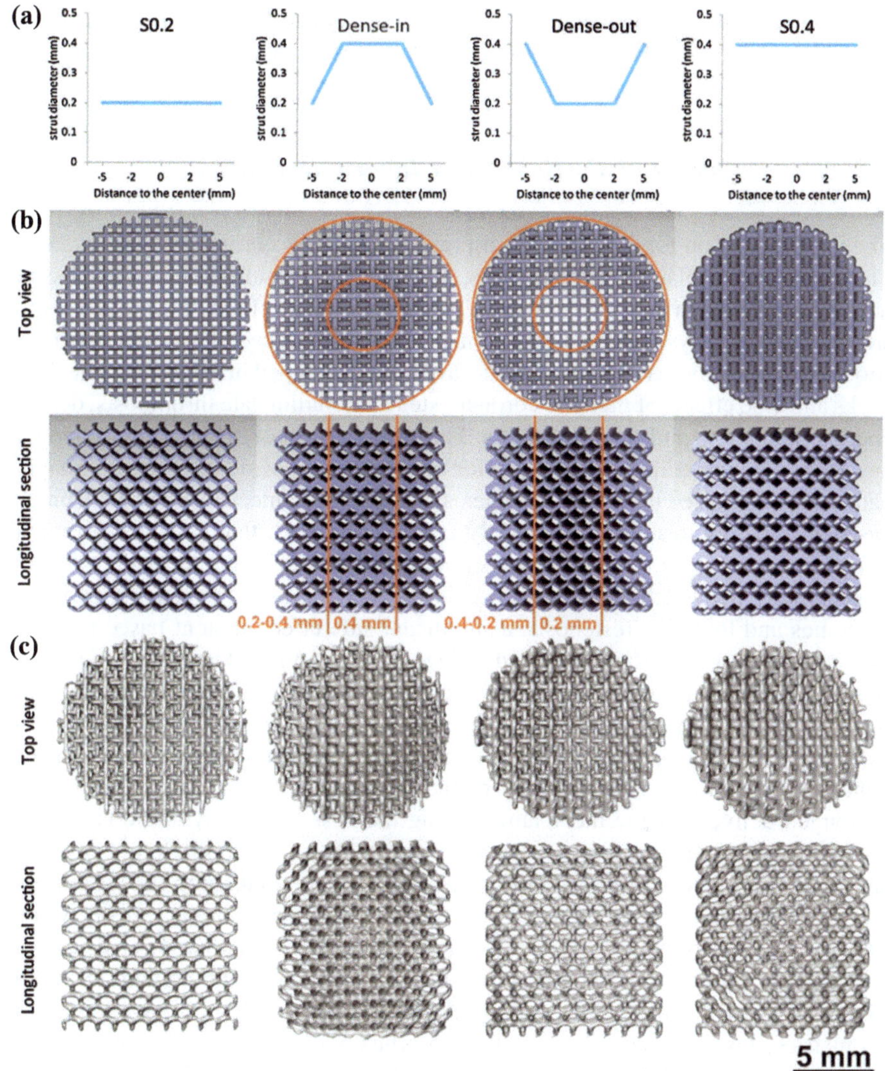

Fig. 4 The specimens from the S0.2, Dense-in, Dense-out, and S0.4 groups had four distinct topological designs: **a** the distribution of strut sizes; **b** the longitudinal cross-section and top view of the CAD models; and **c** the micro-CT reconstructions of the AM porous iron specimens. Reused with permission from Elsevier [39]

the dimensional variance in the OPTS cell wall. The porosity content of the manufactured iron foam was found to have a substantial impact on the macroplastic and microelastic deformation under compressive loading. The primary cause of failure was observed to be the yielding and bending of cell walls. The produced OPTS's compressive mechanical characteristics, including yield strength, Young's modulus, plateau stress, and ultimate compressive strength, ranged from 136 to 257 MPa, 11 to 54 MPa, 12 to 64 MPa, and 11 to 67.3 MPa, respectively.

3.2 Development of Fe-Based Cardiovascular Stent

The primary reason for the global death rate's increase is coronary artery disorders. The formation of plaque in the arteries, known as atherosclerosis, causes obstruction, which results in uneven blood flow. In order to restore blood flow, stenotic blood arteries are typically implanted with stents, which are meshed tubular constructions. Biodegradable polymer coronary stents have been developed in response to the drawbacks and side effects of metallic coronary stents, including late thrombosis, delayed endothelialization, and stent restenosis. The advancement of 3D printing technology has opened up incredible possibilities for the production of stents. Medical imaging data can now be used to modify complex stent designs to meet each patient's unique vascular anatomy, resulting in a highly tailored approach that enhances treatment effectiveness while minimizing side effects. 3D-printed stents have the potential to accurately deliver therapeutic molecules to targeted areas, enhancing treatment outcomes and lowering restenosis, a significant side effect of stent-based therapies [40]. Neointimal overgrowth-induced restenosis is a significant issue with bare metal stents. Drug-eluting stents containing cytotoxic or immunosuppressive medications were created to prevent neointimal hyperplasia in order to address this issue. Drug loading typically occurs from a polymeric coating, which facilitates elution kinetics control. The anti-restenotic efficacy of the device depends on the controlled and prolonged delivery of the medication during the first 10–30 days following transplantation, during the healing of the surgical site [41]. Curcumin-eluting stent OM and SEM images are displayed in Fig. 5a–d. Curcumin-containing PLGA coatings are incredibly smooth, as seen in Figs. 5a, b, and the curcumin is easily visible prior to angioplasty balloon dilatation. The coatings are smooth and consistent. No webbings or fissures exist between the struts. Additionally, the polymeric coating for iron-based BRS needs to be completely biodegradable. By printing several stent geometries, Jasvinder et al. [42] tested the efficacy of a technology created based on the solvent-cast 3D printing technique. Three current stent designs, including ABBOTT BVS1.1, PALMAZSCHATZ, and ART18Z, were printed using carbonyl iron powder (CIP) reinforced polycaprolactone (CIPC). Platelet adhesion, hemolysis, protein adsorption, in vitro cell viability, and live/dead cell viability assays were used to examine the stents' cytocompatible and hemocompatible properties. The findings showed that stents possessed sufficient mechanical qualities to carry out the required tasks in the human heart. The findings showed that the PALMAZ-SCHATZ stent had superior mechanical qualities, superior cell viability, improved blood compatibility, and improved degradation behavior.

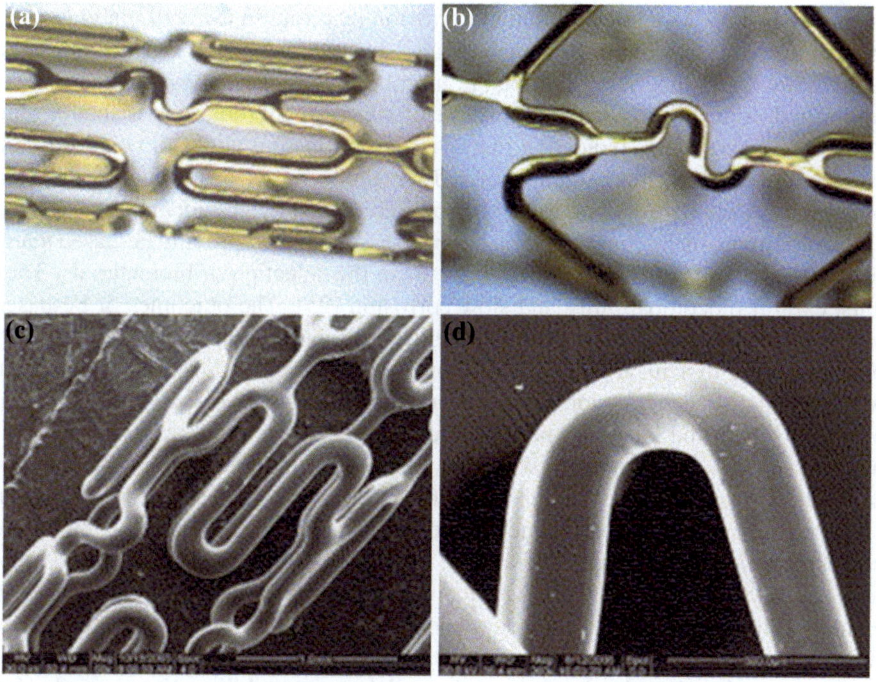

Fig. 5 Metallic stents spray-coated with PLGA and curcumin are shown in optical microscope and SEM micrographs as follows: **a** (×40) and **c** (×50) are pre-expansion images, and **b** (×40) and **d** (×200) are post-dilation images. Reused with permission from Elsevier [41]

4 In Vivo and In Vivo Studies of Fe-Based Biomaterials

In numerous in vitro investigations, iron (Fe) has demonstrated exceptional biocompatibility. It was shown that pure electroformed iron did not restrict the metabolic activity of rat smooth muscle cells (SMCs) or stop cell proliferation when it was utilized in cardiovascular stents [43]. Additionally, Chen et al. [44] observed that pure industrial-grade Fe exhibited good hemocompatibility and biocompatibility, with no cytotoxicity. In their investigation of the in vitro cytotoxicity of nitride iron stents, Lin et al. [45] discovered that the extraction medium and oxygen availability were essential for both incubation and extraction. L929 cells did not exhibit any cytotoxicity from extracts with high iron concentrations. However, alloying elements were found to have an impact on the biocompatibility of Fe alloys [44]. According to studies on alloying cells, the inhibitory effect was highly influenced by the alloy content in the medium. Multicomponent Fe alloys were also studied. In their study of the cytocompatibility of Fe–Mn–C (-Pd) alloy with human umbilical vein endothelial cells (HUVECs), Schinhammer et al. [22] discovered that the content of the extract medium had a significant impact on both metabolic activity and cell viability. Additionally, there was a strong correlation between cell viability and the rates of corrosion of Fe alloys [14], with high corrosion resulting in low cell viability [46]. Porous Fe-based scaffolds were investigated by Orinak et al. [47] in vitro static cell cultures, and they reported notable morphological alterations, enhanced scaffold blood compatibility, and increased cell survival. Fibroblasts perished within 24 hours, nevertheless, indicating that cytotoxicity testing ought to be carried out in an unpredictable environment to avoid the proliferation of too many corrosion byproducts. Cell viability was observed to be enhanced by Fe/bioceramics [48].

Additionally, surface modification has been essential in improving Fe-based materials' biocompatibility, which is a key factor in the selection of biomaterials. They must exhibit biocompatibility and slow corrosion [25]. The outcomes of Fe-based materials' in vivo experiments are compiled in Table 4 below. Peuster et al. [2] performed the first in vivo testing of iron stents by implanting pure Fe stents in white rabbits' descending aortas. There were no thromboembolic adverse effects or problems during the follow-up period. Additionally, there was no noticeable inflammatory response, neointimal growth, or systemic toxicity. The identical stents were then placed in pigs' aortas and observed for more than a year; no Fe-related organ damage or Fe overload was found. There was no evidence of local toxicity from corrosion products. In a brief research, Fe stents were placed in young pigs' coronary arteries and left there for 28 days to evaluate their safety and effectiveness [49]. Histomorphometry showed no fibrin deposition or severe inflammation, and there was no evidence of thrombosis or particle embolization.

Stents made of nitrite iron showed good biocompatibility. Kraus et al. [50] implanted Fe–21Mn–0.7C–1Pd, Fe–10Mn–1Pd, and Fe pins into the rat femur. There were no reports of inflammation or pathological symptoms following a year of follow-up. The wound usually healed as expected, despite the fact that there was some slight edema for one to two days after surgery. Histological research showed that Fe^{2+} and Fe^{3+} were the main degradation products of iron close to the tissue; no toxicity or inflammation was seen, and the implants did not cause any tissue damage. Additionally, Fe–Mn–Si implants demonstrated exceptional biocompatibility. Fe implants were found to be slightly toxic in certain tissue investigations [51], while the surrounding tissues only suffered minor harm. The bioactivity of Fe-bioceramic composites in Indonesian sheep leg bones was evaluated by Ulum et al. [52]. After 35 days of implantation, the stent-induced inflammation subsided, and the tissue showed no reaction as the wound healed. The existence of bioceramic phases with asymmetrical morphologies in the Fe matrix and extensive sintering is demonstrated by the microstructure of the sintered and polished surfaces of pure Fe and Fe-based composites. The in vitro degradation mechanism of gold-sputtered iron was examined by Rabeeh et al. [53], who discovered that the dispersion of gold particles on the iron surface accelerated the degradation process.

5 Conclusion

The prospect of producing CT pictures of metallic absorbable implants, for instance, has long been discussed in conjunction with biodegradable iron. New production methods that enable the creation of porous materials with regulated porosity have given biodegradable iron new significance. Customized porous iron structures made using new manufacturing techniques, like additive manufacturing, will allow for the customization of mechanical and corrosion properties. This is a significant advancement in the field of materials engineering as well as tissue engineering for biomedical applications. It is anticipated that biodegradable iron will find successful uses in the field of tissue engineering, specifically in the areas of cardiovascular devices and temporary implants for non-load-bearing orthopedic applications. This chapter highlights a number of methods to get beyond the limitations of pure iron, including the utilization of porosity and the alloying of specific elements to reduce stiffness and speed up iron corrosion. However, there will still be a number of obstacles to overcome. The in vivo experiments, which are uncommon, will further the clinical application of these kinds of materials. Future directions and research gaps have been noted in order to solve the difficulties in developing iron-based alloys and composites for use as biodegradable medical devices. Future studies on the creation of biodegradable iron actually tend to focus on "multifunctional capabilities." This implies that in order to support structures that have a direct and active relationship with the host, there will be a tendency to combine many elements, such as porosity, alloying of specific elements, composites, and coatings.

References

1. Hermawan H, Dubé D, Mantovani D (2010) Developments in metallic biodegradable stents. Acta Biomater 6:1693–1697. https://doi.org/10.1016/J.ACTBIO.2009.10.006
2. Peuster M, Hesse C, Schloo T, Fink C, Beerbaum P, von Schnakenburg C (2006) Long-term biocompatibility of a corrodible peripheral iron stent in the porcine descending aorta. Biomaterials 27:4955–4962. https://doi.org/10.1016/J.BIOMATERIALS.2006.05.029
3. Zheng YF, Gu XN, Witte F (2014) Biodegradable metals. Mater Sci Eng R: Rep 77:1–34. https://doi.org/10.1016/j.mser.2014.01.001
4. Venezuela J, Dargusch MS (2020) Addressing the slow corrosion rate of biodegradable Fe–Mn: current approaches and future trends. Curr Opin Solid State Mater Sci 24:100822. https://doi.org/10.1016/J.COSSMS.2020.100822
5. Moravej M, Mantovani D (2011) Biodegradable metals for cardiovascular stent application: interests and new opportunities. Int J Mol Sci 12:4250. https://doi.org/10.3390/IJMS12074250
6. Tripathi G, Pandey PM (2023) Experimental investigations into pressureless microwave sintering of iron–hydroxyapatite composite for biodegradable implants. Proc Inst Mech Eng C J Mech Eng Sci 237:5933–5950. https://doi.org/10.1177/09544062231166488
7. Wang S, Xu Y, Zhou J, Li H, Chang J, Huan Z (2017) In vitro degradation and surface bioactivity of iron-matrix composites containing silicate-based bioceramic. Bioact Mater 2:10–18. https://doi.org/10.1016/J.BIOACTMAT.2016.12.001
8. Gambaro S, Paternoster C, Occhionero B, Fiocchi J, Biffi CA, Tuissi A, Mantovani D (2021) Mechanical and degradation behavior of three Fe–Mn–C alloys for potential biomedical

applications. Mater Today Commun 27:102250. https://doi.org/10.1016/J.MTCOMM.2021.102250
9. Mishra DK, Pandey PM (2021) Effects of morphological characteristics on the mechanical behavior of 3D printed ordered pore topological Fe scaffold. Mater Sci Eng: A 804. https://doi.org/10.1016/j.msea.2021.140759.
10. Chen Y, Li T, Jia Z, Scarpa F, Yao CW, Wang L (2018) 3D printed hierarchical honeycombs with shape integrity under large compressive deformations. Mater Des 137:226–234. https://doi.org/10.1016/j.matdes.2017.10.028
11. Gąsior G, Szczepański J, Radtke A (2021) Biodegradable iron-based materials—what was done and what more can be done? https://doi.org/10.3390/ma14123381
12. Kasaeian-Naeini M, Sedighi M, Hashemi R (2022) Severe plastic deformation (SPD) of biodegradable magnesium alloys and composites: a review of developments and prospects. J Magnes Alloys 10:938–955. https://doi.org/10.1016/J.JMA.2021.11.006
13. Sharma P, Jain KG, Pandey PM, Mohanty S (2020) In vitro degradation behaviour, cytocompatibility and hemocompatibility of topologically ordered porous iron scaffold prepared using 3D printing and pressureless microwave sintering. Mater Sci Eng: C 106:110247. https://doi.org/10.1016/J.MSEC.2019.110247
14. Das P, Pathak DK, Sharma P, Pandey PM (2024) A review on the mechanical and biocorrosion behaviour of iron and zinc-based biodegradable materials fabricated using powder metallurgy routes. Corros Rev. https://doi.org/10.1515/CORRREV-2023-0169/ASSET/GRAPHIC/J_CORRREV-2023-0169_CV_004.JPG
15. Sharma P, Pandey PM (2018) Morphological and mechanical characterization of topologically ordered open cell porous iron foam fabricated using 3D printing and pressureless microwave sintering. Mater Des 160:442–454. https://doi.org/10.1016/J.MATDES.2018.09.029
16. Hermawan H (2018) Updates on the research and development of absorbable metals for biomedical applications. Prog Biomater 7:93–110. https://doi.org/10.1007/S40204-018-0091-4
17. He J, He FL, Li DW, Liu YL, Liu YY, Ye YJ, Yin DC (2016) Advances in Fe-based biodegradable metallic materials. RSC Adv 6:112819–112838. https://doi.org/10.1039/C6RA20594A
18. Prakasam M, Locs J, Salma-Ancane K, Loca D, Largeteau A, Berzina-Cimdina L (2017) Biodegradable materials and metallic implants—a review. J Funct Biomater 8:44. https://doi.org/10.3390/JFB8040044
19. Schinhammer M, Hänzi AC, Löffler JF, Uggowitzer PJ (2010) Design strategy for biodegradable Fe-based alloys for medical applications. Acta Biomater 6:1705–1713. https://doi.org/10.1016/J.ACTBIO.2009.07.039
20. Mishra DK, Pandey PM (2020) Mechanical behaviour of 3D printed ordered pore topological iron scaffold. Mater Sci Eng: A 783:139293. https://doi.org/10.1016/J.MSEA.2020.139293
21. Singh J, Singh G, Pandey PM (2021) Additive manufacturing of functionalized nanomaterials for the modern health care industry. In: Additive manufacturing with functionalized nanomaterials. pp 55–85. https://doi.org/10.1016/B978-0-12-823152-4.00003-X
22. Schinhammer M, Steiger P, Moszner F, Löffler JF, Uggowitzer PJ (2013) Degradation performance of biodegradable Fe–Mn–C(-Pd) alloys. Mater Sci Eng C Mater Biol Appl 33:1882–1893. https://doi.org/10.1016/J.MSEC.2012.10.013
23. Liu B, Zheng YF (2011) Effects of alloying elements (Mn Co, Al, W, Sn, B, C and S) on biodegradability and in vitro biocompatibility of pure iron. Acta Biomater 7:1407–1420. https://doi.org/10.1016/J.ACTBIO.2010.11.001
24. Cheng J, Zheng YF (2013) In vitro study on newly designed biodegradable Fe-X composites (X = W, CNT) prepared by spark plasma sintering. J Biomed Mater Res B Appl Biomater 101:485–497. https://doi.org/10.1002/JBM.B.32783
25. Sharma P, Pandey PM (2019) Corrosion behaviour of the porous iron scaffold in simulated body fluid for biodegradable implant application. Mater Sci Eng: C 99:838–852. https://doi.org/10.1016/J.MSEC.2019.01.114

26. Khan AR, Grewal NS, Zhou C, Yuan K, Zhang HJ, Jun Z (2023) Recent advances in biodegradable metals for implant applications: Exploring in vivo and in vitro responses. Results Eng 20:101526. https://doi.org/10.1016/J.RINENG.2023.101526
27. Feng Q, Zhang D, Xin C, Liu X, Lin W, Zhang W, Chen S, Sun K (2013) Characterization and in vivo evaluation of a bio-corrodible nitrided iron stent. J Mater Sci Mater Med 24:713–724. https://doi.org/10.1007/s10856-012-4823-z
28. Li C, Chen J, Hou S, Zhang Q, Li X, Kou S (2023) Effects of spark plasma sintering parameters on properties of Fe-based bulk amorphous alloys. Mater Today Commun 37:107068. https://doi.org/10.1016/J.MTCOMM.2023.107068
29. Hermawan H, Dubé D, Mantovani D (2007) Development of degradable Fe-35Mn alloy for biomedical application. Adv Mat Res 15–17:107–112. https://doi.org/10.4028/WWW.SCIENTIFIC.NET/AMR.15-17.107
30. Hermawan H, Dubé D, Mantovani D (2010) Degradable metallic biomaterials: design and development of Fe-Mn alloys for stents. J Biomed Mater Res A 93:1–11. https://doi.org/10.1002/JBM.A.32224
31. Dargusch MS, Dehghan-Manshadi A, Shahbazi M, Venezuela J, Tran X, Song J, Liu N, Xu C, Ye Q, Wen C (2019) Exploring the role of manganese on the microstructure, mechanical properties, biodegradability, and biocompatibility of porous iron-based scaffolds. ACS Biomater Sci Eng 5:1686–1702. https://doi.org/10.1021/ACSBIOMATERIALS.8B01497/ASSET/IMAGES/MEDIUM/AB-2018-01497C_0016.GIF
32. Kupková M, Hrubovcáková M, Kupka M, Oriňáková R, Turonová AM (2015) Sintering behaviour, graded microstructure and corrosion performance of sintered Fe-Mn biomaterials. Int J Electrochem Sci 10:9256–9268. https://doi.org/10.1016/S1452-3981(23)11175-8
33. Fiocchi J, Biffi CA, Gambaro S, Paternoster C, Mantovani D, Tuissi A (2020) Effect of laser welding on the mechanical and degradation behaviour of Fe–20Mn–0.6C bioabsorbable alloy. J Mater Res Technol 9:13474–13482. https://doi.org/10.1016/J.JMRT.2020.09.104
34. Xu W, Lu X, Tan L, Yang K (2011) Study on properties of a novel biodegradable Fe–30Mn–1C alloy. Jinshu Xuebao/Acta Metall Sin 47:1342–1347. https://doi.org/10.3724/SP.J.1037.2011.00258
35. Mandal S, Ummadi R, Bose M, Balla VK, Roy M (2019) Fe–Mn–Cu alloy as biodegradable material with enhanced antimicrobial properties. Mater Lett 237:323–327. https://doi.org/10.1016/J.MATLET.2018.11.117
36. Rabeeh VPM, Hanas T (2022) Progress in manufacturing and processing of degradable Fe-based implants: a review. Prog Biomater 11:163–191. https://doi.org/10.1007/S40204-022-00189-4
37. Guo Z, Zhou C (2021) Recent advances in ink-based additive manufacturing for porous structures. Addit Manuf 48:102405. https://doi.org/10.1016/J.ADDMA.2021.102405
38. Sharma P, Pandey PM (2019) Rapid manufacturing of biodegradable pure iron scaffold using amalgamation of three-dimensional printing and pressureless microwave sintering. Proc Inst Mech Eng C J Mech Eng Sci 233:1876–1895. https://doi.org/10.1177/0954406218778304
39. Li Y, Jahr H, Pavanram P, Bobbert FSL, Puggi U, Zhang XY, Pouran B, Leeflang MA, Weinans H, Zhou J, Zadpoor AA (2019) Additively manufactured functionally graded biodegradable porous iron. Acta Biomater 96:646–661. https://doi.org/10.1016/J.ACTBIO.2019.07.013
40. Khan MA, Khan N, Ullah M, Hamayun S, Makhmudov NI, MBBS R, Safdar M, Bibi A, Wahab A, Naeem M, Hasan N (2024) 3D printing technology and its revolutionary role in stent implementation in cardiovascular disease. Curr Probl Cardiol 49:102568. https://doi.org/10.1016/J.CPCARDIOL.2024.102568
41. Pan CJ, Tang JJ, Weng YJ, Wang J, Huang N (2006) Preparation, characterization and anticoagulation of curcumin-eluting controlled biodegradable coating stents. J Controlled Release 116:42–49. https://doi.org/10.1016/J.JCONREL.2006.08.023
42. Singh J, Pandey PM, Kaur T, Singh N (2021) A comparative analysis of solvent cast 3D printed carbonyl iron powder reinforced polycaprolactone polymeric stents for intravascular applications. J Biomed Mater Res B Appl Biomater 109:1344–1359. https://doi.org/10.1002/JBM.B.34795

43. Jurgeleit T, Quandt E, Zamponi C (2017) Magnetron sputtering as a fabrication method for a biodegradable Fe32Mn alloy. Materials 10:1196. https://doi.org/10.3390/MA10101196
44. Chen CZ, Shi XH, Zhang PC, Bai B, Leng YX, Huang N (2008) The microstructure and properties of commercial pure iron modified by plasma nitriding. Solid State Ion 179:971–974. https://doi.org/10.1016/J.SSI.2008.03.019
45. Lin W, Qin L, Qi H, Zhang D, Zhang G, Gao R, Qiu H, Xia Y, Cao P, Wang X, Zheng W (2017) Long-term in vivo corrosion behavior, biocompatibility and bioresorption mechanism of a bioresorbable nitrided iron scaffold. Acta Biomater 54:454–468. https://doi.org/10.1016/J.ACTBIO.2017.03.020
46. Walker DE, Wilcox GD (2008) Molybdate based conversion coatings for zinc and zinc alloy surfaces: a review. Trans IMF 86:251–259. https://doi.org/10.1179/174591908X345022
47. Oriňák A, Oriňáková R, Králová ZO, Turoňová AM, Kupková M, Hrubovčáková M, Radoňák J, Džunda R (2014) Sintered metallic foams for biodegradable bone replacement materials. J Porous Mater 21:131–140. https://doi.org/10.1007/S10934-013-9757-4/METRICS
48. Oriňaková R, Gorejová R, Macko J, Oriňak A, Kupková M, Hrubovčáková M, Ševc J, Smith RM (2019) Evaluation of in vitro biocompatibility of open cell iron structures with PEG coating. Appl Surf Sci 475:515–518. https://doi.org/10.1016/J.APSUSC.2019.01.010
49. Waksman R, Pakala R, Baffour R, Seabron R, Hellinga D, Tio FO (2008) Short-term effects of biocorrodible iron stents in porcine coronary arteries. J Interv Cardiol 21:15–20. https://doi.org/10.1111/J.1540-8183.2007.00319.X
50. Kraus T, Moszner F, Fischerauer S, Fiedler M, Martinelli E, Eichler J, Witte F, Willbold E, Schinhammer M, Meischel M, Uggowitzer PJ, Löffler JF, Weinberg A (2014) Biodegradable Fe-based alloys for use in osteosynthesis: outcome of an in vivo study after 52 weeks. Acta Biomater 10:3346–3353. https://doi.org/10.1016/J.ACTBIO.2014.04.007
51. Feng YP, Gaztelumendi N, Fornell J, Zhang HY, Solsona P, Baró MD, Suriñach S, Ibáñez E, Barrios L, Pellicer E, Nogués C, Sort J (2017) Mechanical properties, corrosion performance and cell viability studies on newly developed porous Fe-Mn-Si-Pd alloys. J Alloys Compd 724:1046–1056. https://doi.org/10.1016/J.JALLCOM.2017.07.112
52. Ulum MF, Arafat A, Noviana D, Yusop AH, Nasution AK, Abdul Kadir MR, Hermawan H (2014) In vitro and in vivo degradation evaluation of novel iron-bioceramic composites for bone implant applications. Mater Sci Eng: C 36:336–344. https://doi.org/10.1016/J.MSEC.2013.12.022
53. Rabeeh VPM, Hanas T (2022) Enhancing biointerfacial properties of porous pure iron by gold sputtering for degradable implant applications. Mater Today Commun 31:103492. https://doi.org/10.1016/J.MTCOMM.2022.103492

Chapter 3
Fabrication Techniques and Characterization for Iron-Based Biodegradable Implants

Shubhra Dixit, Ajay Sharma, and Surbhi Gupta

1 Introduction

Biodegradable metal implants are a new type of medical materials that provide temporary support to the body and slowly break down over time, eliminating the need for a second surgery to remove them. Among these metals, iron (Fe) and its alloys are considered promising because they are strong, biocompatible, and capable of supporting heavy loads, making them useful for bone and heart implants [1]. Unlike magnesium-based implants, which dissolve too quickly, iron degrades at a slower rate, allowing it to maintain its strength for a longer time. For achieving an optimized degradation performance and mechanical integrity of iron-based implants, many production techniques have been introduced with each impacting on the material's microstructure, porosity [2], and surface features. Standard procedures such as powder metallurgy can accommodate careful composition control as well as porosity, while cutting-edge techniques such as additive manufacturing (3D printing) provide access to complexly geometric patient-specific implants [3]. In addition, electroforming and surface modification methods have been investigated to increase corrosion rates and enhance biocompatibility [4, 5].

Characterization is of particular importance when assessing the potential of these materials for biomedicine. Microstructure examination, mechanical testing, in vitro and in vivo degradation testing, and biocompatibility tests supply crucial information on how they will perform and interact with biological systems. This chapter discusses fabrication procedures and characterization technologies of iron-based biodegradable implants and recent developments as well as possible approaches to overcome existing limitations in their clinical application.

Emerging fabrication strategies are shifting toward hybrid processing techniques that integrate powder metallurgy with AM to achieve better structural integrity and

S. Dixit · A. Sharma · S. Gupta (✉)
Amity University, Noida, Uttar Pradesh, India
e-mail: sgupta11@amity.edu

A. Dixit et al. (eds.), *Biodegradable Metallic Materials*, Biomedical Materials for Multi-functional Applications, https://doi.org/10.1007/978-981-95-2401-3_3

controlled biodegradation. Nano-engineered surface modifications, such as bioactive and antimicrobial coatings, provide enhanced osteointegration and localized drug release. Machine learning-driven process optimization is being explored to fine-tune AM parameters for better microstructural control. Eco-friendly manufacturing techniques, such as green sintering and solvent-free electrochemical deposition, offer sustainable alternatives for implant fabrication. Additionally, personalized implants tailored to patient-specific anatomical and biomechanical requirements are becoming a reality through AI-assisted 3D printing and computational modelling. These advancements hold the potential to improve implant longevity, reduce complications, and accelerate clinical adoption. Despite advancements in fabrication techniques, several challenges persist in developing iron-based biodegradable implants. Powder metallurgy struggles with achieving uniform porosity, grain refinement, and optimal sintering conditions to balance mechanical strength and degradation rate. Additive manufacturing (AM) techniques like laser powder bed fusion lack process parameter optimization for iron-based materials, leading to inconsistent microstructures and unpredictable corrosion behavior. Casting and sintering methods often result in coarse-grained structures and high residual stresses, which negatively affect mechanical integrity. Electrochemical deposition, though promising for coating applications, suffers from poor adhesion and difficulty in tailoring degradation rates. Mechanical processing techniques such as rolling, extrusion, and forging lack comprehensive studies on their effects on corrosion kinetics and in vivo degradation. Surface modification techniques, including bioactive and drug-eluting coatings, face issues in adhesion durability, long-term stability, and uniformity. Additionally, the scalability and cost-effectiveness of these fabrication methods remain a major hurdle for clinical translation. Research is needed to develop integrated fabrication approaches that combine multiple methods for enhanced performance. Biodegradable implants are medical devices designed to offer temporary mechanical support in the body and break down over time without the need for surgical removal. They are especially useful in orthopedic fixation, cardiovascular stenting, and soft tissue regeneration, where permanent implants can lead to long-term complications [6]. Biocompatible metals such as titanium (Ti) and its alloys, cobalt (Co)-based alloys, stainless steels, and magnesium (Mg)-based alloys are widely utilized in biomedical devices and implants [7]. Figure 1 illustrates the routine use of these metal components in various implantable medical applications.

The following timeline [9] illustrates key milestones in the progress of biomaterials for biomedical applications. Since the 1940s, biomaterials have continued to evolve and advance (Fig. 2). Bio-inertness for close applications was the primary emphasis of the first generation of biomaterials. Further development in the context of resorbable biomaterials, controlled reactions with the surrounding application region, bonding, and continuous releases of the medication follow the second generation, which has bioactivity as its ultimate goal. The primary goal of the third generation of biomaterials, which essentially began in 1980, was to regenerate functional tissues, including those that were interactive toward functionality, integrated, proliferative, and differentiated toward the growth or healing process.

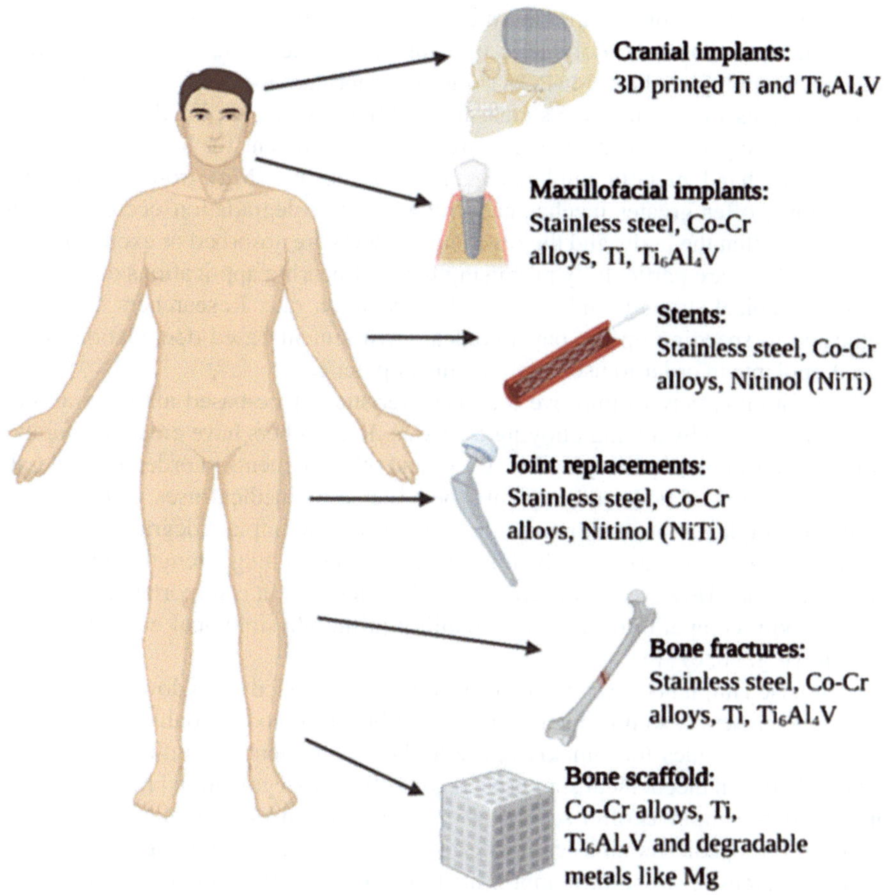

Fig. 1 Classification of metallic implants in the human body [8]

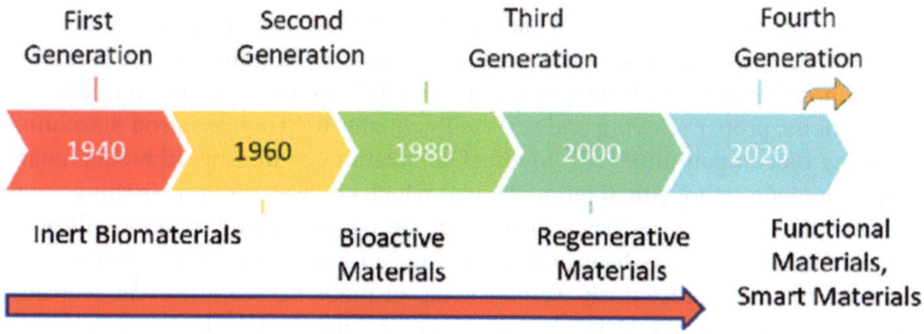

Fig. 2 Progress in biomaterials for biomedical applications [9] with permission from Elsevier

Biodegradable implants are classified based on their material composition and degradation mechanisms. Polymeric implants, including those made of polylactic acid (PLA), polyglycolic acid (PGA), and polycaprolactone (PCL), are commonly used in sutures, drug delivery systems, and soft tissue scaffolds. While they degrade through hydrolysis, their mechanical strength is insufficient for load-bearing [10]. On the other hand, metallic implants, made of magnesium (Mg), iron (Fe), and zinc (Zn) alloys, offer greater mechanical strength. Their degradation occurs through corrosion within the body, and the resulting products are absorbed or excreted. Iron-based implants are particularly promising for load-bearing applications due to their high mechanical strength and controlled degradation rate. Researchers have been dedicated to speeding up the physiological environment-based degradation of Fe-based implants in order to use them in clinical practice.

One common way to improve the characteristics of Fe-based alloys is to alter their composition by adding alloying elements. Researchers have created a number of Fe-based samples by alloying Mn, C, Si, and other elements in order to accelerate the deterioration of Fe-based implants for possible biomedical uses. For example, as in bone implantation, the progress of bone fusion and implant degradation occurs simultaneously (Fig. 3a) [11] which is best for corresponding strength and matches the healing rate. However, while choosing alloy elements, it is important to take into account both compatibility and degradability in the physiological environment of humans (Fig. 3b, c) [11].

Iron-based implants have mechanical strength, controlled breakdown, and natural biological presence, so it is extensively researched as a biodegradable metal. While magnesium degrades too quickly and zinc lacks sufficient strength, iron offers a balance between mechanical support and controlled biodegradation [12]. Iron-based biodegradable implants offer several advantages in biomedical applications, particularly in orthopedic fixation, cardiovascular stenting, and tissue engineering. Iron possesses a tensile strength of approximately 200–300 MPa, making it comparable to traditional stainless-steel implants. This slow degradation rate makes iron ideal for load-bearing applications like bone fixation plates, screws, and cardiovascular stents, preventing premature loss of mechanical integrity and ensuring sustained structural support during tissue healing. Unlike rapidly corroding magnesium, iron degrades gradually. Furthermore, iron is an essential element in the human body, crucial for hemoglobin formation and oxygen transport. While excessive iron ion release can cause toxicity, proper alloying and surface treatments help regulate iron dissolution, improving biocompatibility [13]. Medical applications of iron-based biodegradable implants include orthopedic fixation devices which are used for bone fracture repair, including plates, screws, and rods and provide strong mechanical support while gradually degrading as the bone heals. Unlike permanent stents, biodegradable stents used in arteries to prevent restenosis (re-narrowing of blood vessels) degrade over time. This characteristic lowers the risk of long-term complications such as chronic inflammation [14]. In tissue engineering scaffolds, they support the growth of new tissues and degrade as cells replace the scaffold [15] and are used in soft tissue and bone

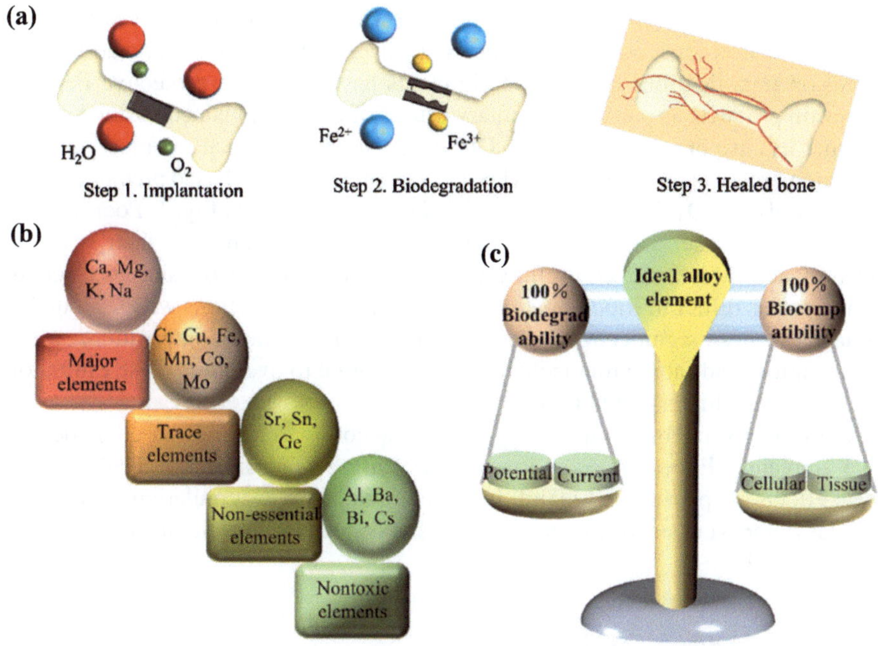

Fig. 3 **a** Schematic representation of Fe-based implant biodegradation [11]; **b** classification of metallic elements in the human body [11]; **c** necessary and sufficient conditions for Fe-based alloys to be biodegradable metals [11] with permission from Elsevier

regeneration applications [16]. Although iron-based implants offer several advantages, challenges related to corrosion control, cytotoxicity, and bioresorption rates must be addressed for widespread clinical adoption [17].

Advantages of iron-based biodegradable implants include their superior mechanical strength which provides structural support like conventional stainless-steel implants [18], and their predictable corrosion rate is unlike that of magnesium; iron degrades in a more controlled manner [12]. Adaptability for various medical applications which can be used in orthopedics, cardiovascular stents, and tissue scaffolds [19], and its potential for surface modifications, which are used for coatings and nano-engineered surfaces for enhancing biodegradability and biocompatibility [20], are key advantages.

1.1 Challenges of Iron-Based Biodegradable Implants

Iron-based biodegradable implants face several challenges despite their promising potential. One major issue is their slow degradation rate, which may not match the tissue healing timeline. Furthermore, the corrosion process can result in the accumulation of iron ions, potentially causing localized toxicity or inflammation. Most corrosion investigations on Fe-based implants have identified protective Fe-based corrosion products as one of the primary causes of their sluggish corrosion kinetics, whether in vitro or in vivo [21–24]. Our observations show that the most prevalent corrosion products produced by in vitro Fe corrosion are Ca/P-based precipitates [25, 26], FeOOH, Fe_2O_3, Fe_3O_4, and $Fe(OH)_3$ [26–28] as shown in Fig. 4. Poor mechanical properties, such as brittleness and lack of ductility, can limit their structural performance. Moreover, controlling uniform degradation in physiological conditions remains difficult. Biocompatibility concerns and interactions with surrounding tissues also pose significant hurdles. Advancements in alloy composition, surface modification, and fabrication techniques are essential to overcome these limitations for successful clinical application.

Cytotoxicity and biocompatibility are critical concerns for iron-based biodegradable implants. Excessive iron ion release during corrosion can lead to cytotoxic effects, impairing cell viability and function. Iron accumulation may disrupt cellular homeostasis and trigger oxidative stress, leading to inflammation or tissue

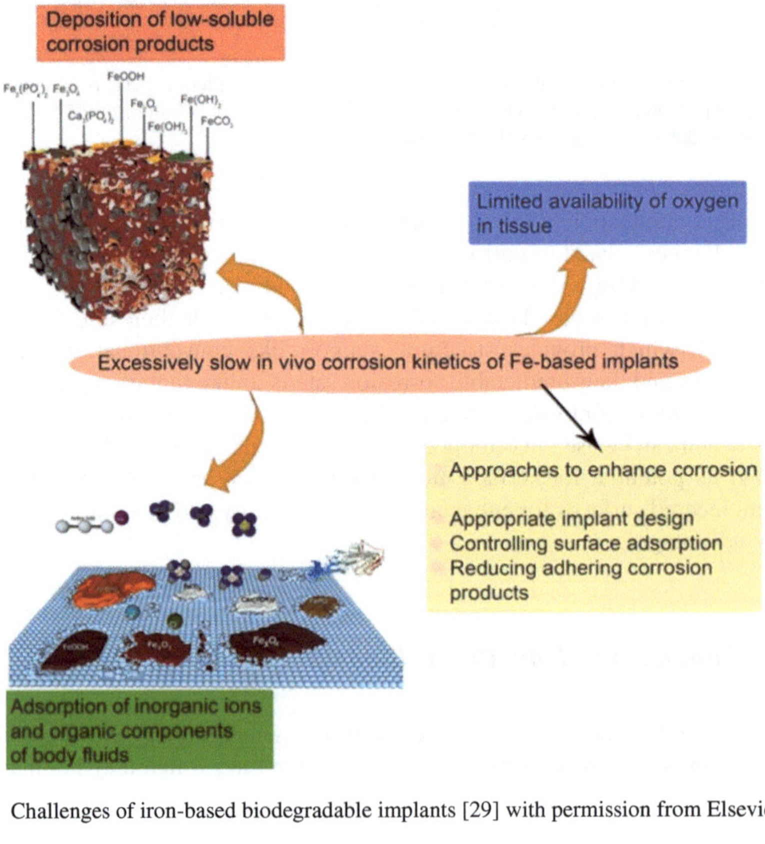

Fig. 4 Challenges of iron-based biodegradable implants [29] with permission from Elsevier

damage. Additionally, non-uniform degradation can result in sharp corrosion by-products that physically irritate surrounding tissues. Biocompatibility also depends on the implant's interaction with blood, immune cells, and local tissue, where adverse responses can hinder healing. Strategies like alloying, surface coatings, and controlled degradation are being explored to improve biocompatibility and minimize cytotoxic effects for safer clinical integration.

2 Fabrication Techniques for Iron-Based Biodegradable Implants

Modern manufacturing uses various processes to create components with specific properties, geometries, and functionalities. Among these, Powder Metallurgy (PM) [30], Additive Manufacturing (AM) [15], and Casting are key manufacturing methods for Fe-based biodegradable metals shown in Fig. 5. Each has unique process characteristics and requires tailored design strategies to optimize performance, cost, and manufacturability. PM involves compacting and sintering metal powders, favoring simple, symmetric designs for uniform density and easy ejection. AM builds parts layer by layer from digital models, enabling complex geometries, lightweight lattice structures, and part consolidation. Casting involves pouring molten metal into molds, needing careful design for flow, solidification, and mold release, with considerations like draft angles and uniform wall thickness. Understanding each process's strengths and limitations allows designers to optimize part geometry, functionality, and manufacturability for performance, cost-efficiency, and application-specific requirements.

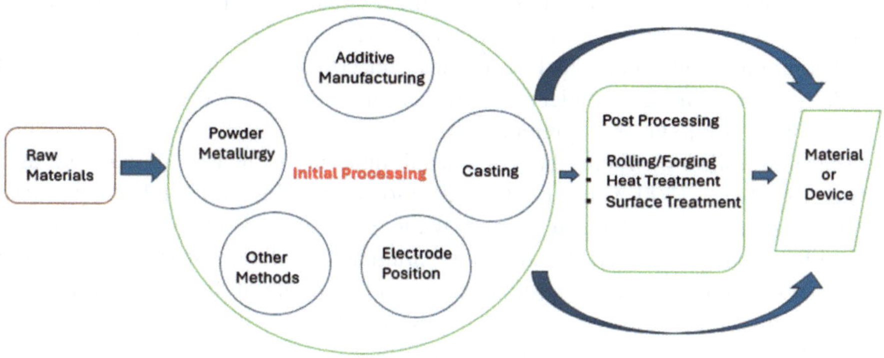

Fig. 5 Manufacturing methods for Fe-based biodegradable metals

2.1 Powder Metallurgy (PM)

It is a widely used method for fabricating iron-based biodegradable metals, particularly for biomedical applications such as temporary implants. This process enables precise control over composition, porosity, and mechanical properties, making it a popular technique for creating Fe-based biomaterials [31] and suitable for designing materials with controlled biodegradation rates. The adaptability of this powder metallurgy (PM) process allows for the creation of components with diverse shapes and characteristics. Using PM [31, 32] fabricated Fe–Mn alloys with Mn content ranging from 20 to 35% (b wt). Their findings indicated that these alloys degraded faster in physiological environments than pure Fe. They also observed that Fe–20Mn and Fe–25Mn alloys exhibited multiphase microstructures, while Fe–30Mn and Fe–35Mn alloys had single-phase microstructures. Notably, multiphase alloys showed a higher degradation rate (1.1–1.3 mmpy) compared to single-phase alloys (0.4–0.7 mmpy). Furthermore, the magnetic susceptibility of all these alloys was comparable to that of SS316L [33]. To create porous samples, the powder metallurgy approach was also successfully applied. The rate of degradation can be of better quality, and the mechanical characteristics can be brought closer to those of human bone by managing and adjusting the porosity [34, 35]. A porous shape lowers the implant's modulus of elasticity and speeds of degradation. Utilizing implants with a modulus of elasticity comparable to bone can prevent stress shielding, a documented problem with conventional devices [36–39]. Furthermore, porosity is a desirable characteristic in implants because it facilitates vascular invasion, osseointegration, and effective oxygen and nutrition transfer. Additionally, it makes better fixation between the implant and surrounding tissues and increases bioactivity by increasing the interfacial surface area [40].

The PM process begins with the production of fine iron powder, typically through gas atomization, water atomization, or chemical reduction. The iron powder is then blended with alloying elements like manganese (Mn), zinc (Zn), or magnesium (Mg), which influence the mechanical properties and corrosion behavior. The blended powder is then compacted under high pressure into a predefined shape using a die, forming what is known as a "green compact."

Sintering follows, where the compacted material is heated to temperatures between 1100 and 1300 °C in a controlled atmosphere. This step bonds the particles together, enhancing mechanical strength while maintaining some porosity, which is crucial for biodegradability. By adjusting parameters such as sintering temperature and time, manufacturers can fine-tune the degradation rate of the final product.

Metal-matrix composites were also developed using the powder metallurgy approach. To improve biocompatibility and biodegradability, numerous research teams created Fe/bioceramic composites [20, 41]. The addition of bioceramic to the Fe matrix, while improving its degradation rate, resulted in composites with lower yield and compressive strength compared to pure Fe. Sharma et al. [42] developed a new hybrid additive manufacturing (AM) process for creating scaffolds made of biodegradable hierarchically controlled porous (HCP) Fe–Mn alloy. The Fe–30Mn

Fig. 6 **a** SEM image **b** microstructure [42]

scaffold's morphological, mechanical, magnetic, and electrochemical characteristics were assessed and compared with those of pure Fe. The sample's austenite (γ) and martensite (ε) phases are visible in the XRD spectrum. The microstructure (Fig. 6b) and SEM image (Fig. 6a) demonstrate the bonding between the particles as well as the presence of some intrinsic porosity. The development of isolated pores results in inherent porosity, which signifies full densification.

The ability of powder metallurgy (PM) to produce porous structures is a key benefit, as it promotes tissue ingrowth and enhances bio-integration. The porosity also accelerates the degradation of iron implants, ensuring they dissolve at a desirable rate within the body. Additionally, PM minimizes material waste, making it a cost-effective and sustainable technique.

However, PM components generally exhibit lower mechanical strength compared to wrought materials, necessitating post-processing treatments like rolling, forging, or heat treatment. Despite this, PM remains a highly effective method for producing bioresorbable implants, including orthopedic scaffolds and cardiovascular stents.

Through careful selection of powder composition and sintering conditions, PM enables the development of biodegradable iron-based materials tailored to specific biomedical applications, offering a promising alternative to traditional permanent implants. Comparison of Powder Metallurgy, Additive Manufacturing, and Casting for manufacturing iron-based biodegradable metals is shown in Table 1.

2.2 Additive Manufacturing (AM)

Additive Manufacturing (AM), often referred to as 3D printing, is an advanced manufacturing technique that permits the precise fabrication of biodegradable metals based on iron. Unlike traditional methods, AM enables the direct creation of complex structures with controlled porosity, making it particularly useful for biomedical implants.

Table 1 Comparison of powder metallurgy, additive manufacturing, and casting for manufacturing iron-based biodegradable metals

Aspect	Powder metallurgy (PM)	Additive manufacturing (AM)	Casting
Process overview	Uses fine iron powders compacted and sintered at high temperatures	Layer-by-layer fabrication using laser or electron beam fusion of metal powders	Melts iron and alloying elements, pours them into a mold, and allows solidification
Material utilization	High material efficiency, minimal waste	Very high material efficiency with precise control	Some material loss due to mold design and machining
Porosity control	Good porosity control via powder selection and sintering parameters	Excellent control over porosity for tuned degradation rates	Difficult to introduce porosity; requires post-processing
Mechanical properties	Moderate strength, tunable via sintering and post-processing	Excellent mechanical properties with precise control over microstructure	High strength but lower control over microstructure
Surface finish	Requires post-processing for smooth finish	High-resolution finish but may need polishing	Often, a rough surface requiring additional finishing
Degradation rate control	Moderate control via sintering and alloying	Excellent control over degradation due to microstructure customization	Limited control; requires secondary processing for degradation tuning
Complex geometry feasibility	Limited; best for simple to moderately complex shapes	Excellent; enables complex, patient-specific designs	Moderate; best for bulk components rather than intricate designs
Production speed	Relatively slow; requires multiple steps	Can be slow due to the layer-by-layer process but allows customization	Fast for bulk production but slower for precision parts
Cost consideration	Cost-effective due to minimal material waste but requires specialized equipment	High cost due to expensive equipment and powder materials	Relatively low cost for large-scale production but requires post-processing
Suitability for medical applications	Well-suited for controlled biodegradation in bone implants and scaffolds	Ideal for customized implants, cardiovascular stents, and orthopedic scaffolds	Suitable for large, strong biodegradable implants but requires additional tuning for bioresorption

(continued)

Table 1 (continued)

Aspect	Powder metallurgy (PM)	Additive manufacturing (AM)	Casting
Post-processing requirements	Sintering, heat treatment, and sometimes machining	Heat treatment, surface modification, and stress relief	Heat treatment, machining, and sometimes coating for better degradation
Scalability	Moderate; suitable for medium-scale production	Limited scalability; best for customized implants	High scalability; good for mass production

The progress in additive manufacturing (AM) technologies [15, 43] has unlocked unparalleled potential for creating porous metallic orthopedic implants that can facilitate bone regeneration. Different AM technologies [15] have been successfully used to produce porous, degradable Fe-based implants with appropriate mechanical characteristics. Notably, [44] employed ink-jet 3D printing to fabricate a Fe–30Mn scaffold with 36.3% open porosity. The resulting scaffold demonstrated mechanical properties similar to human bone and a degradation rate of 0.73 mmpy, indicating its potential for use in degradable Fe-based implants.

Another additive manufacturing method for producing porous Fe-based biodegradable materials is 3D binder jet printing [45, 46]. The method was applied by Hong et al. [45] to create porous scaffolds using a combination of Ca/Mg and Fe–Mn powder. The ultimate compressive strength (UCS) of the 3D-printed Fe–Mn and Fe–Mn–1Ca scaffolds was 228 MPa, while their porosity was 39.3% and 52.9%, respectively [47]. Additionally, compared to sintered pellets of the identical alloys, the alloy showed a degradation rate that was around 12 times higher. The Fe–35Mn scaffold was created by Carluccio et al. [48] using selective laser melting (SLM) shown in Fig. 7. Alloying with Mn led to a degradation rate of 0.42 ± 0.03 mmpy, a fivefold increase over pure Fe, and the scaffold exhibited good cell adhesion and proliferation.

The creation of these implants starts with a digital 3D model designed using Computer-Aided Design (CAD) software, which is then sliced into layers for sequential printing with metal powders. Two common Additive Manufacturing (AM) techniques used for biodegradable iron metals are Selective Laser Melting (SLM) and Electron Beam Melting (EBM). Selective Laser Melting is a metal AM process that utilizes a high-powered laser in an inert gas environment to fully melt and fuse metal powder, layer by layer, resulting in high-density, intricate parts with excellent mechanical properties. Its compatibility with metals like titanium, stainless steel, and aluminum makes it suitable for high-precision, high-strength applications in aerospace, automotive, and medical fields. Electron Beam Melting is another AM process that employs an electron beam in a vacuum to melt and build parts from metal powder layer by layer. The entire powder bed is preheated, reducing residual stress and distortion. EBM typically uses titanium and nickel alloys, producing durable components for aerospace and medical implants. Though less precise than

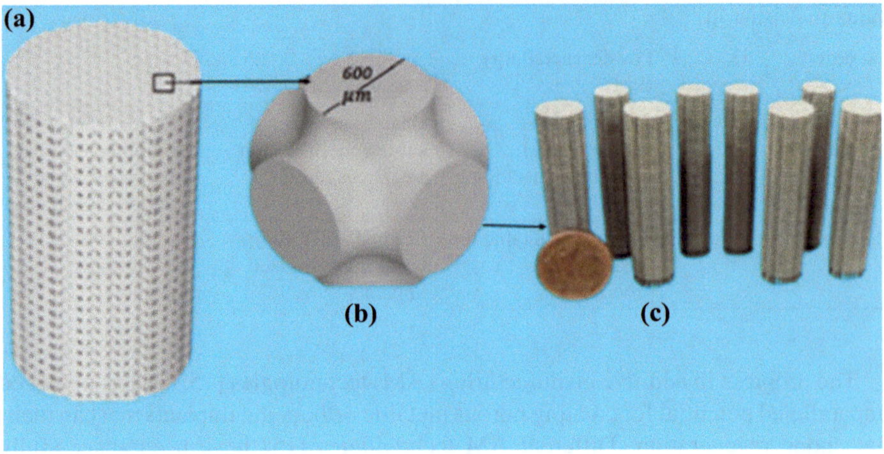

Fig. 7 Design stages of the scaffold: **a** CAD, **b** unit cell, **c** manufactured scaffold [30]

SLM, it excels in high-temperature, performance-critical applications with excellent structural integrity.

A major advantage of Additive Manufacturing (AM) is its capability to create patient-specific implants that precisely match individual anatomy. Furthermore, AM allows for fine control over porosity, a crucial factor in adjusting the degradation rate of biodegradable iron. This porosity increases the surface area exposed to bodily fluids, accelerating bioresorption while simultaneously promoting cell attachment and tissue integration.

Despite its advantages, AM has some limitations. The process requires expensive equipment, extensive post-processing (such as heat treatment or surface modifications), and strict control over printing parameters to avoid defects like residual stress and microcracks. To improve surface quality and mechanical performance, AM-produced components may require additional machining.

In biomedical applications, AM is particularly valuable for manufacturing biodegradable iron-based orthopedic scaffolds, cardiovascular stents, and temporary implants. By offering high design flexibility and the ability to engineer degradation rates, AM represents a breakthrough technology in the field of bioresorbable metal implants, paving the way for highly customized medical solutions.

2.3 Casting

One of the oldest and most cost-effective ways to produce biodegradable iron-based metals is through a process of melting iron with alloying elements, followed by pouring the molten metal into a mold to cool and solidify into a specific form. This process is particularly suitable for producing large implant components or bulk biodegradable metal structures. Because it makes it easier to create complex shapes and offers more possibilities for changing the composition of alloys, casting is the favored method for biodegradable alloys based on iron. Cheng and Zheng [49] examined the use of as-cast Fe, Mn, Mg, Zn, and W for degradable implants, reporting good cytocompatibility, hemocompatibility, and biocompatibility for magnesium, zinc, and iron with specific cell lines. However, He et al. [50] noted that as-cast metals often require post-processing (extrusion, rolling, forging) to improve microstructure, mechanical properties, and corrosion resistance due to defects like segregation, blow holes, and shrinkage.

In dynamic testing, the inclusion of Mn, Al, W, Co, and B greatly decreased the degradation rate by 50–75%, while the addition of C and S had no discernible effect on the degradation rate. When Sn was added, the alloy's yield strength significantly decreased, but Mn, W, B, Co, C, and S somewhat improved the rolled alloys' strength properties; these elements also increased the gap between Fe's yield and ultimate strength.

The casting process typically follows several steps. First, a mold is created using materials like sand, ceramic, or metal, depending on the required precision and surface finish. Next, iron is melted in a furnace, and alloying elements such as manganese (Mn), zinc (Zn), or magnesium (Mg) are added to modify the mechanical and degradation properties. The molten metal is then carefully poured into the mold and left to solidify. After cooling, the cast part is removed, cleaned, and subjected to optional heat treatments or machining for improved mechanical performance.

A major advantage of casting is its ability to produce complex shapes at a relatively low cost. However, controlling porosity and biodegradation behavior is challenging. Unlike Powder Metallurgy or Additive Manufacturing, casting typically results in dense structures, which may lead to slower degradation rates. This can be mitigated by alloying strategies, surface modifications, or post-processing techniques such as oxidation to enhance corrosion.

Despite potential drawbacks like shrinkage defects, inclusions, and lower precision compared to AM, casting remains an effective method for producing iron-based biodegradable materials. It is commonly used in research and development to prototype biodegradable implants before employing more advanced manufacturing methods.

In summary, casting is a practical technique for fabricating biodegradable iron-based medical implants, especially for applications where high mechanical strength is required. When combined with secondary processing, it can provide an affordable and scalable solution for bioresorbable metal components. Each method has its advantages and limitations summarized in Table 2, and the best technique depends on the specific biomedical application.

Table 2 Manufacturing methods for Fe-based biodegradable devices: advantages and disadvantages [30]

Method	Advantages	Disadvantages
Powder metallurgy	• Allows the direct production of relatively complex shapes • An appropriate technique for creating porous materials, composites, and alloys • By making small adjustments to production conditions, the mechanical properties and rate of degradation can be changed over a wide range • It is possible to obtain materials with tuned properties • The simplicity of customizing biodegradability through porosity optimization	• Challenges in preparing powder • Poor mechanical characteristics are caused by the possibility of inhomogeneity during the mixing and milling process
Additive manufacturing	• Best technique for creating porous scaffolds • Permits the mechanical properties of implant materials to resemble those of human bone • Osteoconduction and osseointegration are determined by the controllable shape and geometry of pores • Optimizing porosity makes it simple to customize biodegradability • Precision and flexibility	• Difficulties in powder preparations
Casting	• Easy to make complicated geometry. Affordability of alloy component customization	• Casting flaws like segregation, blowholes, and shrinking are quite likely to occur • After casting, post-processing is necessary • Fe-bioceramic composite cannot be produced • Protection against high stress

2.4 Electrochemical Deposition

Electrochemical deposition, also called electroforming, is a process where electricity is used to coat a metal surface with a thin layer of another metal from a liquid solution. This method allows scientists to precisely control the thickness, structure, and composition of iron-based biodegradable implants. To perfectly replicate the geometric features and dimensions of final parts, an article is electrodeposited onto a mold and then separated from the deposit (Fig. 8a, b).

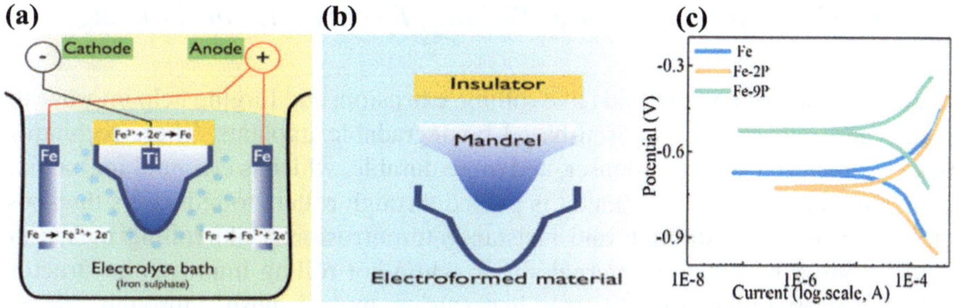

Fig. 8 **a**, **b** Device components of electroforming: electrolyte, anode, and cathode; **c** potentiodynamic polarization curves for Fe, Fe–2P, and Fe–9P samples [11] with permission from Elsevier

By adjusting factors like voltage, solution ingredients, and processing time, researchers can change how fast the implant degrades and improve its mechanical strength. Another big advantage is that this technique can add special bioactive coatings to implants, helping them resist corrosion and making them more compatible with the body [51].

Electrochemical deposition is especially useful for creating thin, high-quality iron structures for medical implants. However, it is mostly used for surface modifications, meaning it improves the outer layer rather than making the entire implant. While it is not ideal for bulk production, it plays an important role in enhancing implant durability and biocompatibility [52]. Fe-based biodegradable materials with superior mechanical qualities and strong corrosion activity can be produced by electroforming; these materials outperform cast iron. Furthermore, electroforming excels at small-scale manufacturing and replicating intricate shapes [53]. All of these offer favorable circumstances for the production of biodegradable materials based on iron.

Mostavan et al. [54] prepared a series of Fe samples by selecting a tin sheet as a support cathode via electroforming. The potentiodynamic polarization curves showed that the lowest corrosion potential (_0.725 V) was the Fee2P sample, and its corrosion rate was 1.46 ± 0.06 mm y^{-1} which the highest of all samples shown in Fig. 8c.

Interestingly, it is particularly suited to producing thin-walled, high-purity iron structures for biomedical devices, although it is mainly used for surface modifications and not bulk fabrication.

Electrochemical deposition enables rigorous control of microstructure, thickness, and composition of iron-based biodegradable implants. One can adjust the degradation rate and strength of the implant by changing operating parameters like electrolyte concentration, voltage, and deposition time. This method also offers the feasibility of introducing bioactive coatings in order to boost biocompatibility and corrosion protection. Most notably, it is well-suited to produce thin-walled, high-purity iron structures in biomedical devices but is largely only appropriate for surface modification rather than bulk fabrication.

2.5 Mechanical Processing (Rolling, Extrusion, and Forging)

Mechanical processing methods like rolling, extrusion, and forging help improve the quality and performance of iron-based biodegradable implants. These techniques make the material stronger, denser, and more durable, which is essential for medical applications [12]. In rolling, metal is passed through rollers to reduce its thickness while increasing its strength and resistance to corrosion. Cold rolling makes the metal stronger by creating internal stress, while hot rolling improves its structure and reduces internal stress. However, in extrusion, metal is pushed through a shaped opening (die) to create uniform structures with precise shapes. This makes the implant strong and consistent, which is ideal for load-bearing implants like bone supports. In forging, metal is shaped by applying strong compressive forces, which makes it tougher and more resistant to fractures. Open-die forging is used for larger implants, while closed-die forging is better for smaller, detailed implants with fewer defects.

These methods produce strong, durable, and defect-free implants [55]. However, they cannot precisely control porosity, which is important for managing how the implant degrades inside the body. Finding ways to improve porosity control is a key challenge for making even better biodegradable implants.

2.6 Fabrication-Performance Relationship

Each fabrication technique presents trade-offs between microstructure, mechanical properties, and corrosion behavior, which influence the implant's suitability for biomedical applications. Table 3 provides well-structured linking fabrication techniques for iron-based biodegradable implants to their impact on implant performance.

3 Characterization of Iron-Based Biodegradable Implants

Mechanical characterization of iron-based biodegradable implants is essential to ensure they provide adequate support during tissue healing. Pure iron typically exhibits a tensile strength (TS) of 200–250 MPa, compressive strength (CS) over 400 MPa, and flexural strength (FS) between 150 and 200 MPa [33, 58]. However, its low ductility and limited fatigue resistance can compromise performance under physiological loads [22]. Alloying with elements like manganese, zinc, and palladium has been shown to enhance mechanical strength and toughness while adjusting degradation rates [59]. Porous structures fabricated via powder metallurgy or additive manufacturing can reduce elastic modulus and prevent stress shielding. Surface treatments and grain refinement through thermo-mechanical processing also improve strength

Table 3 Impact of fabrication techniques on iron-based biodegradable implants

S. No.	Fabrication technique	Principle	Microstructure and porosity	Mechanical properties	Corrosion and degradation behavior	References
1	Powder metallurgy (PM)	Sintering of metal powders under high temperature and pressure	Moderate porosity, homogenous microstructure	Moderate strength, tunable mechanical properties	Controlled corrosion with porosity-dependent degradation rate	[56, 57]
2	Additive manufacturing (AM)	Layer-by-layer deposition using 3D printing techniques	High porosity, customizable structure	Strength depends on build parameters, good mechanical tunability	Accelerated degradation due to interconnected pores	[56, 57]
3	Electroforming	Electrochemical deposition of iron from electrolyte solution	Dense microstructure, low porosity	High strength, ductile properties	Slower corrosion due to low porosity	[56, 57]
4	Mechanical milling and compaction	Powder refinement followed by pressing into shape	Fine microstructure, moderate porosity	Improved mechanical strength compared to PM	Moderate corrosion rate	[56, 57]
5	Laser cladding	Surface melting of powder or wire via laser energy	Dense, refined microstructure	High hardness and wear resistance	Lower corrosion rate due to dense structure	[56, 57]

and corrosion behavior. Thus, optimized mechanical performance is achievable through material and structural engineering.

3.1 In Vitro and in Vivo Performance of Iron and Iron-Based Alloys

In vitro studies of iron and iron-based alloys demonstrate good mechanical strength and biocompatibility, but slow degradation limits their biomedical application [33]. High iron ion concentrations can induce cytotoxicity and oxidative stress in cell cultures [28]. In vivo, these materials show promising tissue compatibility and gradual integration, yet their corrosion rates remain insufficient for timely bioresorption [58]. Alloying with elements like Mn or Zn and surface modifications have been employed to enhance degradation and reduce adverse biological responses [60]. Further development is essential for optimized clinical performance due to the slow degradation of pure iron in physiological environments; scientists have explored various iron alloy designs for biomedical applications. The development of Fe-based alloys for biodegradable applications should focus on maintaining three crucial characteristics: (i) sufficient strength, ideally approaching that of 316L SS alloy; (ii) biocompatibility, ensuring no release of toxic breakdown products; and (iii) a controlled degradation rate within a clinically relevant period of 1–2 years. The mechanical properties of iron and iron-based alloys, such as ductility, yield strength, hardness, and ultimate strength, are detailed in Table 4.

The impact of alloying elements on the performance characteristics of biodegradable iron-based materials was initially documented by Refs. [5, 32, 33]. They created several Fe–Mn alloys with the goal of achieving a degradation behavior more appropriate than that of pure iron and mechanical qualities comparable to stainless steel 316L. da Silva et al. [61] evaluated the corrosion performance of the Fe–35Mn alloy using open circuit potential (OCP) and potentiodynamic polarization tests. Both experiments were carried out in Modified Hank's solution (H1387, Sigma-Aldrich, Darmstadt, Germany), which is comparable to human blood plasma in terms of ionic content and concentration. Figure 9 displays the Fe–35Mn alloy's potentiodynamic polarization curves in Hank's solution.

In vivo research on metallic iron and iron-based alloy implants has gained significant attention in the biomedical area during the past 15 years, particularly for applications in osteosynthesis and cardiovascular disease. An 18-month angiographic follow-up [58] demonstrated that the use of a degradable iron stent in the descending aorta of rabbits did not appear to induce thrombotic events, inflammation, or neointimal proliferation during the study.

Table 4 Mechanical properties: iron and its alloys [19]

Materials	Yield strength, MPa	Ultimate strength, MPa	Elongation % at fracture	Hardness HV/10
99.9% Fe (Armco)	250	300	37.5	85 ± 1
Nanocrystalline Fe:ECAP, 8 passes	470 ± 29	–	–	(Microhardness = 444 ± 31 kgF/mm^{-2})
Fe (carbon steel, Sht)	700	900	9.5	403 ± 2
Fe–10Mn (Sht)	NA	NA	NA	428 ± 6
Fe–10Mn (ht 1)[c]	800	1400	14	384 ± 5
Fe–10Mn (ht 2)[c]	650	1300	14	374 ± 7
Fe–10Mn–1Pd (Sht)[b]	950	1500	2	432 ± 8
Fe–10Mn–1Pd (ht 1)	900	1550	7	437 ± 3
Fe–10Mn–1Pd (ht 2)	850	1450	11	376 ± 6
Fe–20% Mn	421 ± 27	702 ± 11	7.5 ± 1.5	230 ± 1
Fe–25% Mn	361 ± 33	723 ± 19	4.8 ± 0.4	200 ± 1
Fe–30% Mn	239 ± 13	518 ± 14	19 ± 1.4	110 ± 1
Fe–35% Mn	234 ± 7	428 ± 7	32 ± 0.8	106 ± 2
Fe–21Mn–0.7C (TWIP)-ReXX	345 ± 10	980 ± 5	62 ± 4	
TWIP-1Pd(Rexx)[a]	360 ± 4	970 ± 35	64 ± 3	
TWIP-1Pd CW12[a]	690 ± 35	1120 ± 25	38 ± 3	
TWIP-1Pd CW12-ht[a]	505 ± 20	1020 ± 10	53 ± 7	
TWIP-1Pd CW23[a]	1095 ± 35	1320 ± 15	29 ± 2	
TWIP-1Pd CW23-ht[a]	725 ± 20	1255 ± 15	38 ± 2	
Stainless steel (316 L SS)	190	490	40	
Pure iron, annealed plate	150	210	40	119 ± 3 (HV 0.2/10)
Co–Cr alloys (ASTM F90)	310	860	20	
Mg, as cast	20	86	13	
Fe–5% Pd	445 ± 13	754 ± 20	–	184 ± 10 (HV0.2/10)
Fe–5% Pt	503 ± 22	785 ± 18	–	309 ± 8 (HV0.2/10)
Electroformed iron	360 ± 9	423 ± 12	8.3 ± 2	

(continued)

Table 4 (continued)

Materials	Yield strength, MPa	Ultimate strength, MPa	Elongation % at fracture	Hardness HV/10
Electroformed iron annealed at 550 °C	270 ± 6	292 ± 14	18.4 ± 4	
Electroformed iron annealed at 600 °C	130 ± 7	169 ± 9	32.3 ± 5	
Pure iron annealed at 550 °C	140 ± 10	205 ± 6	25.5 ± 3	
WE43 Mg alloy	162	250	2	
Fe–30Mn–6Si alloy: solution treated	180	450	16	
Fe alloyed by different elements (Mn, Co, Al, W, Sn, B, C, and S): as cast	100–220	190–360	12–23	

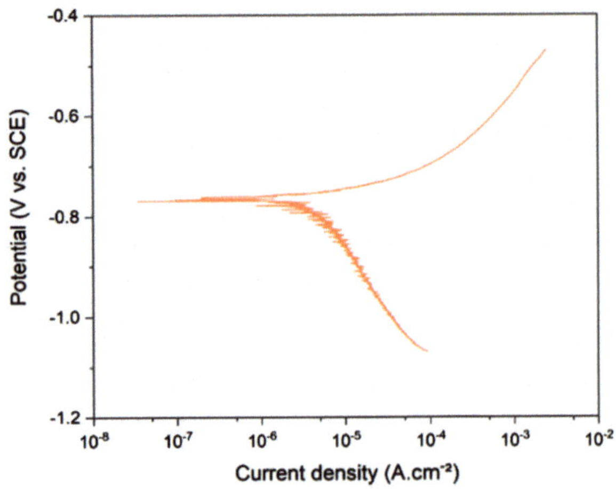

Fig. 9 Average potentiodynamic polarization curve of Fe–35Mn alloy samples in Hank's solution at 37 °C [61]

3.2 *Biological Performance and In vitro/In Vivo Studies*

The biological performance of Fe-based biodegradable implants is critical to evaluate their clinical viability. This includes not only cytocompatibility and biocompatibility but also thrombogenicity and hemocompatibility, especially for cardiovascular applications.

Cytotoxicity is a significant limitation in the development of iron-based biodegradable implants. During corrosion, iron releases Fe^{2+} and Fe^{3+} ions into the surrounding environment. While iron is an essential trace element, elevated local concentrations of these ions can disrupt cellular homeostasis and generate reactive oxygen species

(ROS), leading to oxidative stress, lipid peroxidation, DNA damage, and ultimately cell death [62]. In vitro studies have shown that excessive iron ion release adversely affects the viability, proliferation, and function of key cell types such as endothelial cells, fibroblasts, and smooth muscle cells [22]. These effects are particularly concerning for vascular implants, where proper endothelialization is essential for successful integration and healing.

The severity of cytotoxicity often depends on the corrosion rate and the total surface area of the implant. Although pure iron corrodes slowly, localized degradation or surface imperfections can lead to bursts of ion release. Moreover, iron's poor corrosion uniformity can exacerbate tissue irritation and cytotoxic effects [33]. To address these challenges, alloying iron with biocompatible elements such as manganese, zinc, and calcium has been explored to balance mechanical properties with controlled degradation. Surface modifications, including coatings with biopolymers or bioactive ceramics, also aim to reduce direct ion release and shield tissues from cytotoxic effects.

Ultimately, minimizing cytotoxicity is essential for the safe and effective clinical translation of iron-based implants. Further studies integrating in vitro assays with in vivo models are necessary to evaluate long-term biological responses.

Thrombosis and hemocompatibility are critical factors in the evaluation of iron-based biodegradable implants, especially for cardiovascular applications such as stents. Thrombosis, the formation of blood clots, can be triggered by poor blood-material interactions, leading to implant failure or life-threatening complications [63]. Iron surfaces can activate platelets and promote coagulation cascades, increasing thrombotic risk [64]. Moreover, corrosion products and ion release may exacerbate inflammation and disturb endothelial function, further impairing hemocompatibility [65].

In vitro hemocompatibility assessments have shown that pure iron and some iron-based alloys cause platelet adhesion and activation, although the extent varies with surface topography and corrosion behavior. Alloying elements such as manganese (Mn), zinc (Zn), and palladium (Pd) have been investigated to improve blood compatibility and reduce thrombogenicity. Additionally, surface modification techniques such as passivation, polymer coatings, and endothelial cell pre-seeding have shown promise in minimizing platelet activation and enhancing hemocompatibility.

Animal studies have also demonstrated that iron-based implants can remain thrombus-free under optimized conditions, indicating potential for safe in vivo performance. However, continued research is necessary to fully understand and mitigate the thrombogenic risks of these materials in clinical settings.

4 Prospects in Iron-Based Biodegradable Implants

Iron-based biodegradable implants hold significant promise in the field of biomedical engineering, particularly for orthopedic and cardiovascular applications. Unlike traditional permanent implants, these materials gradually degrade in the body, eliminating the need for secondary surgeries to remove hardware. Iron offers excellent mechanical strength and biocompatibility, closely matching the properties of bone, making it ideal for load-bearing applications. Recent research focuses on overcoming iron's inherently slow degradation rate by alloying it with elements like manganese, zinc, or phosphorus and by using surface modification techniques.

Additionally, innovations in porous structures and nano-engineering are enhancing the corrosion rate and promoting tissue integration. These developments are crucial for achieving timely resorption and ensuring biofunctionality during healing. Iron-based biodegradable implants are also being explored for drug delivery, allowing localized, controlled release of therapeutic agents. Regulatory interest and increasing demand for minimally invasive procedures are further fueling research and commercialization efforts.

Challenges remain, including optimizing degradation rates, ensuring uniform corrosion, and conducting comprehensive long-term biocompatibility studies. However, with ongoing advancements, iron-based biodegradable implants have the potential to revolutionize temporary implant technologies, offering safer, more efficient healing with reduced patient risk and healthcare costs. The field is poised for impactful clinical translation in the near future.

Future research should focus on:

- **Multi-material fabrication**: Combining different metals or coatings to create gradient biodegradability and enhanced mechanical properties
- **Advanced surface modifications**: Enhancing bioactivity and corrosion control through nano-structuring and functional coatings
- **In vivo performance studies**: Expanding long-term degradation and biocompatibility assessments in physiological environments
- **Hybrid manufacturing approaches**: Integrating PM, AM, or electroforming with post-processing techniques for tailored implant performance
- **Computational modelling**: Utilizing AI-driven design and simulations to predict corrosion behavior and mechanical performance before fabrication

5 Conclusion

Iron-based biodegradable implants have introduced new avenues in medical treatment, especially in orthopedic and cardiovascular interventions. They offer temporary mechanical support and slowly dissolve in the body, obviating the need for surgery for removal. However, finding the appropriate balance between mechanical strength, corrosion rate, and biocompatibility is still a challenge that necessitates advancement in fabrication and characterization methods. Several manufacturing techniques have been investigated, such as powder metallurgy, additive manufacturing, casting, sintering, electrochemical deposition, and mechanical processing. Each of these methods affects the microstructure, porosity, and degradation behavior of the implant, which makes it imperative to choose the most appropriate technique for medical applications. Surface modification processes also assist in improving corrosion resistance and promoting biological interactions. The identification of these implants is critical to determine structural integrity, degradation behavior, and biocompatibility. Scanning electron microscopy (SEM), X-ray diffraction (XRD), electrochemical testing, and in vitro/in vivo tests are some of the techniques used to gain an understanding of material performance. Using iron for implants that dissolve inside the body is a big step forward, especially for fixing broken bones and heart problems. These implants give support while healing happens and then disappear, so you do not need another surgery. However, achieving the right balance between mechanical strength, corrosion rate, and biocompatibility remains a challenge, requiring advancements in fabrication and characterization techniques. Scientists are trying different ways to build these implants, like pressing metal powders together, 3D printing, and coating them with special materials. Each method changes how the implant looks and works. Choosing the right method is important for each medical use. We also need to check how these implants are doing. We use microscopes and X-rays to see their structure. We also test them in labs and on living things to see how they break down and how the body reacts. This helps us make them better. While significant progress has been made, further research is needed to optimize degradation control, improve mechanical properties, and ensure regulatory compliance. Future advancements in biodegradable alloys, coatings, and composite materials will help refine implant performance, making them more reliable for clinical use and improving patient outcomes.

References

1. Ravanbakhsh S et al (2023) Study on the mechanical properties of magnetron sputtered W-based degradable radiopaque coatings for tiny biodegradable metallic endovascular implants. Eur J Mech A Solids 101:105072. https://doi.org/10.1016/j.euromechsol.2023.105072
2. Zhai Y, Zhang H, Wang J, Zhao D (2023) Research progress of metal-based additive manufacturing in medical implants. Rev Adv Mater Sci 62(1). https://doi.org/10.1515/rams-2023-0148
3. Ren X et al (2015) Surface modification and endothelialization of biomaterials as potential scaffolds for vascular tissue engineering applications. Chem Soc Rev 44(15):5680–5742. https://doi.org/10.1039/C4CS00483C
4. Bram M, Ebel T, Wolff M, Cysne Barbosa AP, Tuncer N (2013) Applications of powder metallurgy in biomaterials. In: Advances in powder metallurgy. Elsevier, pp 520–554. https://doi.org/10.1533/9780857098900.4.520
5. Moravej M, Prima F, Fiset M, Mantovani D (2010) Electroformed iron as new biomaterial for degradable stents: development process and structure–properties relationship☆. Acta Biomater 6(5):1726–1735. https://doi.org/10.1016/j.actbio.2010.01.010

6. Baburaj A et al (2024) Biodegradable based TENGs for self-sustaining implantable medical devices. Nano Energy 127:109785. https://doi.org/10.1016/j.nanoen.2024.109785
7. Singh J, Kaur T, Singh N, Pandey PM (2020) Biological and mechanical characterization of biodegradable carbonyl iron powder/polycaprolactone composite material fabricated using three-dimensional printing for cardiovascular stent application. Proc Inst Mech Eng H 234(9):975–987. https://doi.org/10.1177/0954411920936055
8. Benčina M, Resnik M, Starič P, Junkar I (2021) Use of plasma technologies for antibacterial surface properties of metals. Molecules 26(5):1418. https://doi.org/10.3390/molecules26051418
9. Prasad A (2021) Bioabsorbable polymeric materials for biofilms and other biomedical applications: recent and future trends. Mater Today Proc 44:2447–2453. https://doi.org/10.1016/j.matpr.2020.12.489
10. Rane AV, Abitha VK, Sisanth KS, Kanny K (2024) Introduction to polymer materials for implants. In: Polymeric materials for biomedical implants. Elsevier, pp 1–29. https://doi.org/10.1016/B978-0-323-99690-7.00009-1
11. Li S et al (2022) Dilemmas and countermeasures of Fe-based biomaterials for next-generation bone implants. J Market Res 20:2034–2050. https://doi.org/10.1016/j.jmrt.2022.07.089
12. Gąsior G, Szczepański J, Radtke A (2021) Biodegradable iron-based materials—what was done and what more can be done? Materials 14(12):3381. https://doi.org/10.3390/ma14123381
13. Bandyopadhyay A, Mitra I, Goodman SB, Kumar M, Bose S (2023) Improving biocompatibility for next generation of metallic implants. Prog Mater Sci 133:101053. https://doi.org/10.1016/j.pmatsci.2022.101053
14. Khan AR, Grewal NS, Zhou C, Yuan K, Zhang H-J, Jun Z (2023) Recent advances in biodegradable metals for implant applications: exploring in vivo and in vitro responses. Results Eng 20:101526. https://doi.org/10.1016/j.rineng.2023.101526
15. Jiang D, Ning F, Wang Y (2021) Additive manufacturing of biodegradable iron-based particle reinforced polylactic acid composite scaffolds for tissue engineering. J Mater Process Technol 289:116952. https://doi.org/10.1016/j.jmatprotec.2020.116952
16. Mishra DK, Pandey PM (2022) Corrosion behavior and degradation mechanism of micro-extruded <scp>3D</scp> printed ordered pore topological Fe scaffolds. J Biomed Mater Res B Appl Biomater 110(6):1439–1459. https://doi.org/10.1002/jbm.b.35011
17. Wang A, Venezuela J, Dargusch MS (2023) Enhancing the corrodibility of biodegradable iron and zinc using poly(lactic) acid (PLA) coating for temporary medical implant applications. Prog Org Coat 174:107301. https://doi.org/10.1016/j.porgcoat.2022.107301
18. Salama M, Vaz MF, Colaço R, Santos C, Carmezim M (2022) Biodegradable iron and porous iron: mechanical properties, degradation behaviour, manufacturing routes and biomedical applications. J Funct Biomater 13(2):72. https://doi.org/10.3390/jfb13020072
19. Francis A, Yang Y, Virtanen S, Boccaccini AR (2015) Iron and iron-based alloys for temporary cardiovascular applications. J Mater Sci Mater Med 26(3):138. https://doi.org/10.1007/s10856-015-5473-8
20. Wang H, Zheng Y, Li Y, Jiang C (2017) Improvement of in vitro corrosion and cytocompatibility of biodegradable Fe surface modified by Zn ion implantation. Appl Surf Sci 403:168–176. https://doi.org/10.1016/j.apsusc.2017.01.158
21. Trincă LC et al (2021) Evaluation of in vitro corrosion resistance and in vivo osseointegration properties of a FeMnSiCa alloy as potential degradable implant biomaterial. Mater Sci Eng, C 118:111436. https://doi.org/10.1016/j.msec.2020.111436
22. Kraus T et al (2014) Biodegradable Fe-based alloys for use in osteosynthesis: outcome of an in vivo study after 52weeks. Acta Biomater 10(7):3346–3353. https://doi.org/10.1016/j.actbio.2014.04.007
23. Wegener B et al (2020) Development of a novel biodegradable porous iron-based implant for bone replacement. Sci Rep 10(1):9141. https://doi.org/10.1038/s41598-020-66289-y
24. Yusop AHM, Alsakkaf A, Kadir MRA, Sukmana I, Nur H (2021) Corrosion of porous Mg and Fe scaffolds: a review of mechanical and biocompatibility responses. Corros Eng, Sci Technol 56(4):310–326. https://doi.org/10.1080/1478422X.2021.1879427

25. Dong H, Lin F, Boccaccini AR, Virtanen S (2021) Corrosion behavior of biodegradable metals in two different simulated physiological solutions: comparison of Mg, Zn and Fe. Corros Sci 182:109278. https://doi.org/10.1016/j.corsci.2021.109278
26. Yusop AHM, Daud NM, Nur H, Kadir MRA, Hermawan H (2015) Controlling the degradation kinetics of porous iron by poly(lactic-co-glycolic acid) infiltration for use as temporary medical implants. Sci Rep 5(1):11194. https://doi.org/10.1038/srep11194
27. Md Yusop AH, Alsakkaf A, Noordin MA, Idris H, Nur H, Szali Januddi F (2021) Degradation-triggered release from biodegradable metallic surfaces. J Biomed Mater Res B Appl Biomater 109(12):2184–2198. https://doi.org/10.1002/jbm.b.34866
28. Zhou J, Yang Y, Alonso Frank M, Detsch R, Boccaccini AR, Virtanen S (2016) Accelerated degradation behavior and cytocompatibility of pure iron treated with sandblasting. ACS Appl Mater Interfaces 8(40):26482–26492. https://doi.org/10.1021/acsami.6b07068
29. Al Sakkaf A, Januddi FS, Yusop AHM, Nur H (2022) Challenges in the use of Fe-based materials for bone scaffolds applications: perspective from in vivo biocorrosion. Mater Today Commun 33:104564. https://doi.org/10.1016/j.mtcomm.2022.104564
30. Rabeeh VPM, Hanas T (2022) Progress in manufacturing and processing of degradable Fe-based implants: a review. Prog Biomater 11(2):163–191. https://doi.org/10.1007/s40204-022-00189-4
31. Hermawan H, Dubé D, Mantovani D (2006) Development of degradable Fe-35Mn alloy for biomedical application. Adv Mat Res 15–17:107–112. https://doi.org/10.4028/www.scientific.net/AMR.15-17.107
32. Hermawan H, Alamdari H, Mantovani D, Dubé D (2008) Iron–manganese: new class of metallic degradable biomaterials prepared by powder metallurgy. Powder Metall 51(1):38–45. https://doi.org/10.1179/174329008X284868
33. Hermawan H, Dubé D, Mantovani D (2010) Degradable metallic biomaterials: design and development of Fe–Mn alloys for stents. J Biomed Mater Res A 93A(1):1–11. https://doi.org/10.1002/jbm.a.32224
34. Zhang Q, Cao P (2015) Degradable porous Fe-35wt%Mn produced via powder sintering from NH_4HCO_3 porogen. Mater Chem Phys 163:394–401. https://doi.org/10.1016/j.matchemphys.2015.07.056
35. Dehghan-Manshadi A, StJohn DH, Dargusch MS (2019) Tensile properties and fracture behaviour of biodegradable iron-manganese scaffolds produced by powder sintering. Materials 12(10):1572. https://doi.org/10.3390/ma12101572
36. Ridzwan MIZ, Shuib S, Hassan AY, Shokri AA, Mohamad Ib MN (2007) Problem of stress shielding and improvement to the hip implant designs: a review. J Med Sci 7(3):460–467. https://doi.org/10.3923/jms.2007.460.467
37. Mour M et al (2010) Advances in porous biomaterials for dental and orthopaedic applications. Materials 3(5):2947–2974. https://doi.org/10.3390/ma3052947
38. Shayesteh Moghaddam N et al (2016) Metals for bone implants: safety, design, and efficacy. Biomanuf Rev 1(1):1. https://doi.org/10.1007/s40898-016-0001-2
39. Prasad K et al (2017) Metallic biomaterials: current challenges and opportunities. Materials 10(8):884. https://doi.org/10.3390/ma10080884
40. Shimko DA, Shimko VF, Sander EA, Dickson KF, Nauman EA (2005) Effect of porosity on the fluid flow characteristics and mechanical properties of tantalum scaffolds. J Biomed Mater Res B Appl Biomater 73B(2):315–324. https://doi.org/10.1002/jbm.b.30229
41. Ulum MF et al (2014) In vitro and in vivo degradation evaluation of novel iron-bioceramic composites for bone implant applications. Mater Sci Eng, C 36:336–344. https://doi.org/10.1016/j.msec.2013.12.022
42. Sharma P, Kumar A, Kumar A (2025) An economically viable hybrid additive manufacturing process for the fabrication of hierarchically controlled porous biodegradable Fe–Mn scaffolds. Mater Lett 380:137701. https://doi.org/10.1016/j.matlet.2024.137701
43. Zadpoor AA, Malda J (2017) Additive manufacturing of biomaterials, tissues, and organs. Ann Biomed Eng 45(1):1–11. https://doi.org/10.1007/s10439-016-1719-y

44. Chou D-T, Wells D, Hong D, Lee B, Kuhn H, Kumta PN (2013) Novel processing of iron–manganese alloy-based biomaterials by inkjet 3-D printing. Acta Biomater 9(10):8593–8603. https://doi.org/10.1016/j.actbio.2013.04.016
45. Hong D et al (2016) Binder-jetting 3D printing and alloy development of new biodegradable Fe–Mn–Ca/Mg alloys. Acta Biomater 45:375–386. https://doi.org/10.1016/j.actbio.2016.08.032
46. Yang C, Huan Z, Wang X, Wu C, Chang J (2018) 3D printed Fe scaffolds with HA nanocoating for bone regeneration. ACS Biomater Sci Eng 4(2):608–616. https://doi.org/10.1021/acsbiomaterials.7b00885
47. Morgan EF, Unnikrisnan GU, Hussein AI (2018) Bone mechanical properties in healthy and diseased states. Annu Rev Biomed Eng 20(1):119–143. https://doi.org/10.1146/annurev-bioeng-062117-121139
48. Carluccio D et al (2020) Additively manufactured iron-manganese for biodegradable porous load-bearing bone scaffold applications. Acta Biomater 103:346–360. https://doi.org/10.1016/j.actbio.2019.12.018
49. Cheng J, Zheng YF (2013) In vitro study on newly designed biodegradable Fe-X composites (X = W, CNT) prepared by spark plasma sintering. J Biomed Mater Res B Appl Biomater 101B(4):485–497. https://doi.org/10.1002/jbm.b.32783
50. He J et al (2016) Advances in Fe-based biodegradable metallic materials. RSC Adv 6(114):112819–112838. https://doi.org/10.1039/C6RA20594A
51. Exner HE, Arzt E (1990) Sintering processes. In: Sintering key papers. Springer Netherlands, pp 157–184. https://doi.org/10.1007/978-94-009-0741-6_10
52. Oriňaková R, Gorejová R, Orságová Králová Z, Oriňak A (2020) Surface modifications of biodegradable metallic foams for medical applications. Coatings 10(9):819. https://doi.org/10.3390/coatings10090819
53. McGeough JA, Leu MC, Rajurkar KP, De Silva AKM, Liu Q (2001) Electroforming process and application to micro/macro manufacturing. CIRP Ann 50(2):499–514. https://doi.org/10.1016/S0007-8506(07)62990-4
54. Mostavan A, Paternoster C, Tolouei R, Ghali E, Dubé D, Mantovani D (2017) Effect of electrolyte composition and deposition current for Fe/Fe-P electroformed bilayers for biodegradable metallic medical applications. Mater Sci Eng, C 70:195–206. https://doi.org/10.1016/j.msec.2016.08.026
55. Chandrasekaran M (2019) Forging of metals and alloys for biomedical applications. In: Metals for biomedical devices. Elsevier, pp 293–310. https://doi.org/10.1016/B978-0-08-102666-3.00010-9
56. Liu J et al (2024) Microstructural evolution, mechanical properties and corrosion mechanisms of additively manufactured biodegradable Zn–Cu alloys. J Mater Sci Technol 186:142–157. https://doi.org/10.1016/j.jmst.2023.10.052
57. Lee M-K et al (2022) Accelerated biodegradation of iron-based implants via tantalum-implanted surface nanostructures. Bioact Mater 9:239–250. https://doi.org/10.1016/j.bioactmat.2021.07.003
58. Peuster M (2001) A novel approach to temporary stenting: degradable cardiovascular stents produced from corrodible metal—results 6–18 months after implantation into New Zealand white rabbits. Heart 86(5):563–569. https://doi.org/10.1136/heart.86.5.563
59. Yang Y et al (2017) Cu-releasing bioactive glass/polycaprolactone coating on Mg with antibacterial and anticorrosive properties for bone tissue engineering. Biomed Mater 13(1):015001. https://doi.org/10.1088/1748-605X/aa87f2
60. Li Y, Wu A, Li Q, Zhao Y, Zhu R, Wang G (2019) Effects of welding parameters on weld shape and residual stresses in electron beam welded Ti2AlNb alloy joints. Trans Nonferrous Metals Soc China 29(1):67–76. https://doi.org/10.1016/S1003-6326(18)64916-7
61. da Silva KB, Carobolante JPA, Nakazato RZ, Caporalli Filho A, Rosifini Alves AP (2025) Influence of a novel thermomechanical processing route on the structural, mechanical, and corrosion properties of a biodegradable Fe-35Mn alloy. Metals (Basel) 15(4):462. https://doi.org/10.3390/met15040462

62. Zhou X, Liu C, Yu L, Liu Y, Li H (2015) Phase transformation behavior and microstructural control of high-Cr martensitic/ferritic heat-resistant steels for power and nuclear plants: a review. J Mater Sci Technol 31(3):235–242. https://doi.org/10.1016/j.jmst.2014.12.001
63. Anderson JM, Rodriguez A, Chang DT (2008) Foreign body reaction to biomaterials. Semin Immunol 20(2):86–100. https://doi.org/10.1016/j.smim.2007.11.004
64. Ferrari G, Khir AW, Fresiello L, Di Molfetta A, Kozarski M (2011) Hybrid model analysis of intra-aortic balloon pump performance as a function of ventricular and circulatory parameters. Artif Organs 35(9):902–911. https://doi.org/10.1111/j.1525-1594.2011.01244.x
65. Frerker C et al (2013) MitraClip® via direct right atrial access in case of a missing inferior vena cava. EuroIntervention 9(5):643–647. https://doi.org/10.4244/EIJV9I5A102

Chapter 4
Clinical Translation, Regulatory Considerations, and Future Perspectives: Challenges in Iron-Based Biodegradable Materials for Biomedical Applications

Vipin Goyal and Girish Verma

1 Introduction

Coronary artery disease, a major cause of death in developed nations, is typified by the narrowing of coronary arteries, which lowers the oxygen flow to the myocardium [1]. The first angioplasty was carried out in 1977, in which an inflatable balloon was inserted into a narrowed blood vessel and then inflated and stored for a prolonged [2]. Although this method was a major breakthrough, it has drawbacks, such as a high rate of restenosis, vascular dissection, and elastic recoil. The late 1980s saw the introduction of coronary stents, which reduced restenosis rates and provided structural support to preserve vessel patency to overcome these restrictions. However, stent thrombosis and neointimal hyperplasia were issues with early stent design that resulted in vasoconstriction [3].

Sigward et al. [4] made the first coronary stent in 1987 using stainless steel for surgical purposes. Stents manufactured with other materials, such as Co–Cr alloys, Ti–Ni alloys, and 316L stainless steel, are also used in clinical practice today. These stents can be broadly divided into two types: balloon-expandable and self-expanding. During delivery, a sheath restricts self-expanding stents, which then expand when the sheath is removed. Although they have certain benefits, such as the capacity to accommodate different vessel diameters, they have also been linked to technical issues and a tendency to cause more neointimal hyperplasia. Because of their good performance characteristics and clinical results, balloon-expandable stents, which are positioned on a balloon catheter and affected by the balloon, have emerged as the preferred option for the majority of coronary procedures [2].

V. Goyal · G. Verma (✉)
Department of Mechanical Engineering, Indian Institute of Technology Indore, Indore, India
e-mail: girish.verma@iiti.ac.in

A. Dixit et al. (eds.), *Biodegradable Metallic Materials*, Biomedical Materials for Multi-functional Applications, https://doi.org/10.1007/978-981-95-2401-3_4

Mani et al. [5] listed nine requirements for the perfect coronary stent, including (i) appropriate expandability ratio, (ii) secure crimping onto the balloon catheter, (iii) adequate flexibility, (iv) low recoil and high radial hoop strength, (v) biocompatibility, (vi) high thromboresistance, (vii) low post-implant restenosis rates, (viii) drug delivery capability, (ix) sufficient radiopacity and reliability for MRI.

The development of biodegradable or bioresorbable stents (BDS) has revolutionized the treatment of coronary artery disease. These stents gradually deteriorate once their purpose is served, usually six months after insertion. They offer blocked vessels temporary mechanical support. Long-term issues with permanent stents, including restenosis, chronic inflammation, and late stent thrombosis, are lessened by this deterioration. Biodegradable stents can lessen bleeding issues associated with long-term anticoagulant therapy, eliminate imaging abnormalities in computed tomography (CT) and magnetic resonance imaging (MRI) evaluations, and lower the likelihood of permanently reduced flow in covered side branches. Moreover, biodegradable stents allow for vascular expansion in juvenile patients, preventing further procedures to remove or replace the implant. Additionally, biodegradable stents enable vascular growth in young individuals, avoiding additional procedures to replace or remove the implant. Therefore, the tenth feature in the criteria for the ideal stent should be complete biodegradability [6].

Metals and polymers are the two main types of materials employed to manufacture bioresorbable stents (BRS). The majority of the early research was on scaffolds made of polymers. The first entirely biodegradable coronary stent, the Igaki-Tamai stent, was implanted in 1998 and was fabricated of high-molecular-weight poly-L-lactic acid (PLLA). The implantation of 25 PLLA scaffolds across 19 lesions in 15 patients was depicted in further research, which was published in 2000 [7]. After three years, the stent scaffold was no longer visible in the vasculature, indicating full bioreabsorption, according to long-term assessments conducted on 50 patients [8].

Although the results were excellent, the device did not advance since it needed a heated contrast, a bigger guide catheter compared to a metallic stent for implantation, and a drug coating. However, the idea of using PLLA in stent design remained, and more research and development were conducted as a result. In 2016, based on the successful implementation of the first PLLA-based absorb drug-eluting device, the absorb drug-eluting device was approved by the US Food and Drug Administration (FDA) 2016. Several biodegradable polymers and copolymers, such as poly(caprolactone) (PCL), phosphorylcholine (ChoP), poly(anhydride ester) (PAE), poly(D, L-lactide-co-glycolic acid) (PLGA), poly(L-lactide-co-caprolactone) (PLCL), poly(vinylidene fluoride-co-hexafluoropropylene) (PVDF-HFP), and poly(desaminotyrosyl-tyrosine ethyl ester) carbonate (PTD-PC) have all been studied in the development of BRS [9]. These polymeric materials have some drawbacks that prevent them from being widely used despite encouraging results. Their poor mechanical properties, such as lower Young modulus value (0.2 and 7.0 GPa) in comparison to their metallic counterparts (54 and 200 GPa) are a major disadvantage [10]. In order to attain the necessary mechanical strength, this discrepancy requires bigger struts in polymer-based stents. This might hinder full expansion during balloon angioplasty, negatively impacting deliverability and

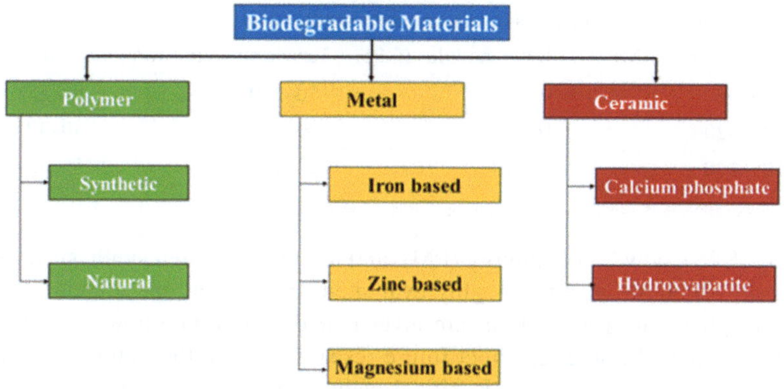

Fig. 1 Classification of biodegradable materials

vascular wall apposition. The manufacturing of biodegradable metal stents is the subject of increased scientific efforts in an attempt to overcome these obstacles.

Corrosion is the main way that metals deteriorate in the physiological environment. As a result, metals like titanium, nitinol, and stainless steel that are typically used in first-generation stents have a strong resistance to corrosion and are inappropriate for uses that call for resorbable materials. Several biodegradable materials are used in different biomedical applications as shown in Fig. 1.

Iron (Fe), magnesium (Mg), and zinc (Zn) are the three main metals that have been discovered by intensive research over the past 20 years as having the potential to be used in biodegradable cardiovascular implants. In contrast, zinc exhibits a corrosion rate that is more in line with the intended stent lifespan, providing a balance between mechanical support and deterioration. However, pure zinc's inherent mechanical qualities are not ideal; thus, improvements are required to increase strength and ductility without negatively affecting corrosion rates. The development of Zn-based alloys with enhanced mechanical performance appropriate for stent applications is still being researched.

Because of their exceptional strength, ductility, and formability, Fe-based materials can be used to create stents with intricate geometries and thinner struts. Pure iron, however, has a very low corrosion rate (CR), which limits its ability to be classified as a truly "biodegradable" material and causes it to remain in the body for a prolonged period of time. Methods to speed up iron's rate of corrosion are being researched in order to make it more appropriate for applications involving temporary implants due to its mechanical benefits and biocompatibility (Table 1).

This article's latter sections go into great detail on iron's physicochemical characteristics, biodegradability, and corrosion behavior, all of which can be assessed using electrochemical and immersion testing methods. The biological performance of Fe-based materials is examined rigorously in terms of their clinical applicability, cellular interactions, tissue biocompatibility, and hemocompatibility. The corrosion behavior of Fe-based alloy depends on the fabrication processes, such as spark plasma

Table 1 The mechanical properties of biodegradable metals [11]

	Yield strength (MPa)	Elastic modulus (GPa)	Tensile strength (MPa)	Hardness (HV)
Mg	50–52	42–46	170–230	36–38
Zn	290–320	16–30	90–180	40–42
Fe	105–120	190–230	220–240	155–157

sintering (SPS), powder metallurgy (PM), and template-assisted synthesis of porous structures, which are also thoroughly evaluated. In addition, the effects of adding secondary phases to iron matrices are investigated to ascertain how they affect the degradation rates. In order to direct future research toward the optimization of Fe-based biodegradable materials, this critical evaluation attempts to separate production techniques that have the potential for biomedical applications from those that have major drawbacks.

2 Biodegradability of Pure Iron

In 2001, Peuster et al. [12] conducted an in vivo study demonstrating that pure iron degrades too slowly and is considered a bioabsorbable substance. This outcome highlighted the need to change iron in order to increase its rate of degradation while maintaining its natural biocompatibility and favorable mechanical characteristics. Subsequent studies have concentrated on understanding the influence of the implant environment on iron's degradation behavior and the types of degradation products that result from it.

In physiological conditions, the cathodic reaction rate and the dissolved oxygen content have a major impact on the rate at which iron corrodes. Iron corrodes more quickly in environments with higher oxygen concentrations. On the other hand, a decreased rate of corrosion results from a decreased oxygen supply. These insights are critical for the development of iron-based biodegradable implants, as they highlight the importance of tailoring the material's degradation characteristics to match the specific oxygen conditions of the implantation site.

$$\mathrm{Fe} \rightarrow \mathrm{Fe}^{2+} + 2\mathrm{e}^{-}(\text{anodic reaction}) \tag{1}$$

Electrochemical reaction for Fe corrosion works as follows [13]:

$$\mathrm{O}_2 + 2\mathrm{H}_2\mathrm{O} + 4\mathrm{e}^{-} \rightarrow 4\mathrm{OH}^{-}(\text{cathodic reaction}) \tag{2}$$

As predicted in the Pourbaix diagram, Fe(OH) is the most common corrosion product due to the reaction:

$$\mathrm{Fe}^{2+} + 2\mathrm{OH}^{-} \rightarrow \mathrm{Fe(OH)}_2(\text{products of the reaction}) \tag{3}$$

In physiological environments, the corrosion of iron yields various degradation products, predominantly iron oxides such as FeO˙ Fe_2O_3˙ The presence of elements like sodium, chloride, carbon, and phosphorus in the human body leads to the formation of additional compounds, including iron carbonates, iron phosphates, and iron chlorides [14]. To account for these complex interactions, Qui et al. [15] have developed the Pourbaix diagram for iron to reflect the specific conditions encountered within the human body, incorporating parameters such as ambient temperature (37 °C), ionic concentrations ($[Fe^{2+}] = 1 \times 10^{-6}$ mol/dm^3, $[Fe^{3+}] = 3 \times 10^{-5}$ mol/dm^3), and partial pressure of oxygen $O_2 = 13.3$ kPa). This adapted diagram serves as a valuable tool for designing and interpreting iron degradation experiments under physiological conditions.

Kraus et al. [16] investigated the corrosion behavior of pure iron and iron-based implant biomaterial Fe–10Mn–1Pd and Fe–21Mn–0.7C–1Pd (Fig. 2). The optical image of the implant material after 4 weeks (Fig. 2a–c) and 52 weeks (Fig. 2d–f) of implantation are shown. A slightly dark degradation was observed on the surface after 4 weeks of implantation. After the 52 weeks (Fig. 2d, e) of implantation, the optical images. Compared with the samples implanted for 4 weeks; however, more voluminous degradation products were observed, and nearly the whole pin surface was covered with degradation products.

After 4 weeks of implantation (Fig. 3a–c), cross-sectional analysis revealed metal at the bottom, degradation products in the middle, and embedding material on top. Degradation layers of pure Fe contained Fe and O, P, and Ca, with traces of Na and K. In Fe–Mn–Pd and Fe–Mn–C–Pd alloys, Mn was present in the degradation products, along with localized Pd accumulation near the metal–matrix interface.

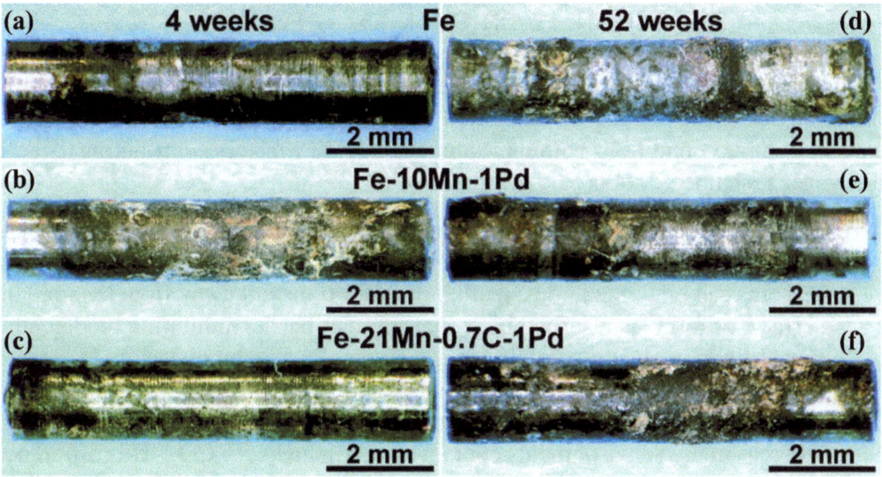

Fig. 2 Optical micrographs of the pins after implantation for (**a**–**c**) 4 weeks and (**d**–**f**) 52 weeks of (**a**, **d**) pure Fe, (**b**, **e**) Fe–10Mn–1Pd, and (**c**, **f**) Fe–21Mn–0.7C

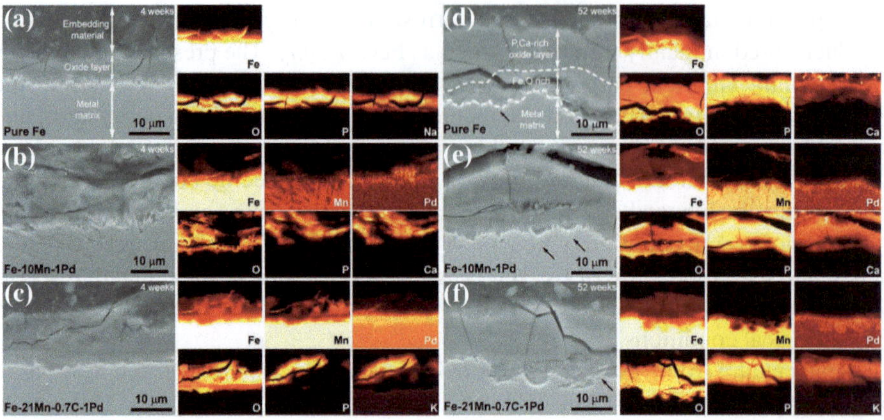

Fig. 3 SEM images of the cross-sections of the pins after implantation for (**a–c**) 4 weeks and (**d–f**) 52 weeks of (**a, d**) pure Fe, (**b, e**) Fe–10Mn–1Pd and (**c, f**) Fe–21Mn–0.7C. The left-hand images show secondary electron contrast; the right-hand images display the corresponding elemental distribution maps

After 52 weeks (Fig. 3d, e), over time, this layer thickened, developing into a bi-layered passivation film. The innermost layer, adjacent to the iron substrate, was primarily composed of Fe and O. The outer oxide layer formed subsequently through the oxidation of initial corrosion products, leading to the formation of stable iron hydroxide (Fe $(OH)_3$). These findings provide insights into the temporal evolution of corrosion layers on iron implants within a physiological environment.

$$Fe(OH)_2 + 1/2H_2O + 1/4O_2 \rightarrow Fe(OH)_3 \quad (4)$$

Differences between in vitro and in vivo CR of iron-based implants are frequently observed due to the inability of simulated physiological environments to fully replicate the complex conditions within the human body. Key factors contributing to these differences include:

(i) **Proteins**: Standard pseudo-physiological fluids mimic the pH and ionic concentrations of blood plasma but lack proteins, which are crucial components. Proteins can bind metal ions, facilitating their transport away from the implant surface, or form an adsorbed layer that acts as a barrier, inhibiting biodegradation. Alternatively, this protein layer may restrict oxygen diffusion, leading to localized corrosion in oxygen-deficient regions.

(ii) **Atmospheric conditions**: Many degradation tests are conducted in ambient air, whereas in vivo environments contain approximately 5% CO_2. This difference can influence the nature and composition of degradation products formed on the implant surface.

(iii) **Cellular interaction**: Post-implantation, the healing process involves cellular coverage of the implant surface within weeks. This cellular layer alters the local environment, affecting corrosion behavior and potentially contributing to the observed disparities between long-term in vitro and in vivo corrosion rates.

(iv) **Dissolved O_2 levels**: Blood contains about 3 cm^3/dm^3 of dissolved O_2 (4.3 mg/dm^3), whereas solutions like Hank's used in cell-based tests may have nearly double this concentration (~ 8 mg/dm^3). This elevated oxygen level can accelerate corrosion rates in vitro, leading to inconsistencies when compared to in vivo conditions.

Corrosion tests

To comprehensively evaluate the corrosion behavior of Fe-based materials intended for physiological applications, in vitro assessments typically encompass two primary methodologies: immersion and electrochemical tests.

(i) **Immersion testing**

As delineated in ASTM G31-21, the standard guide for laboratory immersion corrosion testing of metals, static immersion tests involve suspending the specimen in a simulated physiological fluid that mirrors the ionic composition of the targeted biological environment. The exposure duration is predetermined, typically ranging from 1 to 5 weeks. The mass loss is used to calculate the CR in this test. The following formula relates this mass loss to a time-related unit:

$$\text{CR} = kM/At\rho \tag{5}$$

where CR is measured in mm/year, *k*-conversion constant for computing the degradation rate whose actual value depends on the units of the variables, whose value is equal to 87,600, *M*, mass loss (gm); *A*, specimen surface (cm^2); *t*, exposure time (hr); and ρ, density of the sample in this situation.

In vitro corrosion assessment of iron-based materials for physiological applications encompasses both immersion and electrochemical testing methodologies.

(ii) **Electrochemical Testing**

Electrochemical techniques offer quantitative evaluations of corrosion rates and mechanisms. Potentiodynamic polarization (PDP), as specified in ASTM G59, involves scanning a small potential range to obtain the polarization resistance of the material [17]. This approach facilitates the calculation of current density, which could be employed to estimate the material's CR. Alternatively, electrochemical impedance spectroscopy (EIS), guided by ASTM G106, models material degradation by analyzing impedance data through equivalent electrical circuits. Both methods provide valuable insights into the electrochemical behavior of iron-based materials in simulated physiological environments.

Integrating dynamic immersion testing with electrochemical analyses enhances the predictive accuracy of in vitro corrosion assessments, contributing to developing more durable and biocompatible cardiovascular implants.

3 Iron Metabolism

Fe is an essential element for the human body, playing a pivotal role in numerous biochemical processes, including oxygen transport, electron transfer, and energy production. These functions are facilitated by iron's ability to form diverse and dynamic coordination complexes with organic ligands. The oxidation reaction potential between ferrous (Fe^{2+}) and ferric (Fe^{3+}) ions is critical for driving various biological reactions.

Despite its essentiality, iron's bioavailability is limited. Under aerobic conditions, soluble Fe^{2+} is readily oxidized to Fe^{3+}, which is virtually insoluble at physiological pH levels. This limited solubility poses challenges for iron utilization in biological systems. Moreover, the redox activity of iron, while fundamental to many biochemical reactions, can also contribute to the generation of reactive O_2 species (ROS). Catalytic amounts of Fe can make hydroxyl radicals ($OH^{\cdot}$) from superoxide (O_2^{-}) and hydrogen peroxide (H_2O_2), as well as other reactive species such as thiyl (RS·), thiyl-peroxyl (RSOOS), alkoxyl ($RO^{\cdot}$), and peroxyl (ROOS) radicals. These extremely reactive free radicals have the ability to cause oxidative changes in proteins and nucleic acids. An increase in ROS levels, termed oxidative stress, is associated with various inflammatory, neurodegenerative, and ischemic processes and can accelerate tissue degeneration, contributing to aging.

Fe deficiency can lead to significant health issues. In young children, insufficient Fe levels are related to neurological and psychosomatic developmental disturbances, decrease learning productivity, and could enhance autism. In adults, Fe deficiency can cause fatigue, elevate the risk of depression, and impair thyroid function. Low birth weight, preterm, intrauterine growth restriction, and higher maternal sickness are among the negative consequences linked to low iron levels during pregnancy [18].

Given the similarities in iron metabolism across mammalian species, animal studies provide reliable insights into the long-term behavior of iron-based implants in the human body. A typical adult human body contains approximately 3–5 gm of Fe, with the majority allocated to erythroid precursors for the production of hemoglobin and mature erythrocytes [19]. Each erythrocyte contains over one billion atoms of Fe, contributing to a total erythroid Fe content of about 30 mg/kg body weight. The liver serves as the primary storage site for the body's remaining Fe reserves.

Dietary iron absorption occurs in the duodenum, where enterocytes facilitate its uptake. These cells express ferric reductase enzymes that reduce Fe^{3+} to Fe^{2+}, ensuring iron is in the appropriate oxidation state for transport. The acidic environment of the stomach enhances this process by facilitating the reduction of Fe^{3+} to Fe^{2+}, thereby promoting absorption. Once absorbed, iron can be stored intracellularly as ferritin or exported into the plasma via ferroprotein. In the plasma, iron binds to transfer for delivery to various tissues. The body regulates iron absorption through hepcidin, a peptide hormone that modulates ferroprotein activity, thereby maintaining iron homeostasis.

Iron recycling is a critical component of iron metabolism. Senescent erythrocytes are phagocytosed by macrophages, primarily in the spleen and liver, where hemoglobin is degraded and iron is reclaimed for reuse. This recycling process is highly efficient, with the body conserving most of its iron stores. Daily iron losses occur through mechanisms such as gastrointestinal tract shedding, sweating, and minor bleeding, amounting to approximately 1–2 mg per day. Notably, humans lack a regulated excretory pathway for iron; thus, maintaining iron balance relies predominantly on the regulation of intestinal absorption. When iron levels exceed the body's needs, regulatory mechanisms, such as increased hepcidin expression, reduce dietary iron absorption to prevent overload.

Understanding these pathways is essential when considering the fate of iron released from bioabsorbable implants. Given the body's limited capacity for iron excretion, careful consideration must be given to the potential accumulation of iron from such implants and its subsequent impact on systemic iron homeostasis.

4 Biological Properties of Iron-Based Materials

Biocompatibility is the capacity of a material to carry out its intended function within a host organism without causing negative consequences like immunological rejection, poisonousness to nearby or detached tissues, or mechanical injury. This encompasses attributes including bio-functionality, bio-inertia, bioactivity, and biostability, ensuring that the material integrates seamlessly with the biological environment without provoking undesirable immune responses, allergic reactions, inflammation, or chronic complications. Moreover, biocompatible materials must not exhibit carcinogenic properties. In the context of biodegradable materials, evaluating biocompatibility extends beyond the initial implant to include its degradation products. For iron-based implants, this necessitates a comprehensive assessment at multiple levels.

4.1 Cellular Biocompatibility

Biocompatibility at the cellular level necessitates that the biomaterial does not exhibit toxicity toward endothelial cells, which line the artery walls, or to blood-borne cells. Consequently, in vitro cytotoxicity studies frequently employ endothelial cells, smooth muscle cells (SMCs), and fibroblasts to evaluate the potential adverse effects of these materials [20].

For instance, Huang et al. [21] conducted cytotoxicity assessments on Ag-implanted pure Fe using human umbilical vein endothelial cells (EA.hy-926), human vascular smooth muscle cells (VSMCs), murine fibroblast cells (L-929), and human umbilical vein endothelial cells (EA.hy-926). Similarly, Chang et al. [22] utilized similar cell lines in their study. Additionally, Capek et al. [23] employed the L-929

MG63 cell line, while Liu et al. [24] used human umbilical vein endothelial cells (ECV304) and rodent vascular smooth muscle cells in analogous tests. A different strategy was used by Paim et al. [25], who extracted adipose-derived stem cells from the abdominal adipose tissue of a healthy adult donor undergoing tumescent liposuction and then washed and centrifuged the tissue.

Cytotoxicity evaluations can be performed through direct or indirect contact methods. Direct contact assays involve placing the material in immediate contact with the cells, whereas indirect methods expose cells to the material's extracts. In vitro studies often favor indirect contact to mitigate the effects of localized high concentrations of corrosion products in the confined environment of cell cultures, a scenario less prevalent in vivo due to the continuous washout into the bloodstream [26]. However, it is significant to note that indirect methods may overlook the potential cytotoxic effects of insoluble corrosion products and reactive oxygen intermediates generated during material degradation. Studies have demonstrated that direct contact between iron materials and cells can induce cytotoxicity and oxidative stress, underscoring the necessity of employing multiple evaluation methods to obtain a comprehensive assessment of cytocompatibility.

The half-maximal inhibitory concentration (IC50) is a widely used metric to quantify cytotoxicity, representing the concentration of a substance required to reduce cell viability by 50%. For example, Zhu et al. [13] noticed no toxicity effects on human endothelial cells after 24 h of exposure to a 50 μg/mL concentration of Fe. Considering that a fully degraded stent in a 60 kg healthy person with a circulating blood volume of approximately 2800 mL would result in a blood Fe concentration not exceeding 7 μg/mL, the low risk of toxicity appears minimal. Furthermore, the International Organization for Standardization (ISO) 10993-5 standard stipulates that a material is considered cytotoxic if it reduces cell viability by more than 30%. Adhering to this guideline, Paim et al. [25] demonstrated that Fe does not exhibit toxicity and is suitable for use in biodegradable stents.

4.2 Tissue Biocompatibility

The body starts a series of reactions when a stent is implanted during surgery, including early damage, blood-material interactions, the production of a temporary matrix, acute and chronic inflammation, the development of granulation tissue, foreign-body reaction, and fibrosis. Extrinsic and intrinsic coagulation pathways, the complement system, fibrinolytic processes, kinin-generating systems, and platelets are all activated during this process, which results in thrombus development. In order to enhance the continuous release of bioactive chemicals from the stent and hasten wound healing, a temporary scaffold is manufactured around the implant.

Acute inflammation typically transitions to chronic inflammation, lasting one to two weeks; prolonged inflammation beyond three weeks may indicate infection [27]. Understanding these stages is crucial during in vivo testing to identify and document any abnormal biological responses. Notably, studies have not reported issues with infections or implant rejection due to the presence of iron, suggesting favorable tissue compatibility. However, long-term in vivo animal studies have noted discoloration of surrounding tissues due to iron degradation products accumulating in vessel walls. Despite this discoloration, no infectious responses or tissue damage were observed, and tissue color returned to normal after approximately 53 months. Liu et al. [28] proposed that macrophages facilitate the clearance of degradation products, which are then transported via the lymphatic system to the lymph nodes.

4.3 Hemocompatibility

Ensuring the hemocompatibility of cardiovascular stents necessitates evaluating their interactions with blood components, particularly concerning platelet adhesion and hemolysis. Platelet adhesion assessments, as delineated by ASTM F756-13, quantify the extent of platelet attachment to a material's surface; excessive adherence may indicate thrombogenic potential, raising concerns about thrombus formation. Hemolysis testing evaluates the propensity of a material to induce the rupture of erythrocytes, with a hemolysis rate exceeding 5% relative to a negative control deemed unacceptable according to standardized criteria. These evaluations typically employ human blood from volunteers or animal models, such as rabbit or lamb blood.

Iron has demonstrated commendable hemocompatibility, characterized by minimal platelet adhesion and hemolysis rates well below the critical 5% threshold [25]. However, alloying iron with elements like zinc (Zn) or silver (Ag) can alter these interactions [21]. Notably, certain Zn–Ag-based alloys have exhibited reduced biocompatibility, attributed to the presence of Ag-rich nano-precipitates that promote non-uniform corrosion behavior. Nonetheless, studies have shown that applying thermal treatments to dissolve these precipitates can enhance the uniformity of corrosion and improve biocompatibility. The treated Zn–Ag alloys exhibited reduced neointimal growth and inflammation compared to their untreated counterparts. These findings underscore the complex interplay between alloy composition, microstructural characteristics, and hemocompatibility. They highlight the necessity for meticulous design and processing of iron-based stent materials to ensure optimal performance within the cardiovascular system.

5 Manufacturing of Iron Biodegradable Devices

The selection of manufacturing techniques for biodegradable Fe-based implants significantly influences their mechanical properties and corrosion behavior. Various methods, including spark plasma sintering, powder metallurgy, casting, additive manufacturing, and electroforming, have been employed to fabricate these devices, each offering distinct advantages and limitations (Fig. 4); among all the manufacturing processes, a few have been chosen and explained in this article.

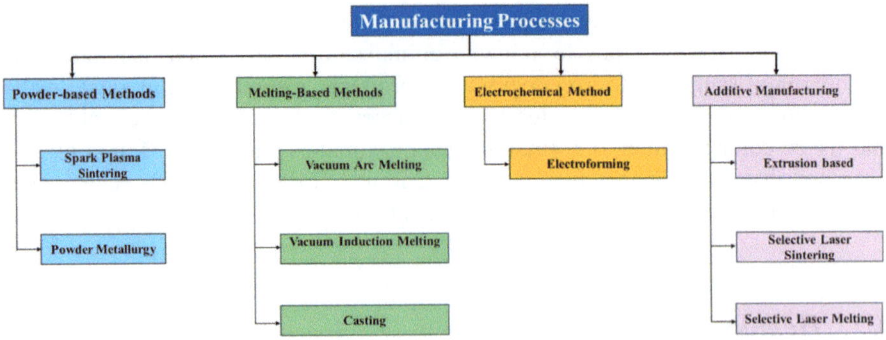

Fig. 4 Classification of manufacturing processes for biodegradable materials

5.1 *Spark Plasma Sintering*

Spark Plasma Sintering (SPS) is an advanced, quick sintering technique for consolidating powdered materials, including metals, ceramics, composites, and intermetallic compounds. Distinct from conventional sintering methods that utilize alternating current for heating, SPS applies periodically repeated pulses of direct high current intensity (ranging from several to tens of thousands of amperes) and low voltage to the consolidated powder. These pulses, lasting from a few to several hundred milliseconds, facilitate efficient heating. The SPS process is characterized by its lower temperature, high efficiency, short processing duration, and requirements compared to traditional sintering techniques.

The application of SPS in fabricating biodegradable Fe-based materials has been explored in various studies. For instance, Huang et al. [29] utilized SPS to prepare Fe–5% Pd and Fe–5% Pt composites by sintering pure elemental powders under vacuum at 1000 °C for a holding time of 5 min. Similarly, Cheng and Zheng [22] produced Fe-based materials with tungsten as the secondary phase, sintering at 950 °C under a pressure of 40 MPa for 5 min.

The corrosion rates (CR) of Fe-based materials fabricated via SPS vary depending on their composition. Table 2 presents the CR for selected Fe-based materials received through SPS, highlighting the influence of alloying elements on degradation behavior.

5.2 *Powder Metallurgy*

Powder Metallurgy (PM) is a manufacturing process that involves four important steps: (1) powder preparation, (2) mixing and blending, (3) compacting, and (4) sintering. Occasionally, secondary operations such as coining, impregnation, and hot forging are employed to enhance the properties of the final product. The PM process offers several advantages, including process efficiency, high purity of the resulting

Table 2 CR for Fe-based materials obtained by SPS

Process	Materials	CR (mm/year)	Ref
SPS	Fe	0.016–0.017	[22]
	Fe2W	0.075–0.080	[22]
	Fe5W	0.138–0.140	[22]
	Fe35Mn5Si	0.025–0.030	[29]
	Fe25Mn10Cu	0.258–0.260	[30]
	Fe34Mn1Cu	0.032–0.035	[30]
	Fe	0.105–0.110	[31]
	Fe3Mn	0.105–0.110	[31]

products, and the capability to mix metals and non-metals in precise proportions. Consequently, PM has been extensively utilized to produce Fe-based biodegradable materials.

However, PM-fabricated materials typically exhibit higher degradation rates compared to their cast or wrought counterparts. This increased degradation is primarily attributed to the inherent porosity in PM materials, which amplifies the surface area exposed to the corrosive environment. Additionally, the presence of oxides formed during processing can further influence corrosion behavior.

For instance, Hermawan et al. [32] employed PM to fabricate a series of Fe–Mn alloys—specifically Fe20Mn, Fe25Mn, Fe30Mn, and Fe35Mn—using highly pure elemental powders. The numerical values denote the normal weight (in %) of Mg. The compacted powder mixtures underwent sintering at 1220 °C for 4 h in a flowing Ar-5% H_2 gas mixture. The incorporation of Mn powder notably altered the degradation properties of the alloys, enhancing their degradation rates. However, it was determined that an optimal manganese content exists; specifically, a 20% addition was deemed most favorable for achieving the desired biodegradability. Increasing the manganese content beyond this threshold results in a reduction of the CR in the Fe–Mn systems (Table 3).

Table 3 CR for Fe-based materials received by PM [33]

Method	Material	CR (mm/year)
PM	Fe20Mn	1.3
	Fe25Mn	1.1
	Fe30Mn	0.7
	Fe35Mn	0.4
	Fe	0.2

Table 4 CR for Fe-based materials achieved by PM followed by cold rolling

Method	Materials	CR (mm/Year)
PM cold rolled	Fe20Mn	0.5
	Fe25Mn	0.5
	Fe30Mn	0.7
	Fe35Mn	0.7

5.3 Mechanical Treatments Inducing Plastic Deformation

Mechanical treatments that cause plastic deformation, like cold rolling or cold drawing, could be used to improve the mechanical properties of pure Fe devices. These processes refine the grain structure, as described by the Hall–Petch relationship, leading to increased strength. However, it is important to note that such treatments may reduce the material's ductility. Additionally, these mechanical treatments influence the corrosion behavior of the material. For instance, studies have shown that cold rolling can affect the corrosion resistance of pure iron, with factors such as rolling temperature playing a significant role. Specifically, cryogenic rolling temperatures have been observed to impact grain refinement and corrosion resistance differently compared to room-temperature rolling. Therefore, while mechanical treatments can enhance certain mechanical properties, their effects on corrosion behavior must be carefully considered in the design and manufacturing of pure iron devices. The corrosion rate of Fe-based material obtained from the PM followed by cold rolling is shown in Table 4 [33, 34].

5.4 Additively Manufactured Fe Materials

Additive manufacturing (AM) is a promising manufacturing technique for developing patient-specific biodegradable metal scaffolds, particularly utilizing materials such as iron. This approach enables the design and production of scaffolds tailored to individual anatomical and physiological requirements. By manipulating the porous architecture of these scaffolds, it is possible to control the degradation rate, which is crucial for applications like bone tissue engineering. Furthermore, medicines or physiologically active compounds can be incorporated into scaffold design to improve therapeutic results.

Among the various AM technologies, electron beam melting (EBM) and selective laser sintering (SLS) are commonly employed for fabricating metallic components. These techniques utilize high-energy beams to selectively fuse metal powders, layer upon layer, to construct complex geometries. However, the application of EBM and SLS in metal printing is constrained by factors such as the high cost of equipment and limited material availability.

Extrusion-based AM has been explored as a more accessible alternative. This method involves the deposition of a continuous flow of material through a micro-sized nozzle, guided by a robotic arm, to build structures layer by layer. It offers flexibility in material selection and cost-effectiveness. For instance, Taylor et al. [35] fabricated the porous Fe scaffolds using extrusion-based AM followed by conventional sintering. In their study, a PLGA-based copolymer served as the binder for the preparation of the metallic-polymer ink. Post-sintering, the resultant porous metallic structures exhibited potential for biomedical applications.

To address the limitations associated with conventional sintering, Sharma and Pandey [36] developed a six-step process that combines AM with pressureless microwave sintering to fabricate porous iron scaffolds. Microwave sintering offers advantages such as reduced energy consumption and shorter processing times compared to traditional methods. Moreover, it has been reported to improve the mechanical properties of the sintered components.

These advancements underscore the potential of AM technologies in creating patient-specific, biodegradable iron-based scaffolds. Ongoing research is focused on optimizing printing parameters, material preparation, and post-processing operations to further improve the performance and applicability of these scaffolds in clinical settings.

6 Iron-Based Material Modifications

6.1 Addition

Alloying is a prevalent strategy to improve the corrosion rate and tailor the mechanical properties of pure iron for biomedical applications. The incorporation of additional elements can either dissolve within the iron matrix or form metal matrix composites, significantly influencing the corrosion behavior of the resultant alloy. The electrochemical characteristics of these alloying elements are pivotal in determining the degradation kinetics of iron-based devices. Two primary approaches are employed to reduce the corrosion resistance of iron alloys:

(i) **Addition of Elements with a Lower Electrochemical Potential than Iron**

Incorporating elements such as gallium (Ga), which possesses a more negative electrochemical potential than iron (− 0.57 V vs. SHE), can enhance the CR through micro-galvanic interactions. Wang et al. [37] observed that adding Ga to pure iron increased the CR by a factor of 5, primarily due to pitting corrosion mechanisms.

(ii) **Addition of Nobler Elements to Induce Galvanic Corrosion**

Introducing elements with higher electrochemical potentials than Fe can produce a second phase that promotes galvanic corrosion, wherein Fe works as the anode, and the nobler metal serves as the cathode.

Palladium (Pd): Pd is completely miscible in iron and can stabilize the austenitic phase, thereby enhancing mechanical properties. Capek et al. [23] demonstrated that alloying iron with Pd increased both mechanical strength and CR, with porous implants fabricated via SPS projected to degrade within approximately five years.

Platinum (Pt): Similar to Pd, Pt has a higher electrochemical potential (+ 1.18 V vs. SHE). Huang et al. [29] compared Fe-5wt%Pt and Fe-5wt%Pd composites and reported that both additions accelerated iron corrosion and improved mechanical properties. Notably, Pt modification resulted in superior enhancements in hardness and yield strength, almost three times those of pure Fe.

Gold (Au): With an electrochemical potential of + 1.69 V versus SHE, Au is partially soluble in iron, allowing for second-phase formation under controlled conditions. Huang et al. [29] manufactured Fe–Au parts using SPS with varying Au content, noting that composites containing 5 wt% Au exhibited the highest corrosion rates after 30 days of immersion, degrading faster than pure iron without increased cytotoxicity or platelet adhesion.

Silver (Ag): Commonly added to medical devices for its biocompatibility and antibacterial properties, Ag has an electrochemical potential of + 0.800 V versus SHE and is insoluble in Fe. Huang et al. [38] reported that adding Ag to pure Fe resulted in a faster CR than Au additions without inducing cytotoxicity or thrombogenic effects. However, Capek et al. [23] observed that incorporating 2 wt% Ag into pure iron decreased the corrosion rate in both long-term immersion and potentiodynamic tests.

In addition to metal additions, non-metallic elements have been explored to modify iron's degradation behavior:

Carbon Nanotubes (CNTs): Cheng and Zheng [22] investigated the incorporation of CNTs into pure iron via SPS, finding that CNTs increased the CR and enhanced mechanical strength.

Nitrogen (N): Nitriding iron has been shown to accelerate CR and enhance mechanical properties. Lin et al. [39] did long-term in vivo studies and reported that nitride iron alloys are fully biocompatible. Conversely, certain elements have been found to have minimal or inconsistent effects on iron's degradation.

Manganese (Mn), Aluminum (Al), and Boron (B): Liu and Zheng [31] observed that small additions of these elements slightly slowed corrosion rates, with Mn often added to Fe biodegradable element to enhance the mechanical properties and biodegradability.

Tin (Sn): The same study reported that Sn additions did not significantly affect corrosion rates or mechanical properties.

Iron Oxide (Fe_2O_3): Cheng and Zheng [22] reported that the effect of adding Fe_2O_3 to pure Fe obtained by SPS and found no substantial differences in corrosion behavior, cytotoxicity, or platelet adhesion between Fe_2O_3–Fe composites and pure Fe.

These findings underscore the complexity of tailoring Fe-based biodegradable materials through alloying. The selection and concentration of alloying elements should be carefully considered to achieve desired degradation rates and mechanical properties suitable for specific biomedical applications.

6.2 *Iron–Manganese Alloys*

The incorporation of manganese (Mn) into iron (Fe) alloys has been extensively investigated to enhance the mechanical properties and biodegradability of materials intended for biomedical applications. This line of research was initiated by Hermawan et al. [32] in 2009, who introduced the Fe-35 wt% Mn (Fe–35Mn) alloy as a promising candidate for biodegradable stents, citing its favorable magnetic susceptibility and corrosion behavior.

Mn is an essential trace element in the human body, predominantly found in the pancreas, liver, bones, kidneys, and adrenal glands. Normal human blood Mn concentrations range between 4 and 15 μg/L, indicating the human body's capacity to tolerate manganese released from implants. However, prolonged exposure to elevated Mn levels can be neurotoxic, potentially leading to neurological disorders such as Parkinson's disease, as well as adverse cardiac effects, including myocardial contractile inhibition and hypotension. Therefore, the design of Fe–Mn biodegradable cardiovascular devices must carefully consider the potential for local accumulation of manganese ions during material degradation. Notably, the presence of Fe in Fe–Mn alloys may mitigate Mn-induced cytotoxicity, as Fe deficiency is known to increase manganese absorption.

Physically, manganese shares similarities with iron, and the two elements are mutually soluble. Mn possesses a lower electrochemical potential than Fe (− 1.18 V vs. SHE), which contributes to the more active corrosion potential observed in Fe–Mn alloys compared to pure iron. Studies have demonstrated that Fe–Mn alloys exhibit higher CR than pure Fe, with the degradation rate increasing with Mn content. For instance, a Fe–30Mn alloy displayed a CR 20 times that of pure iron in PDP tests. The CR of Mn is similar to that described before for Fe degradation:

$$Mn \rightarrow Mn_2^+ + 2e^- (\text{anodic reaction}) \quad (6)$$

$$O_2 + 2H_2O + 4e^- \rightarrow 4OH^- (\text{cathodic reaction}) \quad (7)$$

$$Mn^{2+} + 2OH^- \rightarrow Mn(OH)_2 (\text{product reaction}) \quad (8)$$

The CO_2 content of the atmosphere in human arteries is approximately 5% higher than that of the surrounding air. This contributes significantly to the deterioration of Mn since, according to Mouzou et al. [40] in an environment high in CO_2, two further processes interfere with the alloy corrosion process:

$$CO_2 + H_2O \rightarrow H_2CO_3 \quad (9)$$

$$2H_2CO_3 \rightarrow 2H^+ + 2HCO_3^- \tag{10}$$

$$2HCO_3^- \rightarrow 2CO_3^{2-} + 2H^+ \tag{11}$$

$$Mn \rightarrow Mn^{2+} + 2e^- \tag{12}$$

$$Mn^{2+} + 2HCO_3 \rightarrow Mn(HCO_3)^2 \rightarrow MnCO_3 + CO_2 + H_2O \tag{13}$$

$$Mn^{2+} + CO_3^{2-} \rightarrow MnCO_3 \tag{14}$$

A similar type of reaction also explains the oxidation of the Fe ion present in the alloy:

$$Fe_2^+ + 2HCO_3 \rightarrow Fe(HCO_3)_2 \rightarrow FeCO_3 + CO_2 + H_2O \tag{15}$$

$$Fe^{2+} + CO_3^{2-} \rightarrow FeCO_3 \tag{16}$$

The deterioration process can be stopped by carbonates because $FeCO_3$ and $MnCO_3$ are poorly soluble in water and can provide a stable passivation layer on the device's surface.

Mn plays a vital role in stabilizing the austenitic (γ-phase) structure within iron-based alloys, effectively suppressing ferromagnetic behavior across a broad compositional range. In Fe–Mn alloys containing more than 15% Mn, the formation of ε-martensite (hexagonal close-packed structure) is observed, leading to antiferromagnetic characteristics. Conversely, the presence of α′-martensite (body-centered cubic structure) introduces ferromagnetic properties. Studies have demonstrated that in Fe–Mn–Co alloys with Mn content ranging from 13.2 to 17.3 wt%, both ε and α′ martensite coexist. However, when the Mn content reaches approximately 20.7 wt%, only ε-martensite is present, resulting in the disappearance of ferromagnetic behavior.

To further enhance the properties of Fe–Mn alloys, the introduction of a third element (denoted as X) has been explored, leading to the development of Fe–Mn–X ternary alloys. Several examples of these changes are included in the literature review. Typically, alloy additions fall into one of four fundamental categories:

The addition of molybdenum (Mo) to Fe–Mn alloys has been investigated for its impact on magnetic properties and microstructure. In Fe–Mn–Mo alloys with Mn content between 13.4 and 17.2 wt%, both ε and α′ martensites are observed. However, increasing the Mn content to 20.2 wt% results in the exclusive formation of ε-martensite, accompanied by the complete disappearance of ferromagnetic behavior.

(i) Similarly, the incorporation of elements such as zirconium (Zr) and hafnium (Hf) into Fe–Mn alloys has been studied for their effects on Elinvar and Invar characteristics. These additions influence the Néel temperature and magnetic ordering, thereby affecting the thermal expansion and elastic modulus of the alloys. The specific impacts vary depending on the type and concentration of the added element, underscoring the importance of precise compositional control in tailoring the properties of Fe–Mn–X alloys for specialized applications.

(ii) Due to their high electrochemical potential, noble metals such as Pd, Cu, and Ag, when present in second-phase composites, produce local galvanic corrosion.

(iii) Scientists were interested in the addition of silicon because devices manufactured from the Fe–Mn–Si alloy have a greater rate of corrosion and shape memory.

(iv) The electrochemical potential of the second group's elements, Mg and Ca, is strongly negative (Mg = − 2.36 V vs. SHE, Ca = − 2.78 V vs. SHE). Alloys with active corrosion potentials exist in addition to a Fe–Mn matrix. According to Hong et al. [41] in their analysis of Tafel curves, the Fe–Mn–Ca alloys are more likely to corrode than the Fe–Mn–Mg alloys.

(v) The alloy composition of carbon added to Fe–Mn is like twinning-induced plasticity (TWIP) steels. TWIP steels are incredibly ductile and strong. According to the investigations, TWIP (Fe–Mn–C) alloys deteriorate in a physiological setting.

Venezuela and Dargush lately discussed [42] a more detailed review only about Fe–Mn biodegradable alloys, to which I refer those interested.

6.3 Drug Delivery from Iron-Based Stents

Restenosis due to neointimal hyperplasia remains a significant challenge associated with bare metal stents. To address all these issues, drug-eluting stents (DES) have been developed, incorporating immunosuppressive or cytotoxic agents to inhibit neointimal proliferation. These therapeutic agents are typically delivered via polymeric coatings, which allow for controlled elution kinetics, ensuring a sustained release over the critical 10–30 days post-implantation—an essential period for vascular healing and the prevention of restenosis. For iron-based BRS, it is imperative that the polymeric coatings employed are also fully biodegradable to match the resorption profile of the underlying scaffold.

A variety of pharmacological agents have been explored for use in DES, each targeting specific pathways involved in restenosis:

(i) **Curcumin**: Curcumin is known for its anti-inflammatory properties, and it has been investigated as an inhibitor of platelet aggregation. Yusop et al. [43] examined poly-L-lactic acid (PLLA) coatings incorporating varying concentrations of curcumin on iron scaffolds intended for bone implants. Their findings indicated that the degradation process of iron could influence the release kinetics of curcumin. Mechanically, the composite coating enhanced the scaffold's strength, with a notable reduction observed over 28 days.

(ii) **Sirolimus (Rapamycin)**: This macrolide immunosuppressant effectively inhibits inflammation, cellular proliferation, and migration. Shi et al. [44] performed in vivo studies in rats to assess the biocompatibility and CR of a sirolimus-eluting iron BRS system. After 3 months, the Fe scaffolds exhibited early-stage corrosion, suggesting a favorable degradation profile.

(iii) **Estrogen**: Employed to enhance vascular healing, estrogen's role in DES involves modulating endothelial function and reducing inflammatory responses.

(iv) **Angiopeptin**: This agent functions as a reducer of neointimal hyperplasia, aiming to prevent the excessive proliferation of vascular smooth muscle cells that contribute to restenosis.

(v) **Dexamethasone**: A corticosteroid utilized for its anti-inflammatory effects, dexamethasone aims to mitigate the inflammatory response following stent implantation.

(vi) **Tranilast**: Applied to decrease neointima formation and inflammation, Tranilast interferes with the proliferation of smooth muscle cells and fibroblasts.

Alternative strategies have also been explored to enhance the therapeutic efficacy of Fe-based BRS. For instance, Sharipova et al. [45] investigated a method wherein Fe powder was added with 10% silver and vancomycin hydrochloride as a model drug for treating implant-related infections. The powder mixtures underwent cold sintering at room temperature. In vitro tests demonstrated that the material effectively eradicated pathogens associated with chronic bone infections within five days without exhibiting cytotoxicity.

Additionally, nitric oxide (NO) has garnered attention for its beneficial effects in vascular healing due to its critical role in various biological processes. Nitriding iron stents may not only enhance their mechanical and corrosion properties but also confer therapeutic advantages to the surrounding tissue through the localized release of NO.

In summary, the integration of biodegradable polymer coatings with carefully selected pharmacological agents presents a promising approach to improving the performance of iron-based bioresorbable stents. Such strategies aim to effectively prevent restenosis while ensuring biocompatibility and controlled degradation in alignment with the vascular healing process.

7 In Vitro and in Vivo Trials of Fe-Based Materials

The evaluation of Fe-based biomaterials through in vitro and in vivo studies is critical for understanding their degradation behavior, biocompatibility, and overall suitability for biomedical applications. In vitro experiments are typically conducted under simulated physiological conditions using solutions such as phosphate-buffered saline (PBS) or Hank's balanced salt solution (HBSS) to assess degradation rates, ion release, and surface changes over time [46]. These studies also involve cytocompatibility tests with relevant cell lines (e.g., fibroblasts, osteoblasts, endothelial cells) to examine cellular adhesion, proliferation, and potential cytotoxic effects of Fe degradation products. Results from in vitro assays provide early insight into material performance but may not fully replicate the complexity of biological environments.

To bridge this gap, in vivo animal trials are conducted to evaluate the real-time behavior of Fe-based implants in living systems. Common models include rodents (e.g., rats and mice) and larger animals, such as rabbits or pigs, depending on the implant size and application [47]. These studies assess local tissue response, systemic toxicity, and degradation kinetics under dynamic physiological conditions. Findings often reveal the formation of a corrosion layer enriched with iron oxides, phosphates, and calcium compounds, along with moderate inflammatory responses that tend to diminish over time. While animal trials offer valuable data, eventual translation to humans necessitates rigorous clinical studies under regulatory guidelines. Initial human trials, although limited, have begun to explore Fe-based devices like biodegradable vascular stents, showing promising outcomes in terms of safety and controlled degradation. However, long-term human data are still required to validate the full clinical potential of Fe-based biomaterials [48].

8 Timeline of Clinical Translation of Biodegradable Materials

The clinical translation of biodegradable materials, including Fe-based alloys, follows a multi-phase timeline that spans from fundamental research to regulatory approval and clinical use. The process typically begins with basic research and material development, where novel biodegradable compositions are designed and synthesized. During this stage, focus is placed on optimizing mechanical properties, degradation behavior, and biocompatibility through laboratory testing. This phase can take 2–4 years, depending on the complexity of the material system and the intended biomedical application [49].

Following successful material development, the preclinical phase involves extensive in vitro testing and in vivo animal studies to evaluate degradation kinetics, local and systemic toxicity, and biological performance. This stage often spans 3–5 years, as it includes iterative testing and modifications to improve material performance and meet regulatory expectations [50]. Upon satisfactory preclinical outcomes, developers can proceed to the clinical trial phase, which begins with first-in-human studies under strict regulatory oversight. Clinical trials are conducted in multiple phases (I–III) to assess safety, efficacy, and long-term performance in a controlled patient population, typically lasting 5–7 years. Parallel to these efforts, the regulatory submission and approval process requires comprehensive documentation and validation data, adding another 1–2 years [51]. Altogether, the full clinical translation timeline of biodegradable materials can take 10–15 years, emphasizing the importance of interdisciplinary collaboration, sustained funding, and regulatory engagement to bring innovative biomaterials from bench to bedside [20].

9 Regulatory Considerations for the Use of Fe-Based Biomaterials

Fe-based biomaterials have garnered significant attention in biomedical applications, particularly in biodegradable implants such as stents and orthopedic devices. However, their clinical translation is subject to stringent regulatory considerations to ensure safety, efficacy, and long-term biocompatibility. Regulatory agencies such as the US Food and Drug Administration (FDA) and the European Medicines Agency (EMA) mandate rigorous testing protocols, including mechanical integrity, degradation behavior, and in vitro/in vivo biocompatibility assessments [52]. Since Fe-based materials degrade over time within the body and release iron ions, their corrosion products must be thoroughly evaluated to mitigate potential toxicity and adverse biological responses. Key considerations include mechanical stability, degradation kinetics, and the biological impact of degradation byproducts, which must comply with established standards such as ASTM F2129 and ISO 10993 for corrosion resistance and biocompatibility assessments [53].

Moreover, regulatory approval pathways for Fe-based biomaterials differ based on their intended application and classification under medical device regulations. For instance, biodegradable Fe-based implants are often classified as Class III medical devices, requiring extensive preclinical and clinical evaluations [54]. Manufacturers must provide comprehensive documentation on material composition, processing methods, sterilization techniques, and long-term degradation profiles to gain regulatory approval. Additionally, post-market surveillance is essential to monitor potential late-stage adverse effects and ensure compliance with evolving regulatory frameworks. As research advances in surface modifications, alloying strategies, and composite developments, regulatory bodies are expected to update guidelines to accommodate the unique challenges and opportunities associated with Fe-based biomaterials in clinical applications [55].

10 Conclusions and Future Perspectives

Iron-based biodegradable stents have garnered significant attention due to their exceptional biocompatibility and excellent biomedical properties compared to the Mg and Zn alloys. These properties position iron as a promising material for the development of lighter and smaller stents. However, a notable challenge is iron's slow in vivo corrosion rate, which impedes its full biodegradability. Consequently, research efforts have focused on enhancing the CR of Fe-based materials without compromising their inherent advantages.

Various strategies have been explored to accelerate iron corrosion, including alloying, surface coatings, and manufacturing modifications. Alloying elements such as manganese (Mn), palladium (Pd), and silver (Ag) have been investigated for their potential to increase corrosion rates and improve mechanical properties. For instance, Fe–30Mn alloys have demonstrated corrosion rates more than twice that of pure iron while maintaining mechanical properties comparable to 316L stainless steel. Similarly, the addition of Pd has been shown to enhance both the CR and the strength of Fe alloys.

Surface coatings with bioresorbable polymers like polylactic acid (PLA) have also been employed to expedite iron degradation. These coatings locally acidify the environment around the implant during degradation, thereby accelerating corrosion. Studies have demonstrated that iron stents coated with PLA can achieve complete dissolution of thin struts within six months.

Manufacturing methods that enhance the microstructure and surface property of Fe have shown promise in controlling CR. Methods such as PM and template-based porous system synthesis have been identified as effective in producing Fe-based systems with optimized degradation profiles.

Despite these advancements, achieving the ideal degradation rate for Fe-based materials remains an ongoing challenge. Future research should focus on refining manufacturing processes like vacuum arc melting and electroforming to optimize corrosion behavior. Additionally, while introducing alloying elements can enhance degradation rates, it is crucial to consider the potential for adverse side effects due to multiple elements in degradation products. Therefore, optimizing the degradation rate through morphological adjustments, such as controlling porosity and specific surface area, presents a viable approach when using pure iron.

Advancements in manufacturing and imaging technologies hold the potential to enable the production of patient-specific devices tailored in composition and shape. This personalization could reduce stent size and weight without compromising mechanical strength, thereby minimizing adverse effects on blood vessel walls. Furthermore, optimizing drug release mechanisms in stents offers a promising avenue for enhancing personalized therapy, ensuring controlled and sustained drug delivery during critical post-implantation periods.

In conclusion, while significant progress has been made in developing iron-based biodegradable stents, continued research is essential to address existing challenges and fully realize their clinical potential.

References

1. Lloyd-Jones D, Adams R, Carnethon M (2009) Heart disease and stroke statistics—2009 update. Circulation 119. https://doi.org/10.1161/circulationaha.108.191261

2. Byrne RA, Stone GW, Ormiston J, Kastrati A (2017) Coronary balloon angioplasty, stents, and scaffolds. Lancet 390:781–792. https://doi.org/10.1016/S0140-6736(17)31927-X
3. Tan C, Schatz RA (2016) The history of coronary stenting, Interv. Cardiol Clin 5:271–280. https://doi.org/10.1016/j.iccl.2016.03.001
4. Sigwart U, Puel J, Mirkovitch V, Joffre F, Kappenberger L (1987) Intravascular stents to prevent occlusion and re-stenosis after transluminal angioplasty. N Engl J Med 316:701–706
5. Mani G, Feldman MD, Patel D, Agrawal CM (2007) Coronary stents: a materials perspective. Biomaterials 28:1689–1710. https://doi.org/10.1016/j.biomaterials.2006.11.042
6. Veeram Reddy SR, Welch TR, Nugent AW (2021) Biodegradable stent use for congenital heart disease. Prog Pediatr Cardiol 61:101349. https://doi.org/10.1016/j.ppedcard.2021.101349
7. Tamai H, Igaki K, Kyo E, Kosuga K, Kawashima A, Matsui S, Komori H, Tsuji T, Motohara S, Uehata H (2000) Initial and 6-month results of biodegradable poly-L-lactic acid coronary stents in humans. Circulation 102:399–404. https://doi.org/10.1161/01.CIR.102.4.399
8. Campos CAM, Zhang YJ, Bourantas CV, Muramatsu T, Garcia-Garcia HM, Lemos PA, Iqbal J, Onuma Y, Serruys PW (2013) Bioresorbable vascular scaffolds in the clinical setting. Interv Cardiol 5:639–646. https://doi.org/10.2217/ica.13.72
9. Goyal V, Verma GC, Saxena P (2024) Advancements in 3D printing of functional materials for biomedical applications: challenges, opportunities, and case studies: a critical review. Prog Addit Manuf. https://doi.org/10.1007/s40964-024-00901-9
10. Beshchasna N, Saqib M, Kraskiewicz H, Wasyluk Ł, Kuzmin O, Duta OC, Ficai D, Ghizdavet Z, Marin A, Ficai A, Sun Z, Pichugin VF, Opitz J, Andronescu E (2020) Recent advances in manufacturing innovative stents. Pharmaceutics. https://doi.org/10.3390/pharmaceutics12040349
11. Gasior G, Szczepanski J, Radtke A (2021) Biodegradable iron-based materials—what was done and what more can be done? Materials (Basel) 14. https://doi.org/10.3390/ma14123381
12. Drynda A, Hassel T, Bach FW, Peuster M (2015) In vitro and in vivo corrosion properties of new iron-manganese alloys designed for cardiovascular applications. J Biomed Mater Res—Part B Appl Biomater 103:649–660. https://doi.org/10.1002/jbm.b.33234
13. Zhu S, Huang N, Xu L, Zhang Y, Liu H, Sun H, Leng Y (2009) Biocompatibility of pure iron: in vitro assessment of degradation kinetics and cytotoxicity on endothelial cells. Mater Sci Eng C 29:1589–1592. https://doi.org/10.1016/j.msec.2008.12.019
14. Ulum MF, Arafat A, Noviana D, Yusop AH, Nasution AK, Abdul Kadir MR, Hermawan H (2014) In vitro and in vivo degradation evaluation of novel iron-bioceramic composites for bone implant applications. Mater Sci Eng C 36:336–344. https://doi.org/10.1016/j.msec.2013.12.022
15. Qi Y, Li X, He Y, Zhang D, Ding J (2019) Mechanism of acceleration of iron corrosion by a polylactide coating. ACS Appl Mater Interfaces 11:202–218. https://doi.org/10.1021/acsami.8b17125
16. Kraus T, Moszner F, Fischerauer S, Fiedler M, Martinelli E, Eichler J, Witte F, Willbold E, Schinhammer M, Meischel M, Uggowitzer PJ, Löffler JF, Weinberg A (2014) Biodegradable Fe-based alloys for use in osteosynthesis: outcome of an in vivo study after 52 weeks. Acta Biomater 10:3346–3353. https://doi.org/10.1016/j.actbio.2014.04.007
17. Goyal V, Prasad NK, Verma G (2025) Experimental investigations into corrosion behaviour of DMLS manufactured Ti6Al4V alloy in different biofluids for orthopedic implants. Mater Today Commun 42:1–7. https://doi.org/10.1016/j.mtcomm.2024.111158
18. Georgieff MK (2020) Iron deficiency in pregnancy. Am J Obstet Gynecol 223:516–524. https://doi.org/10.1016/j.ajog.2020.03.006
19. Papanikolaou G, Pantopoulos K (2005) Iron metabolism and toxicity. Toxicol Appl Pharmacol 202:199–211. https://doi.org/10.1016/j.taap.2004.06.021
20. Zivic F, Grujovic N, Pellicer E, Sort J, Mitrovic S, Adamovic D, Vulovic M (2017) Biodegradable metals as biomaterials for clinical practice: iron-based materials.https://doi.org/10.1007/978-3-319-68025-5_9
21. Huang T, Cheng Y, Zheng Y (2016) In vitro studies on silver implanted pure iron by metal vapor vacuum arc technique. Colloids Surfaces B Biointerfaces 142:20–29. https://doi.org/10.1016/j.colsurfb.2016.01.065

22. Cheng J, Zheng YF (2013) In vitro study on newly designed biodegradable Fe-X composites (X = W, CNT) prepared by spark plasma sintering. J Biomed Mater Res—Part B Appl Biomater 101:485–497. https://doi.org/10.1002/jbm.b.32783
23. Capek J, Msallamova S, Jablonska E, Lipov J, Vojtech D (2017) A novel high-strength and highly corrosive biodegradable Fe–Pd alloy: structural, mechanical and in vitro corrosion and cytotoxicity study. Mater Sci Eng C 79:550–562. https://doi.org/10.1016/j.msec.2017.05.100
24. Liu B, Zheng YF, Ruan L (2011) In vitro investigation of Fe30Mn6Si shape memory alloy as potential biodegradable metallic material. Mater Lett 65:540–543. https://doi.org/10.1016/j.matlet.2010.10.068
25. Paim TC, Wermuth DP, Bertaco I, Zanatelli C, Naasani LIS, Slaviero M, Driemeier D, Schaeffer L, Wink MR (2020) Evaluation of in vitro and in vivo biocompatibility of iron produced by powder metallurgy. Mater Sci Eng C 115:111129. https://doi.org/10.1016/j.msec.2020.111129
26. Eliaz N (2019) Corrosion of metallic biomaterials: a review. Materials (Basel) 12. https://doi.org/10.3390/ma12030407
27. Anderson JM, Rodriguez A, Chang DT (2008) Foreign body reaction to biomaterials. Semin Immunol 20:86–100. https://doi.org/10.1016/j.smim.2007.11.004
28. Liu Y, Zheng Y, Chen XH, Yang JA, Pan H, Chen D, Wang L, Zhang J, Zhu D, Wu S, Yeung KWK, Zeng RC, Han Y, Guan S (2019) Fundamental theory of biodegradable metals—definition, criteria, and design. Adv Funct Mater 29:1–21. https://doi.org/10.1002/adfm.201805402
29. Huang T, Cheng J, Zheng YF (2014) In vitro degradation and biocompatibility of Fe–Pd and Fe–Pt composites fabricated by spark plasma sintering. Mater Sci Eng C 35:43–53. https://doi.org/10.1016/j.msec.2013.10.023
30. Mandal S, Ummadi R, Bose M, Balla VK, Roy M (2019) Fe–Mn–Cu alloy as biodegradable material with enhanced antimicrobial properties. Mater Lett 237:323–327. https://doi.org/10.1016/j.matlet.2018.11.117
31. Liu B, Zheng YF (2011) Effects of alloying elements (Mn Co, Al, W, Sn, B, C and S) on biodegradability and in vitro biocompatibility of pure iron. Acta Biomater 7:1407–1420. https://doi.org/10.1016/j.actbio.2010.11.001
32. Hermawan H, Moravej M, Dube D, Fiset M, Mantovani D (2007) Degradation behaviour of metallic biomaterials for degradable stents. Adv Mater Res 15–17:113–118. https://doi.org/10.4028/www.scientific.net/amr.15-17.113
33. Moravej M, Purnama A, Fiset M, Couet J, Mantovani D (2010) Electroformed pure iron as a new biomaterial for degradable stents: in vitro degradation and preliminary cell viability studies. Acta Biomater 6:1843–1851. https://doi.org/10.1016/j.actbio.2010.01.008
34. Drevet R, Zhukova Y, Malikova P, Dubinskiy S, Korotitskiy A, Pustov Y, Prokoshkin S (2018) Martensitic transformations and mechanical and corrosion properties of Fe–Mn–Si alloys for biodegradable medical implants. Metall Mater Trans A Phys Metall Mater Sci 49:1006–1013. https://doi.org/10.1007/s11661-017-4458-2
35. Taylor SL, Jakus AE, Shah RN, Dunand DC (2017) Iron and nickel cellular structures by sintering of 3D-printed oxide or metallic particle inks. Adv Eng Mater 19:1–8. https://doi.org/10.1002/adem.201600365
36. Sharma P, Pandey PM (2019) Rapid manufacturing of biodegradable pure iron scaffold using amalgamation of three-dimensional printing and pressureless microwave sintering. Proc Inst Mech Eng Part C J Mech Eng Sci 233:1876–1895. https://doi.org/10.1177/0954406218778304
37. Wang H, Zheng Y, Liu J, Jiang C, Li Y (2017) In vitro corrosion properties and cytocompatibility of Fe–Ga alloys as potential biodegradable metallic materials. Mater Sci Eng C 71:60–66. https://doi.org/10.1016/j.msec.2016.09.086
38. Huang T, Cheng J, Bian D, Zheng Y (2016) Fe–Au and Fe–Ag composites as candidates for biodegradable stent materials. J Biomed Mater Res Part B Appl Biomater 104:225–240. https://doi.org/10.1002/jbm.b.33389
39. Lin W, Zhang G, Cao P, Zhang D, Zheng Y, Wu R, Qin L, Wang G, Wen T (2015) Cytotoxicity and its test methodology for a bioabsorbable nitrided iron stent. J Biomed Mater Res—Part B Appl. Biomater 103:764–776. https://doi.org/10.1002/jbm.b.33246

40. Mouzou E, Paternoster C, Tolouei R, Chevallier P, Biffi CA, Tuissi A, Mantovani D (2016) CO_2-rich atmosphere strongly affects the degradation of Fe-21Mn-1C for biodegradable metallic implants. Mater Lett 181:362–366. https://doi.org/10.1016/j.matlet.2016.06.017
41. Hong D, Chou DT, Velikokhatnyi OI, Roy A, Lee B, Swink I, Issaev I, Kuhn HA, Kumta PN (2016) Binder-jetting 3D printing and alloy development of new biodegradable Fe–Mn–Ca/Mg alloys. Acta Biomater 45:375–386. https://doi.org/10.1016/j.actbio.2016.08.032
42. Venezuela J, Dargusch MS (2020) Addressing the slow corrosion rate of biodegradable Fe–Mn: current approaches and future trends. Curr Opin Solid State Mater Sci 24:100822. https://doi.org/10.1016/j.cossms.2020.100822
43. Yusop AH, Sarian MN, Januddi FS, Ahmed QU, Kadir MR, Hartanto D, Hermawan H, Nur H (2018) Structure, degradation, drug release and mechanical properties relationships of iron-based drug eluting scaffolds: the effects of PLGA. Mater Des 160:203–217. https://doi.org/10.1016/j.matdes.2018.09.019
44. Shi J, Miao X, Fu H, Jiang A, Liu YF, Shi XL, Zhang D, Wang Z (2020) In vivo biological safety evaluation of an iron-based bioresorbable drug-eluting stent. Biometals 33:217–228. https://doi.org/10.1007/s10534-020-00244-2
45. Sharipova A, Unger RE, Sosnik A, Gutmanas E (2021) Dense drug-eluting biodegradable Fe–Ag nanocomposites. Mater Des 204:109660. https://doi.org/10.1016/j.matdes.2021.109660
46. Rabeeh VPM, Hanas T (2022) Progress in manufacturing and processing of degradable Fe-based implants: a review.https://doi.org/10.1007/s40204-022-00189-4
47. Scarcello E, Lison D (2020) Are Fe-based stenting materials biocompatible? A critical review of in vitro and in vivo studies. J Funct Biomater 11. https://doi.org/10.3390/jfb11010002
48. Francis A, Yang Y, Virtanen S, Boccaccini AR (2015) Iron and iron-based alloys for temporary cardiovascular applications. J Mater Sci Mater Med 26:1–16. https://doi.org/10.1007/s10856-015-5473-8
49. Hermawan H, Dubé D, Mantovani D (2010) Degradable metallic biomaterials: design and development of Fe–Mn alloys for stents. J Biomed Mater Res—Part A 93:1–11. https://doi.org/10.1002/jbm.a.32224
50. Peuster M, Hesse C, Schloo T, Fink C, Beerbaum P, von Schnakenburg C (2006) Long-term biocompatibility of a corrodible peripheral iron stent in the porcine descending aorta. Biomaterials 27:4955–4962. https://doi.org/10.1016/j.biomaterials.2006.05.029
51. Moravej M, Mantovani D (2011) Biodegradable metals for cardiovascular stent application: interests and new opportunities. Int J Mol Sci 12:4250–4270. https://doi.org/10.3390/ijms12074250
52. Han HS, Loffredo S, Jun I, Edwards J, Kim YC, Seok HK, Witte F, Mantovani D, Glyn-Jones S (2019) Current status and outlook on the clinical translation of biodegradable metals. Mater Today 23:57–71. https://doi.org/10.1016/j.mattod.2018.05.018
53. ASTM International (2010) ASTM F2129-19: standard test method for conducting cyclic potentiodynamic polarization measurements to determine the corrosion susceptibility of small implant devices. pp 1–6
54. Prakasam M, Locs J, Salma-Ancane K, Loca D, Largeteau A, Berzina-Cimdina L (2017) Biodegradable materials and metallic implants-a review. J Funct Biomater 8:1–15. https://doi.org/10.3390/jfb8040044
55. Gorejova R, Haverova L, Oriňakova R, Orinak A, Orinak M (2019) Recent advancements in Fe-based biodegradable materials for bone repair. J Mater Sci 54:1913–1947. https://doi.org/10.1007/s10853-018-3011-z

Chapter 5
Composition and Design Strategies for Magnesium-Based Biodegradable Alloys

Priyabrata Das and Pulak Mohan Pandey

1 Introduction

Magnesium-based biodegradable alloys have emerged as a promising category of materials for temporary implants like orthopedic screws, cardiovascular stents, and tissue scaffolds. While magnesium (Mg), zinc (Zn), and iron (Fe) are all potential candidates for biodegradable implants [1], Mg stands out due to its superior biocompatibility and its successful use in human trials [2, 3], making it a leading choice for clinical applications. Furthermore, the mechanical properties of Mg, such as its density and elastic modulus, closely resemble those of human bone, thereby minimizing stress shielding effects and facilitating patient recovery [4]. However, the clinical adoption of magnesium-based alloys faces significant challenges, primarily due to their rapid and often uncontrollable degradation in the human body, which can compromise implant integrity and release excessive hydrogen gas [5]. Resolving these issues necessitates a precise approach to alloy composition and design methodologies. The selection of alloying elements and their concentrations significantly influences the mechanical strength, corrosion resistance, and biocompatibility of Mg-based alloys. Appropriately tailored compositions can enhance performance by refining microstructure, controlling degradation rates, and promoting cellular compatibility. Further, alloying can be broadly categorized into two approaches: the addition of individual alloying elements at concentrations less than five atomic % (at %) and at concentrations equal to or greater than five at %. Conventional magnesium alloys typically involve alloying elements below five at %, adhering to the solute–solvent concept. On the other hand, the addition of alloying elements above five atomic % leads to the development of compositionally complex alloys (CCAs) which form based on the principle of maximizing configurational entropy to stabilize

P. Das · P. M. Pandey (✉)
Department of Mechanical Engineering, Indian Institute of Technology Delhi, New Delhi, India
e-mail: pmpandey@mech.iitd.ac.in

A. Dixit et al. (eds.), *Biodegradable Metallic Materials*, Biomedical Materials for Multi-functional Applications, https://doi.org/10.1007/978-981-95-2401-3_5

a single-phase structure [6], and bulk metallic glasses (BMGs), which rely on multicomponent compositions to promote amorphous structures by following Inoue's rule [7]. In the past two decades, Mg-based bulk metallic glasses (BMGs) have garnered significant attention as promising candidates for biodegradable implants, owing to their superior mechanical properties, exceptional biocompatibility, and favorable degradation behavior both in in vitro and in vivo conditions. This chapter explores both conventional Mg alloy and Mg based compositionally complex alloy (CCA) computational design approaches, with an emphasis on their mechanical, corrosion, and biological properties.

2 Design Strategies for Magnesium-Based Biodegradable Alloy

2.1 Experimental High-Throughput Screening

High-throughput experiments in material library design involve the rapid synthesis and characterization of multiple material compositions simultaneously, enabling efficient exploration of large compositional spaces. This approach accelerates the discovery of materials with optimal properties by using techniques like combinatorial deposition (continuous or discrete) and automated testing. Datye et al. [8] prepared Mg–Zn–Ca and Fe–Mg–Zn ternary amorphous material library by utilizing DC magnetron co-sputtering. The combinatorial sputtering resulted in 177 unique specimens both in Mg–Zn–Ca and Fe–Mg–Zn alloy systems, covering a wide range of compositions. Automated X-ray diffraction (XRD), automated energy dispersive spectroscopy (EDS), and nanoindentation technique were used to determine the phase, composition, and mechanical properties of the individual alloy films, respectively. The high-throughput experimental results for the Fe–Mg–Zn system can be seen from Fig. 1a–d.

For Mg–Zn–Ca system, a strong dependence of hardness on Zn content was observed. Further, both alloy systems exhibited high hardness, elastic modulus, and yield strength when the Mg content was within the range of 40–50 at %. The corrosion properties of Mg alloy material libraries are also assessed by high-throughput characterization techniques which utilize optical detection methods. Kotoka et al. [9] designed an optical corrosion setup to estimate the corrosion rate of Mg–Zn alloy films (as deposited and annealed) by measuring transmittance before and after corrosion, as illustrated in Fig. 2a, b. The study reported a higher corrosion rate associated with annealed samples and compositions having higher at % of Mg (testing media: phosphate-buffered saline). In summary, this alloy design approach leverages automation and parallelization in both synthesis and characterization, paving a streamlined and highly efficient pathway for Mg alloy selection.

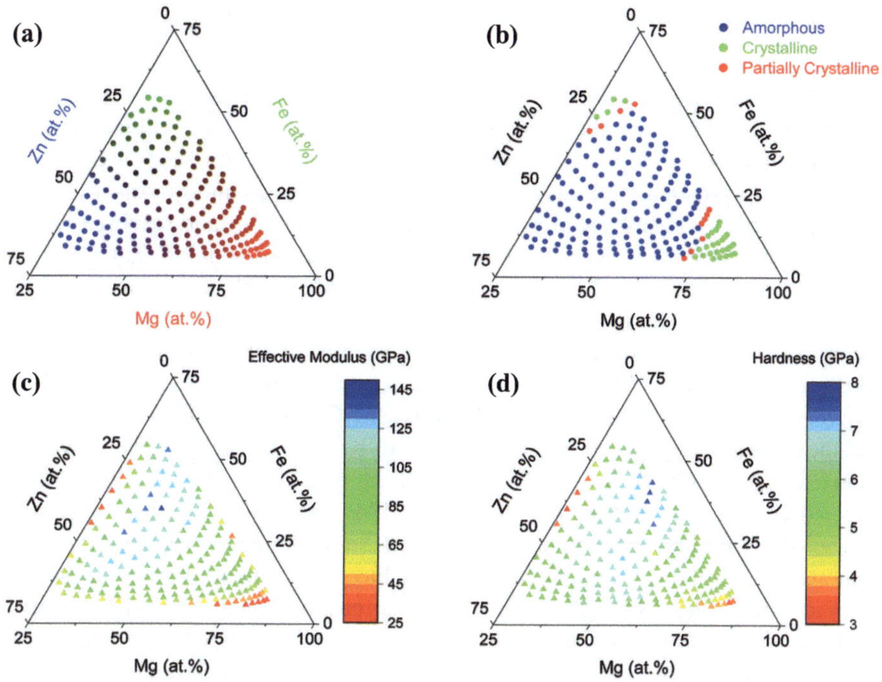

Fig. 1 Fe–Mg–Zn alloy material libraries showing **a** compositions obtained from EDS, **b** phase information obtained from XRD, **c** effective modulus, and **d** hardness obtained from nanoindentation. Reprinted from [8] with permission from the Royal Society of Chemistry

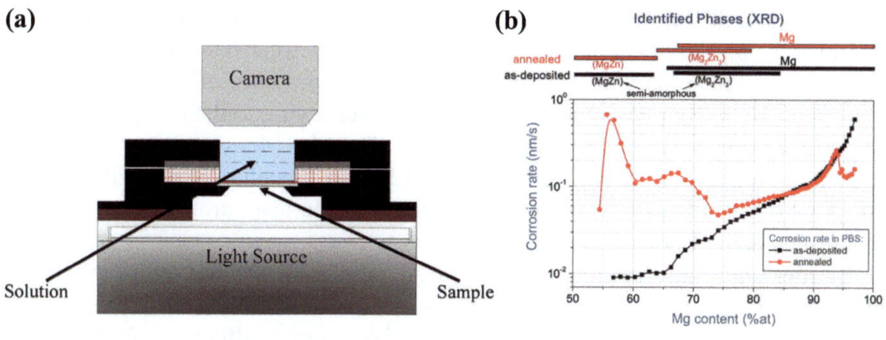

Fig. 2 **a** Optical corrosion setup and **b** estimated corrosion rate of as-deposited and annealed Mg–Zn alloy library in PBS solution. Reprinted from [9] with permission from Elsevier

2.2 Computational Thermodynamic Approach (CALPHAD)

The CALPHAD (CALculation of PHAse Diagrams) method minimizes total Gibbs-free energy to determine chemical equilibrium, using models for Gibbs energy as functions of temperature, pressure, and composition in pure compounds and binaries. By fitting experimental and theoretical data, it generates thermodynamic databases for higher-order systems, enabling accurate phase predictions and solving practical technological challenges. By using CALPHAD approach, Rajendran et al. [10] identified a composition in the Mg–Zn–Ca system capable of exhibiting superior glass-forming ability. The study evaluated the nucleation driving force (NDFs) of the crystalline phases in the abovementioned alloy system at 613 K by using the TCFE11 database and found that $Mg_{65}Zn_{31}Ca_4$ exhibited a minimum value of NDFs, implying a high glass-forming ability. Further, the $Mg_{65}Zn_{31}Ca_4$ ternary alloy was fabricated, and the diameter of the sample was reported to be 30% higher than the previously reported Mg–Zn–Ca alloys which confirmed the high glass-forming ability of the selected composition in line with the CALPHAD prediction. Avraham et al. [11] utilized the CALPHAD method to select Mg–Sn–Zn alloy composition and heat treatment to develop an age-hardenable alloy having excellent microhardness. The study identified Mg–8Sn–1.25Zn alloy composition which showed maximum microhardness after heat treatment due to the precipitation of homogenous Mg_2Sn in α-Mg matrix and metastable Mg–Zn precipitates. Phase prediction through CALPHAD approach is crucial as the phase information can enable the designing of high-strength and corrosion-resistant Mg alloys. To this end, Gandel et al. [12] evaluated the phase information in Mg–Fe–Zr, Mg–Mn–Fe, Mg–Mn–Zr, and Mg–Mn–Zr–Fe alloy systems by using PANDAT (CALPHAD based program). For the ternary alloy systems, an HCP Mg phase was predicted along with one other solid solution and intermetallic phases in minor mass fraction ($< 1\%$) which can be seen from Fig. 3a–c. However, the quaternary alloy system exhibited the presence of two solid solution phases, an HCP Mg phase and intermetallic phases owing to the complexity of the Mg–Mn–Zr–Fe system, as seen from Fig. 3d. The Mg–Mn–Fe alloy exhibited the least mass loss ($3.4 + 1.5$ mg/cm^2 day) in 0.1 M NaCl medium among all the tested alloys which can be ascribed to the formation of a Mn layer around the Fe particle (confirmed using EDS).

Besides phase information, the CALPHAD approach can be used to investigate the thermodynamics of chemical reactions in Mg-based alloy systems in aqueous environments. Wu et al. [13] calculated the free energy difference ($\delta \Delta G^R_{\text{dop}}$) and enthalpy difference ($\delta \Delta H^R_{\text{dop}}$) using Thermo-Calc software to determine whether the addition of certain alloying element promotes or mitigates the corrosion process. The $\delta \Delta G^R_{\text{dop}}$ and $\delta \Delta H^R_{\text{dop}}$ were calculated for Zn, Y, and Ca doped Mg alloys in pure water and human blood fluid (HBF). The calculations indicated that the effect of doping elements on the stability of Mg alloys is consistent in both HBF and water. Further, the reaction products of Mg alloys in HBF were given in Table 1 which demonstrated the presence of ZnO in high concentration in Mg–0.1Zn alloy. The results of the study indicated that the addition of Zn mitigates the hydrolysis process,

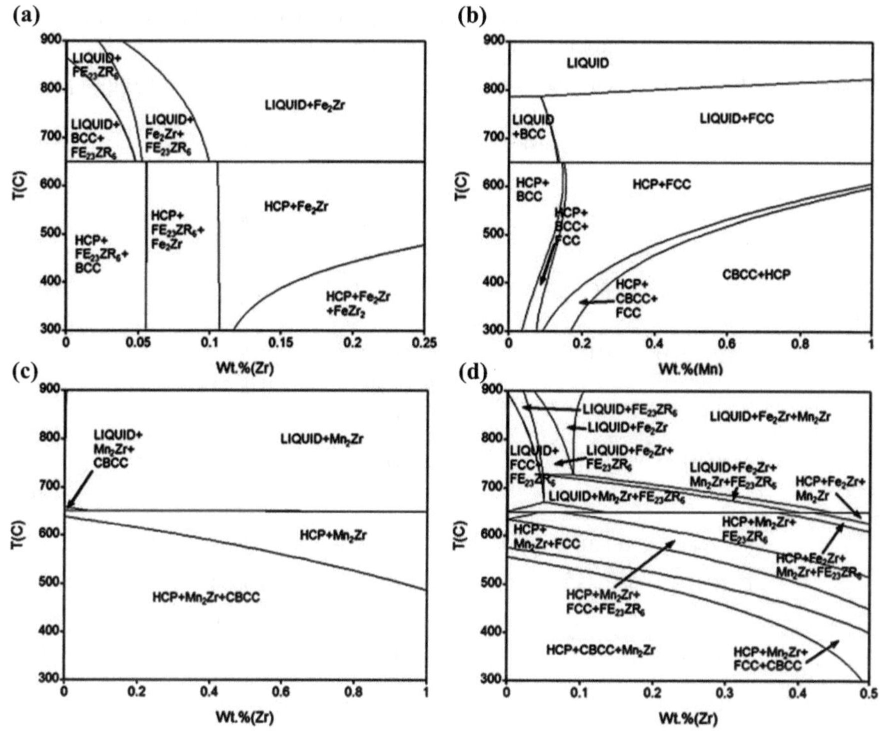

Fig. 3 Isopleth diagrams calculated using PANDAT showing phase transitions in **a** Mg–Fe–Zr alloy as Zr content increases, **b** Mg–Mn–Fe alloy with varying Mn content, **c** Mg–Mn–Zr alloy with increasing Zr content, and **d** Mg–Mn–Zr–Fe alloy as Zr content changes. Reprinted from [12] with permission from Elsevier

whereas the incorporation of Ca or Y enhances the hydrolysis process. Similarly, Avey et al. [14] utilized CALPHAD calculations to gain insight on the dependency of Mg_2Ca brittle phase formation with Zn/Ca ratio in Mg–0.5Ca–xZn–0.5Mn alloys. The critical value of Zn/Ca above which the formation of Mg_2Ca can be eliminated was found to be 2.77. However, the optimum combination of strength, ductility, and corrosion resistance was achieved in an alloy having a Zn/Ca ratio of 1.82 as seen from Fig. 4a. The experimental investigation on phases clearly showed suppression of the Mg_2Ca phase with an increase in the Zn/Ca ratio (Fig. 4b–f) in line with the CALPHAD prediction.

Table 1 Products formed from the reaction of Mg alloys with human blood fluid (concentrations measured in moles) Reprinted from [13] with permission from Elsevier

System	Precipitates	Concentration (moles)	Gas constituents	Concentration (moles)	pH
HBF[a]	$Ca_3(PO_4)_2$	1.045×10^{-4}	H_2	0.023	7.4
			H_2O	0.002	
			CH_4	0.008	
			H_2S	1.487×10^{-4}	
1 Mg + HBF	$Mg(OH)_2$	1.00	H_2	1.0257	9.4
	$Ca_3(PO_4)_2$	1.045×10^{-4}	H_2O	0.06639	
			CH_4	0.00545	
			H_2S	1×10^{-4}	
1 (Mg–0.1Ca) + HBF	$Mg(OH)_2$	0.9015	H_2	1.0257	11.9
	$Ca_3(PO_4)_2$	1.045×10^{-4}	H_2O	0.0666	
	$Ca(OH)_2$	0.0887	CH_4	0.00545	
			H_2S	1×10^{-4}	
1 (Mg–0.1Zn) + HBF	$Mg(OH)_2$	0.9	H_2	1.0256	9.4
	$Ca_3(PO_4)_2$	1.045×10^{-4}	H_2O	0.06625	
	ZnO	0.1	CH_4	0.00545	
	ZnS	1×10^{-4}	H_2S	3×10^{-14}	
1 (Mg–0.1Y) + HBF	$Mg(OH)_2$	0.9	H_2	1.0757	9.4
	$Ca_3(PO_4)_2$	1.045×10^{-4}	H_2O	0.06986	
	Y_2O_3	0.05	CH_4	0.00545	
			H_2S	1×10^{-4}	

[a]HBF = 55.8424 mol human blood fluid

2.3 *Molecular Dynamic (MD) Simulation and Density Functional Theory (DFT) Calculations*

Molecular Dynamics (MD) simulation and Density Functional Theory (DFT) calculations are essential tools for designing magnesium alloys, providing a detailed understanding of atomic-scale interactions and properties. MD simulates the motion of atoms over time by solving Newton's equations of motion using force fields, enabling the study of microstructural evolution, grain boundary behavior, and diffusion processes in Mg alloys [15]. On the other hand, DFT, grounded in quantum

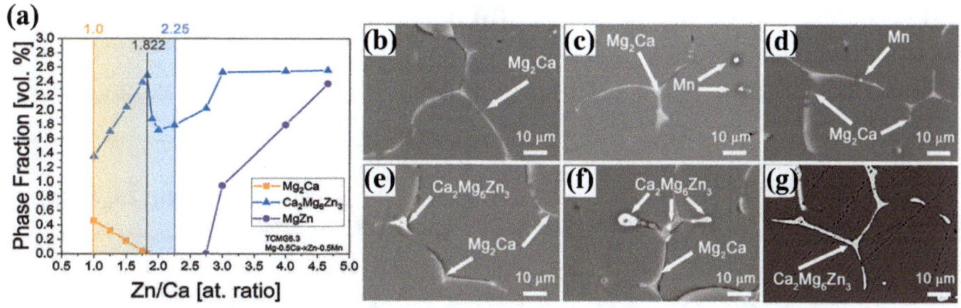

Fig. 4 **a** Phase fractions determined from Scheil calculations (shaded portion represents area of interest), SEM micrographs of Mg–0.5Ca–xZn–0.5Mn alloy having Zn/Ca ratio of **b** 1.18, **c** 1.4, **d** 1.8, **e** 2.07, **f** 2.36, and **g** 4.66, following heat treatment at 390 °C for 24 h. Reprinted from [14] with permission from Elsevier

mechanics, uses electron density to solve the Schrödinger equation and provides insights into the electronic structure, phase stability, and bonding characteristics of Mg alloys. By predicting thermodynamic properties and the influence of alloying elements at an atomic level, DFT aids in optimizing alloy compositions. Together, these methods enable a comprehensive understanding of both macroscopic properties and atomic-scale phenomena, advancing the development of Mg alloys with enhanced mechanical performance and corrosion resistance. Sun et al. [16] studied the effect of Sr incorporation in $Mg_{66}Zn_{30}Ca_4$ bulk metallic glass (BMG) by combining MD simulation and DFT calculations. The MD simulation conducted with a unit cell containing 8000 atoms (Fig. 5a) predicted the formation of BMG structures at the local level for both $Mg_{66}Zn_{30}Ca_4$ and $Mg_{66}Zn_{30}Ca_3Sr_1$ which was in line with the experimental XRD data (Fig. 5b). The stress–strain curve of both the alloys was obtained by simulating the uniaxial tensile test on a 3 × 2 × 2 supercell comprising 96,000 atoms with open boundary condition on the BMG surface and periodic boundary condition on the other two surfaces as seen from Fig. 5c. The predicted elastic modulus and tensile strength for $Mg_{66}Zn_{30}Ca_4$ and $Mg_{66}Zn_{30}Ca_3Sr_1$ alloys were found to be in close agreement with the experimentally determined values. The engineering stress–strain curve obtained from MD simulation can be seen in Fig. 5d. The DFT calculations showed that incorporation of Sr into $Mg_{66}Zn_{30}Ca_4$ alloys reduces electron transfer and enhances surface corrosion resistance.

Further, Yuan et al. [17] utilized the MD studies to evaluate the effect of electroplating temperature on the bonding of Ca-P coatings to Mg substrate by calculating the binding energy ($\Delta E_{\text{bonding}}$) of $CaHPO_4.2H_2O$ with Mg at different plating temperatures (298, 323, 333, 343, and 353 K). The $\Delta E_{\text{bonding}}$ showed the least value at 333 K plating temperature implying the excellent binding force and corrosion-resistant properties. MD simulations are also useful to evaluate the adsorption of fibrinogen (Fg) on Mg alloy surfaces which was undertaken by the study reported by Wang et al. [18]. The effect of eight different alloying elements, i.e., Zn, Li, Ce, Cu, Ca, Mn, Zr, and Y, on Fg adsorption behavior was evaluated. The results of the study indicated that the electronegativity of the alloying element directly influences

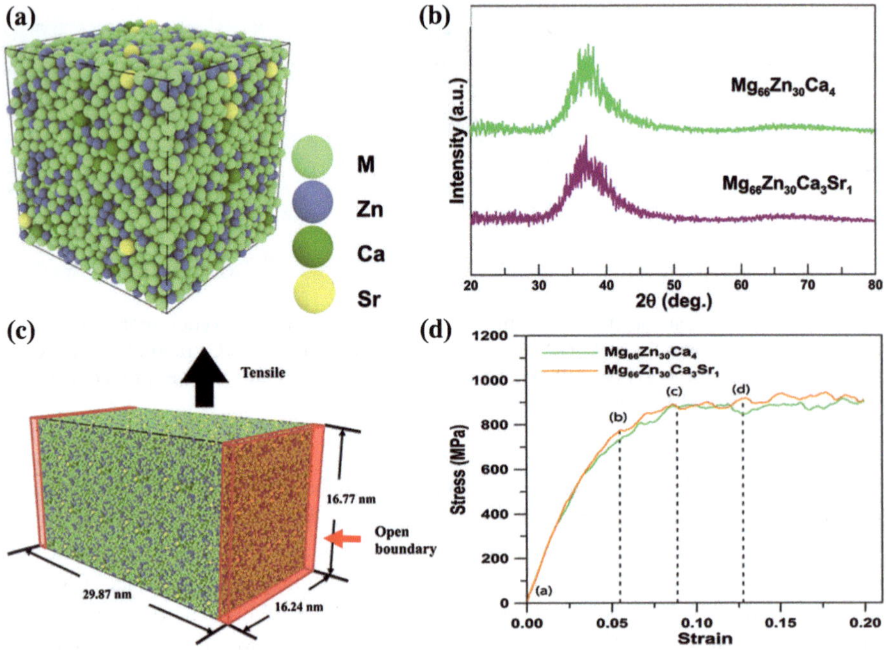

Fig. 5 **a** $Mg_{66}Zn_{30}Ca_3Sr_1$ unit cell containing 8000 atoms, **b** XRD plot of $Mg_{66}Zn_{30}Ca_4$ and $Mg_{66}Zn_{30}Ca_3Sr_1$ alloys, **c** tensile simulation on 3 × 2 × 2 supercell and **d** stress–strain graph of $Mg_{66}Zn_{30}Ca_4$ and $Mg_{66}Zn_{30}Ca_3Sr_1$ alloys [16]

the Fg adsorption. Incorporation of alloying elements having relatively lower electronegativity than Mg such as Ca, Li, Ce, and Y enhanced the protein adsorption with more residues anchored to the alloy surface which was also confirmed from the experimental studies.

2.4 Machine Learning

Machine learning (ML) is a powerful approach for advancing the design of magnesium alloys by enabling data-driven predictions of critical properties such as mechanical strength, corrosion resistance, and phase stability. Through the use of experimental and computational datasets, ML models can uncover complex relationships between alloy composition, processing conditions, and resulting properties, significantly accelerating the alloy development process. The process begins with data collection and preprocessing to ensure consistency, followed by feature selection to identify key input variables that significantly affect the output response, i.e., properties in focus. Subsequently, supervised/unsupervised ML algorithms are employed to establish predictive models. A recent development in this domain of ML-assisted Mg alloy design was reported by Das et al. [19, 20]. The reported studies proposed a novel design strategy involving ML and a multi-objective evolutionary algorithm to design Mg-based non-equiatomic medium entropy alloy (MEA) compositions which can be visualized in Fig. 6. During the feature selection step, multiple physical and thermodynamic parameters [21] which affect the phase formation in CCAs were considered and further screened using appropriate methods. Supervised ML models (artificial neural network and k-nearest neighbor) were employed for the phase prediction, and composition optimization was carried out using a multi-objective genetic algorithm. Finally, a quaternary Mg MEA ($Mg_{60}Ti_{24.24}Zn_{9.66}Nb_{6.1}$) was identified and fabricated using a powder metallurgy approach which showed the presence of a solid solution phase in the fabricated alloy in line with the ML phase prediction. The fabricated alloy exhibited excellent cytocompatibility with appropriate mechanical properties and biocorrosion properties [22] and was successfully used to fabricate Fe-MgMEA composite scaffolds with improved biodegradation and cytocompatibility [23]. Similarly, Dong et al. [24] utilized the ML approach to predict ultimate tensile strength, yield strength, elongation, and hardness of Mg alloy. Experimental parameters such as alloy composition, process parameters, and testing conditions were used as features for the prediction of targets. Further, the incorporation of the SHapley Additive exPlanations (SHAP) method quantified the interaction of features with target mechanical properties.

Chen et al. [25] identified the material processing parameter for ZE62 alloy, i.e., Mg-6wt%Zn-2wt% RE (RE:Y, Gd, Ce, Nd) to bring simultaneous improvement in strength and ductility by using ML-assisted two different multi-o.jective optimization strategy. Both strategies provided sets of heat treatment parameters iteratively by taking the ML predictions of unexplored space as the initial population. Strategy I determines the angle (θ^p) between the target vector (ω^t) and the Pareto front vector

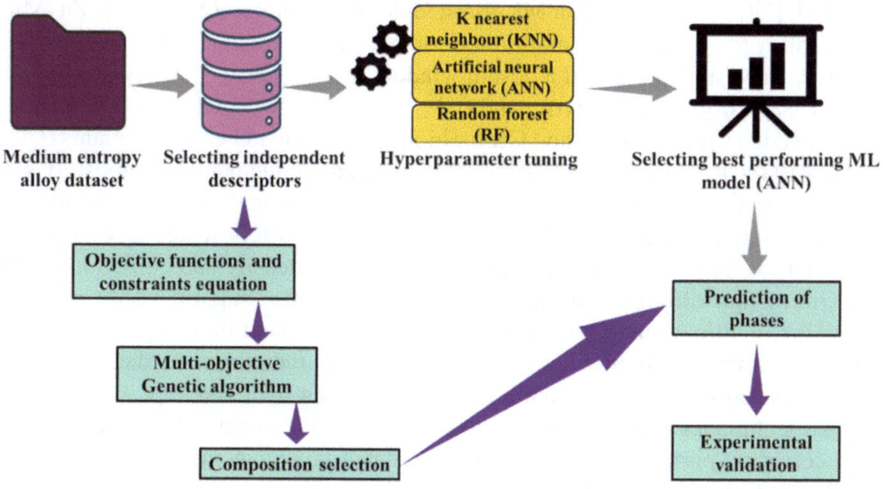

Fig. 6 Schematic diagram of computational framework adopted in the works of Das et al. [19, 20]

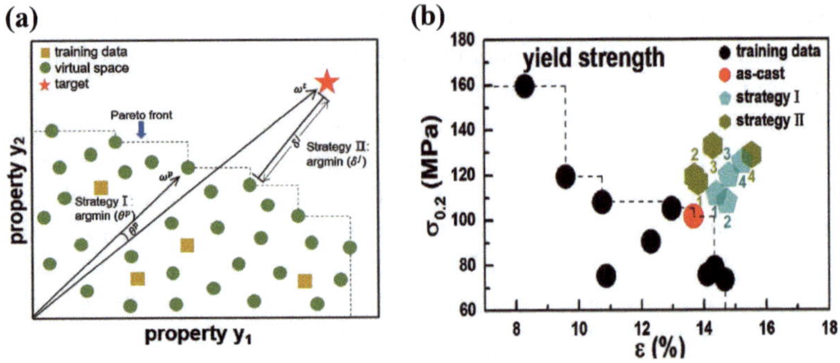

Fig. 7 **a** Schematic representation of strategy I and II, **b** the training data, as-cast, and different iterations of strategy I and II distributed in the yield strength-elongation plane. Reprinted from [25] with permission from Elsevier

(ω^p), while Strategy II computes the distance (δ^j) from a point in the virtual space (y_i^j) to the target (y_i^t) which can be seen from Fig. 7a (property y1 and y2 represent the properties of interest). The consecutive iterations were obtained by minimizing the θ^p or δ^j. As per the heat treatment parameters obtained from different iterations of two strategies, the Mg alloy was fabricated, and its mechanical properties were determined. From Fig. 7b, it can be conferred that both the strategies simultaneously improved the strength and ductility of Mg alloys with consecutive iterations. Notably, the strength and ductility of the optimal alloy improved by 27% and 13.5%, respectively, with just four additional successive experiments, indicating the efficacy of the proposed strategy.

Besides predicting Mg alloy compositions and processing conditions, several studies [26–28] have focused on the ML-assisted corrosion prediction in these alloys. Suh et al. [28] also provided an ML-assisted design guide for ZMJX (Mg–Zn–Mn–Sr–Ca) alloy to obtain optimum ultimate compressive strength (UCS) and corrosion rate (CR). The study provided a design window of $0.5 \leq Zn \leq 1.5$ wt%, $0.6 \leq Mn \leq 1.2$ wt%, $0.0 \leq Sr \leq 0.6$ wt%, and $0.0 \leq Ca \leq 0.15$ wt% which adhered to the recommended values of UCS and CR for biodegradable load bearing implants (UCS ≥ 240 MPa, CR ≤ 1 mm/year). Lu et al. [26] used the elemental compositions of alloying elements, chloride ion concentration, and 16 physical and atomic properties as the features (calculated from Mg alloy compositions by following rule of mixture) to predict two target properties, i.e., corrosion current density and corrosion potential. Out of four different ML algorithms (random forest, support vector machine regression, multiple linear regression, and extreme gradient boosting), the random forest algorithm was found to provide most accurate predictions. As data collection and processing technologies continue to advance, big data and sophisticated data analysis techniques can be leveraged to extract valuable insights for predicting Mg alloy corrosion performance from extensive datasets, enhancing the accuracy and reliability of predictive models.

2.5 *Integrated Computational Materials Engineering (ICME)*

The Integrated Computational Materials Engineering (ICME) approach is revolutionizing Mg alloy design by integrating computational simulations, experimental data, and data-driven methods to accelerate alloy development and optimize their performance. Computational tools such as CALPHAD are used for thermodynamic predictions and phase stability analysis, while molecular dynamics (MD) and density functional theory (DFT) offer insights into atomic-scale interactions and diffusion mechanisms. Machine learning (ML) complements these by analyzing large datasets and providing rapid property predictions to refine alloy design. These computational models are supported by experimental validation to ensure accuracy and improve predictions. ICME enables the efficient exploration of compositional and processing spaces, facilitating the selection of alloying elements and manufacturing conditions to achieve desired properties such as mechanical strength, ductility, and corrosion resistance. This integration reduces the reliance on costly and time-intensive trial-and-error experiments, streamlining alloy development while minimizing material waste and associated costs. The holistic use of the ICME approach for Mg alloy design can be seen in plenty of literature [29–33].

3 Conclusion

In conclusion, the integration of advanced computational tools in the design of Mg alloys for biomedical applications offers significant advantages in terms of accelerating development and improving material performance. By using methods like CALPHAD, MD, DFT, and ML, researchers can precisely predict and optimize key properties such as mechanical strength, corrosion resistance, and biocompatibility. The application of the ICME framework facilitates a comprehensive approach to alloy design, reducing reliance on trial-and-error experiments and enabling more efficient exploration of material design spaces. This computationally driven approach enhances the understanding of the underlying mechanisms governing alloy behavior in biological environments, thereby ensuring the design of high-performance Mg alloys. As these methodologies continue to mature, they promise to provide more sustainable, cost-effective, and targeted solutions for developing biodegradable Mg alloys with improved reliability and performance in clinical applications. The continued development and integration of these tools will further streamline the design process, reducing the time to market and allowing for more personalized and effective biomedical implants.

References

1. Das P, Pathak DK, Sharma P, Pandey PM (2025) A review on the mechanical and biocorrosion behaviour of iron and zinc-based biodegradable materials fabricated using powder metallurgy routes. Corros Rev.https://doi.org/10.1515/CORRREV-2023-0169
2. Holweg P, Herber V, Ornig M, Hohenberger G, Donohue N, Puchwein P, Leithner A, Seibert F (2020) A lean bioabsorbable magnesium-zinc-calcium alloy ZX00 used for operative treatment of medial malleolus fractures. Bone Joint Res 9:477–483. https://doi.org/10.1302/2046-3758.98.BJR-2020-0017.R2
3. Haude M, Ince H, Abizaid A, Toelg R, Lemos PA, Von Birgelen C, Christiansen EH, Wijns W, Neumann FJ, Kaiser C, Eeckhout E, Lim ST, Escaned J, Garcia-Garcia HM, Waksman R (2016) Safety and performance of the second-generation drug-eluting absorbable metal scaffold in patients with de-novo coronary artery lesions (BIOSOLVE-II): 6 month results of a prospective, multicentre, non-randomised, first-in-man trial. Lancet 387:31–39. https://doi.org/10.1016/S0140-6736(15)00447-X
4. Akbarzadeh FZ, Sarraf M, Ghomi ER, Kumar VV, Salehi M, Ramakrishna S, Bae S (2024) A state-of-the-art review on recent advances in the fabrication and characteristics of magnesium-based alloys in biomedical applications. J Magnes Alloy 12:2569–2594. https://doi.org/10.1016/J.JMA.2024.06.015
5. Das P, Kumar TSS, Sahu KK, Gollapudi S (2022) Corrosion, stress corrosion cracking and corrosion fatigue behavior of magnesium alloy bioimplants. Corros Rev 40:289–333. https://doi.org/10.1515/CORRREV-2021-0088
6. Das P, Nandan R, Pandey PM (2022) A review on corrosion properties of high entropy alloys fabricated by additive manufacturing. Trans Indian Inst Met 75:2465–2476. https://doi.org/10.1007/S12666-022-02610-9
7. Inoue A (2000) Stabilization of metallic supercooled liquid and bulk amorphous alloys. Acta Mater 48:279–306. https://doi.org/10.1016/S1359-6454(99)00300-6
8. Datye A, Alexander Kube S, Verma D, Schroers J, Schwarz UD (2019) Accelerated discovery and mechanical property characterization of bioresorbable amorphous alloys in the Mg–Zn–Ca and the Fe–Mg–Zn systems using high-throughput methods. J Mater Chem B 7:5392–5400. https://doi.org/10.1039/C9TB01302D
9. Kotoka R, Konchady M, Ramakrishnan G, Yarmolenko S, Pai D, Sankar J (2016) High throughput corrosion screening of Mg–Zn combinatorial material libraries. Mater Des 108:42–50. https://doi.org/10.1016/J.MATDES.2016.06.078
10. Kumari Rajendran R, Aggarwal D, Bonvalet Rolland M, Gruescu C, Shabadi R (2024) Design and development of large-diameter Mg–Zn–Ca bulk metallic glass for biomedical applications: a mechanical and corrosion perspective. Intermetallics 175:108520. https://doi.org/10.1016/J.INTERMET.2024.108520
11. Avraham S, Katsman A, Bamberger M (2011) Optimization of composition and heat treatment design of Mg–Sn–Zn alloys via the CALPHAD method. J Mater Sci 46:6941–6951. https://doi.org/10.1007/S10853-011-5660-Z/TABLES/3
12. Gandel DS, Easton MA, Gibson MA, Birbilis N (2014) CALPHAD simulation of the Mg–(Mn, Zr)–Fe system and experimental comparison with as-cast alloy microstructures as relevant to impurity driven corrosion of Mg-alloys. Mater Chem Phys 143:1082–1091. https://doi.org/10.1016/J.MATCHEMPHYS.2013.11.008
13. Wu K, Doan ÖN, Velikokhatnyi OI, Kumta PN (2011) A CALPHAD study on the thermodynamic stability of calcium-, zinc-, and yttrium-doped magnesium in aqueous environments. Mater Sci Eng B 176:1727–1732. https://doi.org/10.1016/J.MSEB.2011.04.016
14. Avey T, Cho D, Zhang J, Miao J, Dean D, Luo AA (2024) Determining critical Zn/Ca atomic ratio and its role in mechanical and corrosion properties of biodegradable Mg–Ca–Zn–Mn alloys. Materialia 37:102203. https://doi.org/10.1016/J.MTLA.2024.102203
15. Wei Q, Ramesh KT, Hufnagel TC, Wilkerson J, El-Awady JA, Kimberley J, Ravaji B, Joshi SP (2021) Insights from the MEDE program: an overview of microstructure–property linkages

in the dynamic behaviors of magnesium alloys. Mech Mater 163:104084. https://doi.org/10.1016/J.MECHMAT.2021.104084
16. Sun SJ, Ju SP, Yang CC, Chang KC, Lee IJ (2020) Effects of strontium incorporation to Mg–Zn–Ca biodegradable bulk metallic glass investigated by molecular dynamics simulation and density functional theory calculation. Sci Rep 101(10):1–14. https://doi.org/10.1038/s41598-020-58789-8
17. Yuan J, Dai B, Cui X, Li P (2023) The effects of electrodeposition temperature on morphology and corrosion resistance of calcium phosphorus coatings on magnesium alloy: comparative experimental and molecular dynamics simulation studies. RSC Adv 13:34145–34156. https://doi.org/10.1039/D3RA04162J
18. Wang H, Fang Z, Zhao Y, Yao S, Li J, Wang J, Zhu S, Niu C, Jia Y, Guan S (2020) Effects of alloy elements on adsorption of fibrinogen on biodegradable magnesium alloys surfaces: the MD simulations and experimental studies. Appl Surf Sci 512:145725. https://doi.org/10.1016/J.APSUSC.2020.145725
19. Das P, Pandey PM (2023) Machine learning based phase prediction and powder metallurgy assisted experimental validation of medium entropy compositionally complex alloys. Model Simul Mater Sci Eng 31:085015. https://doi.org/10.1088/1361-651X/AD04F4
20. Das P, Pandey PM (2024) Multi-objective optimization and machine learning assisted design and synthesis of magnesium based novel non-equiatomic medium entropy alloy. J Alloys Compd 985:174066. https://doi.org/10.1016/J.JALLCOM.2024.174066
21. Rai N, Das P, Gollapudi S (2021) Can an amorphous alloy crystallize into a high entropy alloy? Model Simul Mater Sci Eng 30:025007. https://doi.org/10.1088/1361-651X/AC2D9F
22. Das P, Sethy C, Pandey PM, Patel AK (2025) Evaluation of in vitro degradation behavior and biocompatibility assessment of Mg60Ti24.24Zn9.66Nb6.1 medium entropy alloy. J Alloys Compd 1010:178003. https://doi.org/10.1016/J.JALLCOM.2024.178003
23. Das P, Sethy C, Pandey PM, Patel AK (2025) Rapid fabrication of Mg60Ti24.24Zn9.66Nb6.1 reinforced iron composite bone scaffolds with improved biodegradation and biocompatibility. Mater Manuf Process 1–12. https://doi.org/10.1080/10426914.2025.2476464
24. Dong S, Wang Y, Li J, Li Y, Wang L, Zhang J (2024) Machine learning aided prediction and design for the mechanical properties of magnesium alloys. Met Mater Int 30:593–606. https://doi.org/10.1007/S12540-023-01531-6
25. Chen Y, Tian Y, Zhou Y, Fang D, Ding X, Sun J, Xue D (2020) Machine learning assisted multi-objective optimization for materials processing parameters: a case study in Mg alloy. J Alloys Compd 844:156159. https://doi.org/10.1016/J.JALLCOM.2020.156159
26. Lu Z, Si S, He K, Ren Y, Li S, Zhang S, Fu Y, Jia Q, Jiang HB, Song H, Hao M (2022) Prediction of Mg alloy corrosion based on machine learning models. Adv Mater Sci Eng 2022:9597155. https://doi.org/10.1155/2022/9597155
27. Zhang D, Ma S, Markert B (2023) Machine learning based prediction of the corrosion behaviour of biodegradable magnesium alloys with PEO coatings. PAMM 22:e202200319. https://doi.org/10.1002/PAMM.202200319
28. Suh JS, Suh BC, Bae JH, Kim YM (2023) Machine learning-based design of biodegradable Mg alloys for load-bearing implants. Mater Des 225:111442. https://doi.org/10.1016/J.MATDES.2022.111442
29. Wang Y, Xie T, Tang Q, Wang M, Ying T, Zhu H, Zeng X (2024) High-throughput calculations combining machine learning to investigate the corrosion properties of binary Mg alloys. J Magnes Alloy 12:1406–1418. https://doi.org/10.1016/J.JMA.2021.12.007
30. Yao Z, Berman T, Allison J (2020) An ICME method for predicting phase dissolution during solution treatment in advanced super vacuum die cast magnesium alloys. Integr Mater Manuf Innov 9:301–313. https://doi.org/10.1007/S40192-020-00186-0
31. Eswarappa Prameela S, Yi P, Hollenweger Y, Liu B, Chen J, Kecskes L, Kochmann DM, Falk ML, Weihs TP (2022) Strengthening magnesium by design: integrating alloying and dynamic processing. Mech Mater 167:104203. https://doi.org/10.1016/J.MECHMAT.2021.104203
32. Wang Y, Tang Q, Xu X, Weng P, Ying T, Yang Y, Zeng X, Zhu H (2023) Accelerated discovery of magnesium intermetallic compounds with sluggish corrosion cathodic reactions through

active learning and DFT calculations. Acta Mater 255:119063. https://doi.org/10.1016/J.ACTAMAT.2023.119063

33. He X, Liu J, Yang C, Jiang G (2023) Predicting thermodynamic stability of magnesium alloys in machine learning. Comput Mater Sci 223:112111. https://doi.org/10.1016/J.COMMATSCI.2023.112111

Chapter 6
Fabrication Techniques and Characterisation for Magnesium-Based Biodegradable Implants

Siddharth Tevatia and Abhishek Tevatia

1 Introduction

Magnesium (Mg) and its alloys are used in biomedical applications owing to their mechanical characteristics that closely mimic those of human bone, in addition to their biocompatibility and biodegradability [1, 2]. Mg-based materials are advantageous for orthopaedic [3–6], cardiovascular applications [7, 8], and medical implants [9, 10]. Permanent metal implants may induce stress shielding, wherein the implant absorbs stress rather than the bone, resulting in its deterioration [11, 12]. Mg implants undergo gradual degradation [2, 6, 10], eliminating the need for additional procedures, reducing patient anxiety, improving recovery duration, and reducing healthcare expenses [13].

The increasing interest in Mg alloys for medicinal applications stems from their capacity to facilitate bone regeneration [14–16]. Mg ions are essential for bone metabolism and osteogenesis (the development of bones), thus facilitating healing [17]. Mg implants have been studied for several biological applications. The distinct features of Mg may be used for precise applications in antitumour therapies [10], surgical wires [18–20], cardiovascular instruments such as stents [7, 8], and orthopaedic devices including screws, pins, and bone plates [3–6] as shown in Fig. 1. The natural degradation of the implants mitigates the long-term complications associated with conventional permanent implants. The concerns of Mg-based implants

S. Tevatia
Department of Periodontology, I.T.S Dental College, Greater Noida, Uttar Pradesh, India

A. Tevatia (✉)
Department of Mechanical Engineering, Netaji Subhas University of Technology, Dwarka, New Delhi, India
e-mail: abhishek.tevatia@nsut.ac.in

A. Dixit et al. (eds.), *Biodegradable Metallic Materials*, Biomedical Materials for Multi-functional Applications, https://doi.org/10.1007/978-981-95-2401-3_6

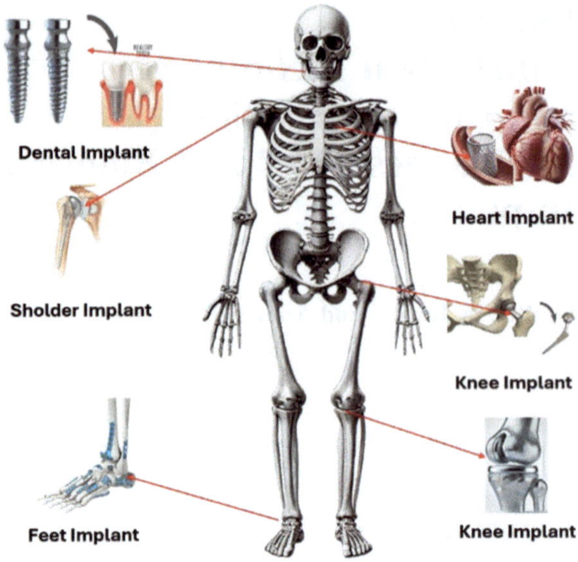

Fig. 1 A schematic depicting the applications of medical implants within the human body [1]

include metal ion buildup, inflammatory reactions, and sustained responsiveness to foreign substances from the implant [6, 10].

The direct application of Mg implants is limited in clinical settings until several difficulties are addressed. Under physiological circumstances, Mg rapidly corrodes and deteriorates. The fast deterioration may jeopardise the implant's mechanical integrity prior to complete tissue recovery, potentially resulting in implant failure [2, 9, 21]. Rusting generates hydrogen gas, perhaps forming gas pockets at the implant location [9]. Gas pockets may obstruct tissue integration, resulting in localised inflammation and diminished implant efficacy [21]. To use Mg-based implants safely and efficiently, the problems such as fast degradation, corrosion, and gas pocket formation must be resolved.

Mechanical engineers and biomedical researchers have explored a variety of techniques to improve the corrosion resistance and mechanical characteristics of Mg-based implants. This chapter examines, in detail, the application of Mg-based implants in the biomedical industry. Aluminium (Al), calcium (Ca), zinc (Zn), manganese (Mn), strontium (Sr), lithium (Li), zirconium (Zr), and rare earth elements (REEs) in Mg alloys may reduce degradation rates and improve mechanical properties [22, 23]. Surface modification methods, including micro-arc oxidation, plasma spraying, and polymeric coatings, allow the creation of protective layers that prevent corrosion and demonstrate biocompatibility [24–26]. Composite coatings and surface treatments using hydroxyapatite have markedly improved corrosion resistance and environmental compatibility [24].

Recent advancements in manufacturing methods have enhanced the optimisation of Mg-based implants. Casting [27, 28], powder metallurgy [29, 30], and 3D printing [31–34] permit the exact control of implant microstructure, porosity, and degradation rates. The fabrication processes facilitate the development of patient-specific implants, enhancing their effectiveness and therapeutic results. Additive printing produces intricate geometries and porous materials that might enhance osseointegration and degradation [32–34].

The creation of Mg-based biodegradable implants necessitates several characterisation processes [1, 5, 35]. The dependability and safety of implants are contingent upon characterisation procedures. Microstructural analyses are enhanced by optical and scanning electron microscopy to examine grain structure, phase distribution, and alloying effects [1, 5, 21, 35]. Tensile, compressive, and fatigue testing assess the strength, toughness, and durability of Mg implants [10]. The dynamics of physiological deterioration are examined by immersion studies in simulated biological fluids and electrochemical methods [1]. All of the mentioned assessments measure corrosion [2, 9, 21, 24]. Biocompatibility studies, including in vitro cell culture and in vivo animal model tests, are essential to evaluate the biological response to Mg implants and ensure their safety for human use [19, 20, 33, 36].

Current research is being conducted on the hybrid materials and composite designs to enhance Mg-based implants [9, 28, 30, 32, 35]. Examples include bioactive glass, polymer–Mg composites, and nano-structured coatings and such coatings may enhance osseointegration and mitigate degradation [24, 37]. The incorporation of Mg into bioresorbable polymers seems to be promising, as it facilitates regulated disintegration while preserving mechanical properties [10].

Standardisation of testing methodologies, prolonged clinical studies, and regulatory endorsement are essential to guarantee the safety and efficacy of Mg-based materials. Additional research is required to develop surface treatments and alloy compositions that may precisely alter degradation rates and mechanical properties for therapeutic applications.

This chapter provides a comprehensive analysis of the various manufacturing processes used in the fabrication of Mg-based biodegradable implants. Furthermore, this chapter examines sophisticated characterisation techniques used to assess the microstructure, mechanical properties, degradation behaviour, and biocompatibility of Mg-based implants. Researchers and medical practitioners may enhance Mg-based biomaterials by gaining a more profound comprehension of related factors, enabling them to enhance their clinical efficacy and dependability as next-generation implant materials. Mg-based implants provide considerable potential to transform the field of medical implants and enhance patient outcomes. The mentioned commitment is favoured by the continuous advancements in the fields of materials science and biomedical engineering.

2 Notable Advancement of Mg-Based Materials in Biomedical Applications

Mg-based materials have been employed in biomedical applications since the late nineteenth century. Figure 2a depicts the use of Mg-based materials from historical to contemporary times. Extensive research and advancements in materials science and biomedical engineering have led to significant improvements in the use of Mg alloys for medical applications [38]. The data suggest that the twentieth century has significant Mg-based research in the medical field; however, further advancement is picking up in the early twenty-first century. Figure 2b illustrates the significant milestones in the development and application of Mg-based materials in biological contexts.

During the nineteenth and early twentieth centuries, Mg was used in biomedical applications owing to its lightweight characteristics and mechanical resemblance to bone. Initially, it was used as ligature wires to regulate haemorrhaging during surgical procedures [39]. Subsequently, pure Mg was evaluated for orthopaedic fixation devices, such as pins, spheres, wires, plates, and nails. Fast corrosion remains a significant problem [40–42]. Mg plates with steel screws were used for fracture repair; nevertheless, uncontrolled deterioration undermined mechanical stability [43]. Magnesium–manganese (Mg–Mn) alloys were developed to mitigate corrosion problems, enhancing osteosynthesis [44]. Subsequent developments included

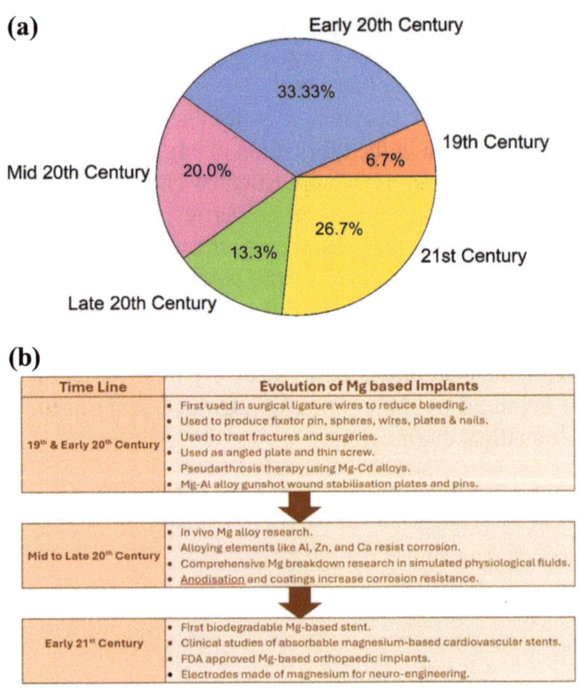

Fig. 2 Biomedical applications of Mg-based materials **a** distribution over time, and **b** notable advancements [10, 38]

the use of magnesium–cadmium (Mg–Cd) alloys for the treatment of pseudarthrosis, providing improved mechanical qualities while eliciting toxicity apprehensions [10, 38]. Magnesium–aluminium (Mg–Al) alloy plates and pins were used for the stabilisation of gunshot wounds, deteriorating within weeks to align with the anticipated bone healing timeframe [10, 38].

In the mid-to-late twentieth century, research focused on enhancing the corrosion resistance and mechanical qualities of Mg implants by alloying with other metals. Comprehensive in vivo investigations were performed on Mg-based alloys, integrating components like Al, Zn, and Ca to improve corrosion resistance [45]. Researchers concentrated on elucidating Mg breakdown in simulated biological fluids, enhancing comprehension of its electrochemical properties [46]. Investigations broadened to include surface modification methods, such as anodisation and coatings, to enhance the corrosion resistance of Mg alloys in biomedical applications [47–49]. The mid-to-late twentieth century's initiatives laid the groundwork for further advancements in Mg-based medical devices.

In the early twenty-first century, substantial advancements were achieved in the commercialisation and clinical use of Mg-based implants. The first biodegradable Mg-based stent was successfully placed in animal models, demonstrating its promise for cardiovascular applications [50]. Clinical studies assessed absorbable Mg cardiovascular stents, demonstrating encouraging biocompatibility and mechanical integrity in the treatment of coronary artery disease [51]. The approval of the first biodegradable Mg-based orthopaedic implants marked a significant advancement [4]. The Drug-Eluting Absorbable Metal Scaffold (DREAMS) clinical trial showed decreased restenosis rates in second-generation Mg-based stents [7]. Mg-based brain electrodes were created for possible neuro-engineering applications, extending beyond orthopaedics and cardiovascular uses [10, 38]. The early twenty-first century has seen a growing recognition of magnesium as a suitable biomaterial for medical implants.

Recent research has enhanced the characteristics of Mg-based biomaterials by sophisticated alloying techniques, nano-structured coatings, and additive manufacturing methods. During the 2020s, 3D printing facilitated the regulation of porosity and degradation rates for individualised Mg-based implants [32–34]. The comprehensive historical progression, together with a timeline of Mg-based medicinal applications, is shown in Table 1.

Contemporary research emphasises hybrid biomaterials that integrate Mg with bioresorbable polymers and bioactive coatings to enhance efficacy [28, 30, 32, 35]. The standardisation of testing protocols and regulatory approvals is essential for the extensive clinical use of Mg-based implants [52]. Recent advancements in alloying techniques, surface modifications, and additive manufacturing have expanded their medicinal uses. The standardisation of regulations and the development of hybrid biomaterials will enhance their safety and efficacy, ensuring that Mg-based implants continue to revolutionise biological applications in the future.

Table 1 Chronological progression of Mg-based medical applications

Time period	Key developments	Description	Reference
1878	First biomedical use	Mg employed as ligature wires to manage haemorrhage during surgical procedures	[39]
1892–1905	Orthopaedic fixation	Pure Mg evaluated for fixator pins, spheres, wires, plates, and nails demonstrated biocompatibility but exhibited significant corrosion	[40–42]
1906	Fracture repair	Mg plates with steel screws employed for fracture repair experienced rapid degradation, compromising stability	[43]
1938	Mg–Mn alloys	Mg–Mn alloys are implemented to mitigate deterioration and enhance osteosynthesis	[44]
1944	Mg–Cd alloys	Utilised in the treatment of pseudarthrosis, providing strength while elevating toxicity concerns	[10, 38]
1945	Mg–Al alloys	Utilised for the stabilisation of gunshot wounds, it degrades between 4 and 6 weeks, coinciding with the period of bone recovery	[10, 38]
1950s–1960s	Biocompatibility studies	Comprehensive in vivo investigations on Mg alloys incorporating Al, Zn, and Ca to improve corrosion resistance	[45]
1970s	Corrosion studies	Comprehensive investigation of Mg breakdown in biological fluids, enhancing electrochemical comprehension	[46]
1980s	Surface modification	Methods such as anodisation and coatings examined to improve corrosion resistance	[47–49]
2003	First Mg-based stent	Implanted in animal models, indicating promise for cardiovascular uses	[50]
2005–2007	Clinical trials	Absorbable Mg stents evaluated for the treatment of coronary artery disease, demonstrating biocompatibility	[51]
2010	FDA-approved implants	Initial FDA-approved biodegradable Mg orthopaedic implants launched	[4]
2013	DREAMS study	Second-generation bioresorbable Mg stents demonstrated decreased restenosis rates	[7]

(continued)

Table 1 (continued)

Time period	Key developments	Description	Reference
2018	Mg-based brain electrodes	Engineered for neuro-engineering applications extending beyond orthopaedic and cardiovascular domains	[10, 38]
2020s	3D printing and hybrid biomaterials	3D printing enhanced porosity and degradation rates; hybrid biomaterials integrated Mg with bioresorbable polymers	[32–34]

3 Mg Biomaterial Alloys: Metallurgical and Biological Characterisation and Their Toxicity

The historical development of Mg-based implants dates back to 1878, when pure Mg was initially employed in biomedical applications for surgical ligatures [39]. Researchers have investigated its potential in orthopaedic fixation and fracture treatment throughout the years; nevertheless, quick degradation has constrained its clinical efficacy [40–42]. Mg alloys have been produced for several medical uses, such as orthopaedic screws [3–6], cardiovascular stents [7, 8], and bioresorbable scaffolds [18–20]. Mg is often alloyed with additional elements to improve its mechanical properties, corrosion resistance, and biocompatibility [22, 23]. Although pure Mg is lightweight and exhibits good biocompatibility, its mechanical strength and corrosion resistance are inadequate for certain applications necessitating load-bearing capabilities. Consequently, Mg is often alloyed with other metals to enhance its properties, such as Al, Ca, Zn, Mn, Sr, Li, Zr, and REEs, which are commonly utilised in alloying processes [1, 10, 26, 53].

Mg alloys, including Ca, have attracted significant interest in recent years. Ca is crucial for bone mineralisation, rendering Mg–Ca alloys very biocompatible [54]. The incorporation of Ca into Mg alloys enhances their mechanical strength while concurrently preserving good degrading resistance [10, 53]. Zinc is an element commonly included in Mg alloys. Zinc enhances the mechanical properties of the alloy and its corrosion resistance. Furthermore, it has been ascertained that Zn exerts beneficial impacts on the human body, encompassing the facilitation of tissue regeneration and osteogenesis (the process of bone formation) [1, 10, 26, 53, 55]. Moreover, Mg can be alloyed with various elements, including Mn, Sr, Li, Zr, and REEs, to modify its properties to meet the specific demands of certain applications. When compared to pure Mg, Mg-based alloys exhibit enhanced mechanical properties, reduced degradation rates, and improved biocompatibility. Table 2 provides a comprehensive analysis of the biological, metallurgical, and toxicological profiles of commonly used alloying elements in Mg biomaterials found in the human body.

Table 2 Metallurgical and biological characterisation of Mg biomaterial alloying elements and toxicity

Alloying elements	Alloying elements characterisation		Toxicology	References
	Metallurgical	Biological		
Calcium (Ca)	• Enhances elongation, creep resistance, hardness, and strength • Enhances corrosion resistance • Recommends Ca below 1%	• Crucial for bone growth and teeth maintenance • Essential for cell function, muscle contraction, and heartbeat regulation • Deficit can cause osteoporosis	Excessive Ca levels in blood • Leads to kidney stones • Causes bone weakness • Disrupts heart and brain function	[1, 10, 26, 53, 54]
Zinc (Zn)	• Zn content below 5% enhances corrosion resistance, strength, and mechanical properties • Up to 3% decreases grain size, reduces H_2 gas accumulation	• Crucial for enzyme function, wound healing, neurotransmission, and protein synthesis • Deficiency can delay the T-cell response	• Zn^{2+} ions generate $ZnCl_2$, which may harm the stomach cell lining • Excessive absorption can be neurotoxic • Impede bone formation	[1, 10, 26, 53, 55]
Manganese (Mn)	• Enhances corrosion resistance • Reduces elongation • Reduces yield strength • Typically, below 1% wt	• Supports blood coagulation, energy synthesis, and antioxidant defence inside cells • Supports digestion, immunity, and neuroregulation • Supports the brain, neurological system, cell growth, and homeostasis	• Mg overexposure risks neurological disorders	[1, 10, 26, 53, 56]

(continued)

Table 2 (continued)

Alloying elements	Alloying elements characterisation		Toxicology	References
	Metallurgical	Biological		
Strontium (Sr)	• Refines grain size in Mg alloys • Forms Mg–Sr phases along grain boundaries • Reduces strength and corrosion resistance • Excessive Sr weakens mechanical properties and increases corrosion rate	• Decreases bone resorption • Stimulates osteoblast proliferation • Improves bone strength and mineral density • The degradation of Mg–Sr alloys facilitates the incorporation of Sr into hydroxyapatite (HA)	• Increases risk of blood clots • May cause skeletal abnormalities	[1, 10, 26, 53, 57]
Lithium (Li)	• Reduces grain size • Forms Mg–Li phases with a bcc structure • Increases deformability • Improves corrosion resistance with less than 9% Li concentration	• Li used in treating psychoses • Low Li intake linked to increased suicides, homicides, and drug use arrests • Definition of Li deficiency diseases unknown	• Excessive Li intake may cause central nervous system issues	[1, 10, 26, 53, 58]

(continued)

Table 2 (continued)

Alloying elements	Alloying elements characterisation		Toxicology	References
	Metallurgical	Biological		
Zirconium (Zr)	• The addition of Zr (less than 2%) to Mg alloys improves mechanical qualities, corrosion resistance, and the ability to absorb stress and vibration	• Crucial element in skeleton	• Low systemic toxicity	[1, 10, 26, 53, 59]
Rare earth element (REEs)	• They are efficient grain refiners and improve mechanical qualities, corrosion resistance, electrochemical behaviour, and creep resistance	• Y and REEs, found in low abundance in organisms and water bodies, are not essential, but some have potential for cancer treatment	• Ce and La exhibit the highest cytotoxicity. Ce and Pr cause hepatotoxicity	[1, 10, 23, 26, 53]

4 Fabrication Techniques for Mg-Based Implants

Various Mg-based implant fabrication procedures, as shown in Fig. 3, such as casting, wrought processing, powder metallurgy, and laser additive manufacturing, have been established over the past two decades [10]. The casting of Mg bone implants is favoured because of its ease of alloy composition management, economic efficiency, and user-friendliness. Casting needs argon due to the rapid oxidation of Mg [27, 28]. Extrusion, rolling, and forging enhance the mechanical and corrosion properties of Mg alloys [60]. Powder metallurgy enhances the quality of Mg alloys. Sintering should not be prolonged to prevent coarse grains and suboptimal performance. Self-heating spark plasma sintering lowers grain size in advanced powder metallurgy [29, 30].

Mg-based biomaterial additive manufacturing has garnered interest for its capacity to fabricate intricate geometries and biodegradable implants. Materials based on Mg have been produced using powder bed fusion, wire arc additive manufacturing, friction stir additive manufacturing, paste extrusion deposition, and jetting [32]. Additive manufacturing reduces production time and costs while creating customised implants with complex geometries. Mg's pronounced oxygen reactivity, low boiling point, and elevated vapour pressure complicate additive manufacturing. Additive manufacturing materials, particularly wire, powder, and liquid resin, increase surface energy and the potential for oxidation [32, 34]. Table 3 presents the diverse fabrication techniques along with applications in medical implants.

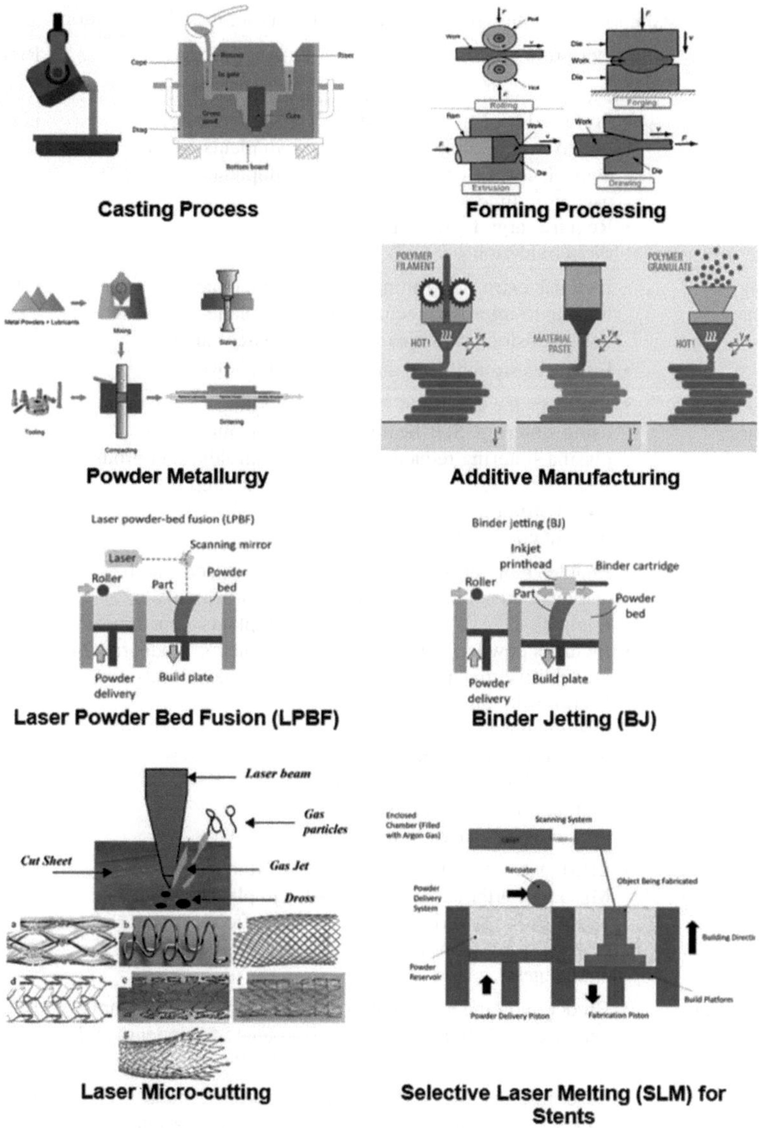

Fig. 3 Various Mg-based implant fabrication procedures [10, 27, 28, 60]

The use of laser powder bed fusion (LPBF) [61], selective laser melting (SLM) [34], and direct metal laser melting for 3D printing Mg-based *orthopaedic implants* [3–6, 16, 18] has shown promise. Research on laser powder bed fusion (LPBF) in an inert environment has shown that biodegradable WE43 Mg-based alloys meet the requirements for orthopaedic biodegradation, biocompatibility, and mechanical integrity [62, 63]. Binder and binderless jetting provide enhanced production

Table 3 Mg-based implants include several methods of fabrication and applications

Fabrication technique	Description	Applications	References
Casting	• Preferred for Mg bone implants due to ease of alloy composition control, cost-effectiveness, and user-friendliness • Requires argon (Ar) to prevent rapid oxidation	Orthopaedic implants—bone implants	[27, 28]
Forming processing	• Includes extrusion, rolling, and forging to improve mechanical and corrosion properties • Enhances strength and durability	Orthopaedic implants—bone screws, plates, and joint implants	[60]
Powder metallurgy	• Improves the quality of Mg alloys using sintering. Self-heating spark plasma sintering reduces grain size • Avoid prolonged sintering to prevent coarse grains	Orthopaedic implants—bone scaffolds and porous structures	[29, 30]
Additive manufacturing	• Enables fabrication of complex and customised biodegradable implants • Includes powder bed fusion, wire arc additive manufacturing, friction stir additive manufacturing, paste extrusion deposition, and jetting • Mg's high oxygen reactivity, low boiling point, and high vapour pressure can cause oxidation	Orthopaedic and cardiovascular implants—customised implants, biodegradable scaffolds, stents	[32, 34]
Laser powder bed fusion (LPBF)	• LPBF used for 3D printing of Mg-based orthopaedic implants • High-resolution printing and material limitations pose challenges	Orthopaedic implants—bone scaffolds, biodegradable structures	[61]
Binder jetting (BJ)	• Binder and binderless jetting enhance scaffold properties	Orthopaedic implants generate biodegradable MgP scaffolds	[62, 63]
Laser microcutting	• Laser microcutting is the primary industrial method for fabricating Mg stents due to its ability to create intricate designs without the need for moulds • Essential for cardiovascular stents due to the need for microtube processing • Mg's poor ductility complicates seamless tube production. High-efficiency fibre lasers improve precision	Cardiovascular implants—stents and microtubes	[65]

(continued)

Table 3 (continued)

Fabrication technique	Description	Applications	References
Selective laser melting (SLM) for stents	• Integrates microtube processing and laser cutting in one step • Laser cutting can cause defects, but improved laser technologies enhance precision	Cardiovascular implants—vascular stents	[66]
3D printing and photochemical etching for surgical wires and staplers	• Used in the fabrication of Mg-based staples with corrosion-resistant coatings • Coatings improve corrosion resistance. Research focuses on failure mechanisms • Enhances biocompatibility and longevity	Surgical wires and staplers—biodegradable staples and medical wires	[67]

options. Binder jetting fuses particles inside a powder bed using a liquid binding agent to include organic or bioactive substances [63]. Binder jetting technique generates biodegradable MgP scaffolds with bone-like tensile and compressive properties [64]. Binderless jetting employs capillary forces among powder particles and soaking duration during sintering to influence material strength, which imparts compressive strength and flexibility akin to bone. Three-dimensional porous scaffolds for bone implant fabrication seem promising. Nonetheless, material limitations and high-resolution printing continue to pose significant challenges [32].

Permanent metallic *stents* [7, 68] need continuous tubing. Tube drawing and laser cutting are essential; however, the inadequate ductility of Mg at ambient temperature makes the production of seamless, precisely diameter Mg-based tubes challenging [66]. Stent applications need microtube processing due to coronary stents having outer diameters under 4.0 mm and wall thicknesses of 0.10–0.20 mm. The efficacy of a stent is contingent upon dimensional precision [66, 69]. Laser microcutting is the primary industrial method for fabricating Mg stents, since it allows intricate designs without the need for moulds. The laser microcutting technique administers energy locally and accurately without physical contact. Demir et al. [65] fabricated a pulsed fibre laser micromachined AZ31 Mg alloy cardiovascular stent with a distinctive mesh configuration. Photochemical machining may produce Mg alloy stents; nevertheless, laser cutting often produces defects. High-efficiency fibre lasers with superior beam quality may address the issue [68].

A novel vascular stent fabrication technique, selective laser melting (SLM), integrates microtube processing and laser cutting into a single step [66]. Recent advancements in laser technology have improved the precision and efficiency of Mg stent manufacturing. To optimise performance, the microstructure of Mg-based stents must be meticulously preserved throughout manufacturing, laser cutting, annealing, and polishing [70].

In comparison to powder-bed fusion technologies [32], the 3D printing of Mg *surgical wires and staplers* [19] for gastrointestinal and anastomotic procedures presents a viable alternative due to its customisation potential, reduced manufacturing time, enhanced performance, biocompatibility, and cost-effectiveness [32, 34].

The enhancement of the mechanical and functional properties of medical equipment is achieved by the manufacture of Mg-based surgical wires and staples using SLM, wire arc additive manufacturing, fused deposition modelling, fused filament fabrication, stereolithography, and composite-based additive manufacturing technology techniques [67]. Post-processing techniques for Mg materials were used in the fabrication of biodegradable Mg AZ31 staples. Mg-based staples were created via photochemical etching and 3D-printed polymer moulds, with coatings that ensured corrosion resistance and biocompatibility [19, 67].

5 Mechanical Characterisation of Mg-Based Implants

Mechanical characterisation is essential to assess the strength, durability, and biocompatibility of Mg-based implants. Tensile and compression tests conducted using universal testing machines (UTMs) assess yield strength, ultimate strength, and elongation. Microhardness tests, such as Vickers and Knoop, evaluate surface hardness and resistance to deformation. Fatigue testing assesses the cyclic stress endurance of load-bearing implants. Nanoindentation reveals hardness and elastic modulus at the microscale. Fracture toughness testing assesses fracture propagation, whereas impact testing evaluates shock resistance, hence validating reliability and performance under physiological conditions.

5.1 Mechanical Properties

The mechanical properties of Mg alloys have been extensively studied in vitro. The incorporation of alloying elements will improve mechanical properties, corrosion resistance, and biocompatibility, as detailed in Table 4. Al-based alloys with 12.7 wt% Mg solubility exhibits excellent castability, moderate mechanical strength, and corrosion resistance. Enhanced Al content augments the corrosion resistance of Mg–Al alloys [45, 47]. Alloys of Zn and Mn enhance strength and ductility at room temperature, making AZ31, AZ61, AZ91, and AM60 attractive biodegradable materials [2, 21]. The incorporation of Ce, Ca, Y, Nd, and Sr enhances the microstructure, hence augmenting mechanical performance [1, 10, 53].

Table 5 provides a comparative analysis of different Mg alloys, emphasising mechanical properties. Aluminium alloys such as AZ91 and AM60 provide moderate strength and corrosion resistance, making them suitable for orthopaedic applications, such as bone plates [45, 47]. Zinc-containing alloys, such as Mg–4Zn, ZK30, and ZK60, demonstrate fine-grained microstructures with improved strength, frequently

Table 4 Influence of alloying elements on mechanical properties, corrosion resistance, and biocompatibility of Mg alloys

Element	Mechanical properties	Corrosion resistance	Biocompatibility	References
Aluminium (Al)	Enhances castability and moderate strength	Improves corrosion resistance in Mg–Al alloys	Limited biocompatibility due to the possible toxicity of Al ions	[1, 10, 45, 47, 53]
Zinc (Zn)	Improves yield strength and ductility	Contributes to corrosion resistance in limited amounts	Essential for cell growth and tissue regeneration	[1, 10, 26, 53, 55]
Zirconium (Zr)	Acts as a superior grain refiner	Enhances corrosion resistance in aluminium-free alloys	Highly biocompatible, with minimal cytotoxicity	[1, 10, 26, 53, 59]
Rare earth elements (REEs)	Improve creep resistance and high-temperature strength	Enhance corrosion resistance (e.g. Nd, Gd)	Some have anti-carcinogenic properties and improve biocompatibility	[1, 10, 23, 26, 53]
Calcium (Ca)	Refines grains but has limited solubility, affecting strength and ductility	Controlled alloying and surface treatments enhance corrosion resistance, maintaining structural integrity	Crucial for bone integration but has limited solubility	[1, 10, 26, 53, 54]
Strontium (Sr)	Strontium enhancing bone regeneration when alloyed with Mg	Improves corrosion resistance, but excessive $Mg_{17}Sr_2$ at grain boundaries may reduce it	Promotes osteoblast maturation and bone formation	[1, 10, 26, 53, 57]

utilised in biodegradable implants and cardiovascular stents [2, 21, 55]. WE43, infused with rare earth elements (REEs), enhances mechanical qualities and is extensively utilised in orthopaedic applications [63]. High-strength Mg–Zn–Y alloys, such as Mg97Zn1Y2 and Mg96Zn2Y2, are very efficacious in biodegradable orthopaedic devices [26, 53, 55].

Mg–Si (0.8%) alloys, despite modest strength, exhibit commendable ductility and are utilised in temporary implants [1, 10, 26, 53, 57]. Zirconium- and strontium-based alloys, shown as Mg–1Zr–2Sr, exhibit exceptional corrosion resistance, rendering them suitable for biodegradable cardiovascular stents [1, 10, 26, 53, 59]. Alloys based on neodymium and gadolinium, such as Mg–3Nd–0.2Zn–0.4Zr [20] and Mg–11.3Gd–2.5Zn–0.7Zr, provide improved mechanical properties and are employed in orthopaedic fixation devices and bone plates [1, 10, 23, 26, 53].

Patented alloys such as Mg–YHRE–Zr and Mg–Gd–Zn–Zr–Nd–Y are engineered for biomedical purposes, providing superior strength, corrosion resistance,

Table 5 Mechanical properties of different Mg alloys

Alloy type	YS (MPa)	UTS (MPa)	Elongation (%)	References
AZ91 (as-cast)	145	275	6	[45, 47]
AM60 (as-cast)	160	3	–	[45, 47]
Mg-4Zn	–	216.8	15.8	[2, 21, 55]
ZK30	215	300	9	[2, 21, 55]
ZK60	235	315	8	[2, 21, 55, 71]
WE43	160	260	6	[63]
Mg97Zn1Y2	610	–	5	[26, 53, 55]
Mg96Zn2Y2	390	–	5	[26, 53, 55]
Mg–Si (0.8%)	52	152	9.5	[1, 10]
Mg–1Zr–2Sr	65–125	200–290	14–38	[10, 53, 59]
Mg–3Nd–0.2Zn–0.4Zr	90	194	12	[20]
Mg–11.3Gd–2.5Zn–0.7Zr	281	341	13.5	[1, 10, 26, 53]
Mg–7.25Y–0.31Zn	149	246	–	[1, 10, 23, 53]
Mg–8Y–1Er–2Zn	275	359	–	[1, 23, 26, 53]
Patented Mg–YHRE–Zr	150–254	244–333	17.5–28	[1, 10, 53]
Patented Mg–Gd–Zn–Zr–Nd–Y	Up to 269	Up to 325	–	[1, 10, 53]

and biocompatibility [1, 10, 53]. Mg-based alloys are essential in medical applications, especially in orthopaedic implants, biodegradable screws, cardiovascular stents, and medical fixation devices. The biodegradability facilitates progressive absorption within the body, reducing the necessity for subsequent procedures. The inclusion of rare earth elements, zinc, and zirconium improves corrosion resistance and biocompatibility, rendering them essential materials for advanced medical implants.

5.2 *Corrosion*

The corrosion behaviour of multi-elemental Mg-based alloys is difficult to predict due to microgalvanic corrosion, which depends on differences between intermetallic phases and the matrix [2, 9, 21, 24, 46, 53, 61, 63]. Mg hydroxide provides corrosion protection in aqueous environments, but chloride concentrations over 30 mmol L^{-1} diminish its effectiveness [24]. Hydrogen gas release from Mg-based orthopaedic implants may cause gas pockets inside bone tissue [46]. Enhancing alloy composition, thermal treatment, processing, and surface modifications increases corrosion resistance. Rolling and equal channel angular pressing (ECAP) enhances grain boundary density and second-phase distributions. Mg-based alloys have reduced

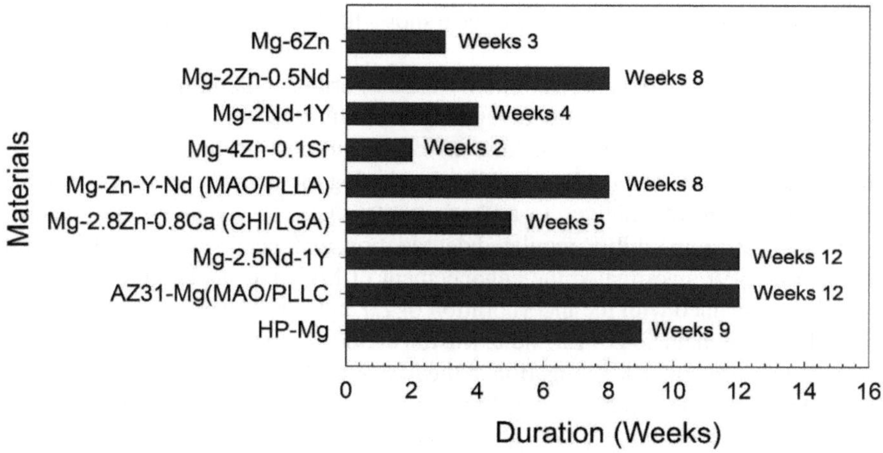

Fig. 4 Duration of in vivo degradation of Mg-based surgical implants [1, 10, 23, 26, 53]

corrosion fatigue limits in vivo compared to air. Research on Mg-based biodegradable implants is necessary to improve durability and longevity in clinical applications [9, 24, 61].

The in vivo degradation properties [21, 46] of Mg-based surgical implants vary depending on alloy composition and surface modifications, as depicted in Fig. 4. Materials with enhanced corrosion resistance, such as AZ31–Mg and Mg–2.5Nd–1Y, are suitable for applications requiring extended mechanical support, such as surgical staples and stents [9, 21, 24]. However, Mg–4Zn–0.1Sr exhibits rapid structural deterioration within 2 weeks, making it unsuitable for prolonged use. Materials with moderate durability, such as HP–Mg and Mg–2Zn–0.5Nd, suggest a balance between degradation and mechanical integrity, potentially beneficial for short- to mid-term applications [24]. Surface coatings like MAO, PLLA, and CHI/LGA can modulate degradation rates, with coated materials showing diminished corrosion rates and improved mechanical retention [21]. However, the high corrosion rate poses a challenge, leading to premature loss of mechanical functioning in some cases. Future research should focus on improving Mg-alloy compositions and developing multifunctional coatings for controlled degradation tailored for specific surgical applications.

6 Applications: Mg Alloy-Based Implants

Medical uses for Mg alloys are developing due to their bone regeneration properties. Mg implants have been studied for surgical wires [18–20], anticancer medications [10], cardiovascular devices like stents [7, 8], and orthopaedic instruments, including screws, pins, and bone plates [3–6]. Table 6 shows the Mg-based implant with various applications.

Mg-based biomaterials are commonly used in orthopaedic implants due to their mechanical properties and ability to repair bone fractures [3–6, 16, 18]. Autografting aids bone regeneration, but financial constraints and post-healing implant removal limit its application. The medical field is developing biocompatible, biodegradable, bone-like implant alternatives to address limitations. Mg–Zn alloys are superior due to their biocompatibility, regulated degradation, and mechanical properties [36]. Zhang et al. [19] investigated the enhancement of Mg's mechanical properties and corrosion resistance with the incorporation of Zn and Mn. Microstructural analyses showed that the inclusion of Zn and extrusion at 300 °C resulted in grain size refinement, tensile strength, and elongation in the Mg–Zn–Mn alloy [19]. Historically, Mg was unpopular in medicine due to its high corrosion rate, but advancements in metallurgy have improved its mechanical properties and corrosion resistance. Mg-based biomaterials are being engineered to replace metallic implants in musculoskeletal and orthopaedic procedures [3–5, 18].

Cardiovascular implants have the potential to save millions of lives by treating coronary stenosis. However, stent materials must meet strict safety and effectiveness criteria, including cellular activity exceeding 80%, yield strength above 200 MPa, and elongation rate above 15% [7, 8, 20, 51, 65, 66, 68–70]. Stents can protect vessels for 6–12 months before decaying or being absorbed, mitigating long-term complications [7]. Biodegradable Mg implants can be substituted with permanent metallic and polymeric implants, providing transient mechanical support during dissolution. Mg's biodegradability reduces thrombogenesis, restenosis, endothelial dysfunction, and other cardiovascular complications. The use of Mg-based materials in cardiovascular applications is in the experimental and preclinical stages [7, 68, 69].

Wound closure devices must be non-toxic, sterile, easy to use, effective, comfortable, and economical. Surgical wires and staplers need high strength-to-diameter ratios, tensile strength, and flexibility for optimal wound adaptation [18, 19, 45, 47]. Materials must be non-electrolytic, non-ferromagnetic, stable, easily sterilisable, biocompatible, and biodegradable. The demand for surgical staplers has increased since the 1980s, with projections showing a CAGR of 4.76% [47]. Anastomotic staples reduce surgical complications and improve patient recovery. Titanium alloys are used for metallic staples due to their biocompatibility, radiopacity, and low infection risk. However, allergic reactions, foreign body responses, inflammation, adhesions, and CT imaging challenges limit their use [18, 19].

7 Challenges and Future Perspectives

Mg-based implants have considerable potential in biomedical applications owing to their biocompatibility and biodegradability. However, before achieving widespread adoption, many crucial difficulties must be resolved. Primary issues include accelerated corrosion, hydrogen gas production, and inadequate mechanical strength, all of

Table 6 Mg-alloy implants and various applications

Category of implant	Description and key considerations	Applications	References
Orthopaedic implants	• Mg-based biomaterials are extensively used for fracture healing owing to their mechanical strength and biodegradability • Faster dissolution rate than biodegradable polymers • Mg–Al and Mg–RE alloys offer strength and controlled corrosion • Mg–Zn alloys offer biocompatibility, degradation control, and mechanical properties • Advances in metallurgy enhance Mg's corrosion resistance and properties	Bone screws, used in fracture fixation, include plates and scaffolds	[3–6, 16, 18]

(continued)

Table 6 (continued)

Category of implant	Description and key considerations	Applications	References
Cardiovascular implants	• Biodegradable, eliminating the need for removal • Reduce thrombosis, restenosis, and endothelial dysfunction • Must meet strict requirements: yield strength > 200 MPa, elongation > 15%, biocompatibility > 80% • In experimental and preclinical phases for use in artificial valves, stents, pacemakers, and grafts	Biodegradable vascular stents, artificial heart valves, pacemaker leads, and grafts	[7, 8, 20, 51, 65, 66, 68–70]
Biodegradable surgical wires and staplers	• Surgical closure devices: strength-to-diameter ratios, tensile strength, flexibility, sterility, and non-toxicity • Traditional titanium staples: risky, inflammation, and imaging issues • Mg-based materials: biocompatibility, biodegradability, and improved healing post-surgery • Market growth projected	Biodegradable surgical wires, staples for gastrointestinal and anastomotic procedures, and haemostasis applications	[18, 19, 45, 47]

which undermine long-term efficacy. The literature, especially the works by Walker et al. [3], has thoroughly recognised Mg alloy implant difficulties.

A key issue is the rapid breakdown rate of Mg in physiological settings. Mg has significant reactivity, resulting in pronounced corrosion when in contact with physiological fluids. The fast degradation not only diminishes implant durability but also produces hydrogen gas as a byproduct, which may collect in tissues and lead to problems. Moreover, the comparatively low mechanical strength of Mg, in contrast to other implant materials like titanium or stainless steel, restricts its use in load-bearing orthopaedic implants [2, 9, 21, 24, 46, 53, 61, 63].

Researchers are concentrating on novel methods to improve the characteristics of Mg implants. Progress in alloy design has resulted in the creation of Mg-based composites exhibiting enhanced strength and corrosion resistance. Researchers seek to enhance mechanical qualities and regulate degradation rates by including components such as Al, Ca, and REEs [1, 10, 26, 53].

Another potential strategy is the use of bioactive coatings. Surface changes, like hydroxyapatite coatings or polymeric layers, have shown efficacy in mitigating corrosion and improving biocompatibility. The coatings serve as protective barriers, minimising direct exposure of the Mg implant to biological fluids while facilitating osseointegration [24, 37].

Alongside material alterations, innovative manufacturing methods are being investigated to enhance the microstructure of Mg implants. Advanced fabrication techniques, including additive manufacturing and extreme plastic deformation, have shown promise in improving mechanical characteristics and corrosion resistance. By customising the microstructural characteristics, researchers may develop implants that disintegrate at a regulated pace, in accordance with the body's natural healing process [32, 34, 62, 70].

Notwithstanding the developments, further study is required to ascertain the long-term durability of Mg implants. Subsequent studies need to concentrate on refining alloy compositions, augmenting surface treatments, and innovating biodegradable coatings that enhance mechanical strength and biocompatibility. Furthermore, in vivo investigations and clinical trials are essential for evaluating the safety and usefulness of Mg implants in practical medical applications [19, 20, 33, 36, 46].

In conclusion, whereas Mg-based implants provide a feasible alternative to conventional materials, several obstacles persist. Researchers want to overcome constraints and fully realise the promise of Mg implants for biomedical applications by ongoing enhancements in alloy composition, bioactive coatings, and production procedures. Through persistent endeavours, Mg-based implants may soon emerge as a dependable and efficacious answer for several medical applications.

8 Conclusion

Biodegradable implants made of Mg provide several advantages over traditional metallic implants, including better biocompatibility, better mechanical properties, and controlled disintegration under physiological settings. Because they improve bone regeneration, reduce the need to remove implants, and alleviate stress shielding, Mg alloys show great potential in cardiovascular and orthopaedic applications. However, in order to ensure its long-term clinical viability, problems including fast corrosion, fluctuating mechanical performance, and hydrogen gas generation must be addressed.

The effectiveness of implants has been enhanced by advancements in alloying processes, surface treatments, and manufacturing techniques, such as powder metallurgy and 3D printing. Researchers can control the pace of deterioration and increase corrosion resistance with coatings such as polymer-based coatings, plasma spraying, and micro-arc oxidation. Bioactive coatings and polymer–Mg composites further improve the biomedical use of Mg-based implants.

To ensure the safety and efficacy of implants, future research should centre on improving alloy formulas, optimising production procedures, and standardising assessment approaches. In order to move Mg implants from the lab to the clinic, long in vivo clinical trials and approval from regulators are required. By continuously researching and collaborating across disciplines, biodegradable implants made of Mg have the potential to revolutionise medical treatments, leading to better patient outcomes and advancements in biomedical engineering.

References

1. Chen Y, Xu Z, Smith C, Sankar J (2014) Recent advances on the development of magnesium alloys for biodegradable implants. Acta Biomater 10:4561–4573. https://doi.org/10.1016/j.actbio.2014.07.005
2. Akbarzadeh FZ, Ghomi ER, Ramakrishna S (2022) Improving the corrosion behavior of magnesium alloys with a focus on AZ91 Mg alloy intended for biomedical application by microstructure modification and coating. Proc Inst Mech Eng H 236:1188–1208. https://doi.org/10.1177/09544119221105705
3. Walker J, Shadanbaz S, Woodfield TBF et al (2014) Magnesium biomaterials for orthopedic application: a review from a biological perspective. J Biomed Mater Res B Appl Biomater 102:1316–1331. https://doi.org/10.1002/jbm.b.33113
4. Staiger MP, Pietak AM, Huadmai J, Dias G (2006) Magnesium and its alloys as orthopedic biomaterials: a review. Biomaterials 27:1728–1734. https://doi.org/10.1016/j.biomaterials.2005.10.003
5. Tsakiris V, Tardei C, Clicinschi FM (2021) Biodegradable Mg alloys for orthopedic implants—a review. J Magnes Alloys 9:1884–1905. https://doi.org/10.1016/j.jma.2021.06.024
6. Antoniac I, Miculescu M, Mănescu (Păltânea) V et al (2022) Magnesium-based alloys used in orthopedic surgery. Materials 15:1148. https://doi.org/10.3390/ma15031148
7. Zhao N, Zhu D (2013) Application of Mg-based alloys for cardiovascular stents. Int J Biomed Eng Technol 12:382. https://doi.org/10.1504/IJBET.2013.057677
8. Prasadh S, Raguraman S, Wong R, Gupta M (2022) Current status and outlook of temporary implants (magnesium/zinc) in cardiovascular applications. Metals (Basel) 12:999. https://doi.org/10.3390/met12060999
9. Ali M, Hussein MA, Al-Aqeeli N (2019) Magnesium-based composites and alloys for medical applications: a review of mechanical and corrosion properties. J Alloys Compd 792:1162–1190. https://doi.org/10.1016/j.jallcom.2019.04.080
10. Akbarzadeh FZ, Sarraf M, Ghomi ER et al (2024) A state-of-the-art review on recent advances in the fabrication and characteristics of magnesium-based alloys in biomedical applications. J Magnes Alloys 12:2569–2594. https://doi.org/10.1016/j.jma.2024.06.015
11. Ridzwan MIZ, Shuib S, Hassan AY et al (2007) Problem of stress shielding and improvement to the hip implant designs: a review. J Med Sci 7:460–467. https://doi.org/10.3923/jms.2007.460.467
12. Niinomi M, Nakai M (2011) Titanium-based biomaterials for preventing stress shielding between implant devices and bone. Int J Biomater 2011:1–10. https://doi.org/10.1155/2011/836587
13. Insua A, Galindo-Moreno P, Miron RJ et al (2024) Emerging factors affecting peri-implant bone metabolism. Periodontol 2000(94):27–78. https://doi.org/10.1111/prd.12532

14. Zhou H, Liang B, Jiang H et al (2021) Magnesium-based biomaterials as emerging agents for bone repair and regeneration: from mechanism to application. J Magnes Alloys 9:779–804. https://doi.org/10.1016/j.jma.2021.03.004
15. Uppal G, Thakur A, Chauhan A, Bala S (2022) Magnesium based implants for functional bone tissue regeneration—a review. J Magnes Alloys 10:356–386. https://doi.org/10.1016/j.jma.2021.08.017
16. Wang N, Yang S, Shi H et al (2022) Magnesium alloys for orthopedic applications: a review on the mechanisms driving bone healing. J Magnes Alloys 10:3327–3353. https://doi.org/10.1016/j.jma.2022.11.014
17. Qi T, Weng J, Yu F et al (2021) Insights into the role of magnesium ions in affecting osteogenic differentiation of mesenchymal stem cells. Biol Trace Elem Res 199:559–567. https://doi.org/10.1007/s12011-020-02183-y
18. Asgari M, Hang R, Wang C et al (2018) Biodegradable metallic wires in dental and orthopedic applications: a review. Metals (Basel) 8:212. https://doi.org/10.3390/met8040212
19. Zhang Y, Cao J, Lu M et al (2023) A biodegradable magnesium surgical staple for colonic anastomosis: in vitro and in vivo evaluation. Bioact Mater 22:225–238. https://doi.org/10.1016/j.bioactmat.2022.09.023
20. Wang Z, Zheng Q, Guan S et al (2019) In vitro and in vivo assessment of the biocompatibility of an paclitaxel-eluting poly-L-lactide-coated Mg–Zn–Y–Nd alloy stent in the intestine. Mater Sci Eng, C 105:110087. https://doi.org/10.1016/j.msec.2019.110087
21. Razavi M, Fathi MH, Meratian M (2010) Microstructure, mechanical properties and bio-corrosion evaluation of biodegradable AZ91-FA nanocomposites for biomedical applications. Mater Sci Eng, A 527:6938–6944. https://doi.org/10.1016/j.msea.2010.07.063
22. Kirkland NT, Birbilis N (2014) Magnesium biomaterials. Springer International Publishing, Cham
23. Tekumalla S, Seetharaman S, Almajid A, Gupta M (2014) Mechanical properties of magnesium-rare earth alloy systems: a review. Metals (Basel) 5:1–39. https://doi.org/10.3390/met5010001
24. Thakur A, Kumar A, Kaya S et al (2022) Recent advancements in surface modification, characterization and functionalization for enhancing the biocompatibility and corrosion resistance of biomedical implants. Coatings 12:1459. https://doi.org/10.3390/coatings12101459
25. Fotovvati B, Namdari N, Dehghanghadikolaei A (2019) On coating techniques for surface protection: a review. J Manuf Mater Process 3:28. https://doi.org/10.3390/jmmp3010028
26. Pesode P, Barve S (2024) A review-recent advances on antibacterial coating on magnesium alloys by micro arc oxidation for biomedical applications. Adv Mater Process Technol 1–43.https://doi.org/10.1080/2374068X.2024.2391192
27. Vahidgolpayegani A, Wen C, Hodgson P, Li Y (2017) Production methods and characterization of porous Mg and Mg alloys for biomedical applications. In: Metallic foam bone. Elsevier, pp 25–82
28. Chandra Kandpal B, Johri N, Kumar L et al (2022) Stir casting technology for magnesium-based metal matrix composites for bio-implants—a review. Mater Today Proc 62:4519–4525. https://doi.org/10.1016/j.matpr.2022.04.957
29. Andani MT, Shayesteh Moghaddam N, Haberland C et al (2014) Metals for bone implants. Part 1. Powder metallurgy and implant rendering. Acta Biomater 10:4058–4070. https://doi.org/10.1016/j.actbio.2014.06.025
30. Singh Arora G, Rao US, Kumar Saxena K (2022) Critical review of Mg matrix composite for bio-implants through powder metallurgy. Mater Today Proc 57:902–907. https://doi.org/10.1016/j.matpr.2022.03.036
31. Attarilar S, Ebrahimi M, Djavanroodi F et al (2024) 3D printing technologies in metallic implants: a thematic review on the techniques and procedures. Int J Bioprint 7:306. https://doi.org/10.18063/ijb.v7i1.306
32. Zamani Y, Ghazanfari H, Erabi G et al (2021) A review of additive manufacturing of Mg-based alloys and composite implants. J Compos Compd 2:71–83. https://doi.org/10.52547/jcc.3.1.7

33. Xie K, Wang N, Guo Y et al (2022) Additively manufactured biodegradable porous magnesium implants for elimination of implant-related infections: an in vitro and in vivo study. Bioact Mater 8:140–152. https://doi.org/10.1016/j.bioactmat.2021.06.032
34. Chua K, Khan I, Malhotra R, Zhu D (2021) Additive manufacturing and 3D printing of metallic biomaterials. Eng Regeneration 2:288–299. https://doi.org/10.1016/j.engreg.2021.11.002
35. Haghshenas M (2017) Mechanical characteristics of biodegradable magnesium matrix composites: a review. J Magnes Alloys 5:189–201
36. Chang Y-H, Tseng CC, Chao C-Y et al (2020) Mg–Zn–Ca alloys for hemostasis clips for vessel ligation: in vitro and in vivo studies of their degradation and response. Materials 13:3039. https://doi.org/10.3390/ma13133039
37. Chen L, Pei Z, Xiao J et al (2017) TiAlN/Cu nanocomposite coatings deposited by filtered cathodic arc ion plating. J Mater Sci Technol 33:111–116. https://doi.org/10.1016/j.jmst.2016.07.018
38. Witte F (2015) Reprint of: the history of biodegradable magnesium implants: a review. Acta Biomater 23:S28–S40. https://doi.org/10.1016/j.actbio.2015.07.017
39. Huse EC (1878) A new ligature. Chicago Med J Examiner 37:171–172
40. PAYR E (1900) Beitrage zur Technik der Blutgesfass und Nervennaht nebst Mittheilungen die Verwendung eines Resorbierharen Metalles in der Chirurgie. Arch Klin Chir 62:67–71
41. Payr E Ueber Yerwendung yon Magnesium zur Behandlung yon Blmgef'~sserkrankungen
42. Höpfner E (Karl HE (1903) Ueber Gefässnaht, Gefässtransplantationen und Replantation von amputirten Extremitäten. L. Schumacher
43. Lambotte A (1909) Technique et indication des protheses dans le traitement des fractures. Presse Med 17:321
44. McBride ED (1938) Magnesium screw and nail transfixion in fractures. South Med J 31:508–514. https://doi.org/10.1097/00007611-193805000-00010
45. Stone P, Lord JW (1951) An experimental study of the thrombogenic properties of magnesium and magnesium-aluminum wire in the dog's aorta. Surgery 30:987–993
46. Fontenier G, Freschard R, Mourot M (1975) Study of the corrosionin vitro andin vivo of magnesium anodes involved in an implantable bioelectric battery. Med Biol Eng 13:683–689. https://doi.org/10.1007/BF02477326
47. Wexler BC (1980) Pathophysiologic responses of spontaneously hypertensive rats to arterial magnesium—aluminum wire implants. Atherosclerosis 36:575–587. https://doi.org/10.1016/0021-9150(80)90250-6
48. Hussl H, Papp C, Höpfel-Kreiner I et al (1981) Resorption time and tissue reactions with magnesium rods in rats and rabbits. Chir Plast 6:117–126. https://doi.org/10.1007/BF00289636
49. Wilflingseder P, Martin R, Papp C (1981) Magnesium seeds in the treatment of lymph-and haemangiomata. Chir Plast 6:105–116. https://doi.org/10.1007/BF00289635
50. Heublein B (2003) Biocorrosion of magnesium alloys: a new principle in cardiovascular implant technology? Heart 89:651–656. https://doi.org/10.1136/heart.89.6.651
51. Erbel R, Di Mario C, Bartunek J et al (2007) Temporary scaffolding of coronary arteries with bioabsorbable magnesium stents: a prospective, non-randomised multicentre trial. The Lancet 369:1869–1875
52. Singh J, Wahab Hashmi A, Ahmad S, Tian Y (2024) Critical review on biodegradable and biocompatibility magnesium alloys: progress and prospects in bio-implant applications. Inorg Chem Commun 169:113111. https://doi.org/10.1016/j.inoche.2024.113111
53. Ding Y, Wen C, Hodgson P, Li Y (2014) Effects of alloying elements on the corrosion behavior and biocompatibility of biodegradable magnesium alloys: a review. J Mater Chem B 2:1912–1933. https://doi.org/10.1039/C3TB21746A
54. Pu F, Chen N, Xue S (2016) Calcium intake, calcium homeostasis and health. Food Sci Hum Wellness 5:8–16. https://doi.org/10.1016/j.fshw.2016.01.001
55. Katarivas Levy G, Goldman J, Aghion E (2017) The prospects of zinc as a structural material for biodegradable implants—a review paper. Metals (Basel) 7:402. https://doi.org/10.3390/met7100402

56. Miah MR, Ijomone OM, Okoh COA et al (2020) The effects of manganese overexposure on brain health. Neurochem Int 135:104688. https://doi.org/10.1016/j.neuint.2020.104688
57. Bornapour M, Muja N, Shum-Tim D et al (2013) Biocompatibility and biodegradability of Mg–Sr alloys: the formation of Sr-substituted hydroxyapatite. Acta Biomater 9:5319–5330. https://doi.org/10.1016/j.actbio.2012.07.045
58. McKnight RF, Adida M, Budge K et al (2012) Lithium toxicity profile: a systematic review and meta-analysis. The Lancet 379:721–728. https://doi.org/10.1016/S0140-6736(11)61516-X
59. Wang C, Sun M, Zheng F et al (2014) Improvement in grain refinement efficiency of Mg–Zr master alloy for magnesium alloy by friction stir processing. J Magnes Alloys 2:239–244. https://doi.org/10.1016/j.jma.2014.09.001
60. Li YQ, Li F, Kang FW et al (2023) Recent research and advances in extrusion forming of magnesium alloys: a review. J Alloys Compd 953:170080. https://doi.org/10.1016/j.jallcom.2023.170080
61. Wang C, Shuai Y, Yang Y et al (2022) Amorphous magnesium alloy with high corrosion resistance fabricated by laser powder bed fusion. J Alloys Compd 897:163247. https://doi.org/10.1016/j.jallcom.2021.163247
62. Putra NE, Mirzaali MJ, Apachitei I et al (2020) Multi-material additive manufacturing technologies for Ti-, Mg-, and Fe-based biomaterials for bone substitution. Acta Biomater 109:1–20. https://doi.org/10.1016/j.actbio.2020.03.037
63. Ling C, Li Q, Zhang Z et al (2024) Influence of heat treatment on microstructure, mechanical and corrosion behavior of WE43 alloy fabricated by laser-beam powder bed fusion. Int J Extreme Manuf 6:015001. https://doi.org/10.1088/2631-7990/acfad5
64. Meininger S, Mandal S, Kumar A et al (2016) Strength reliability and in vitro degradation of three-dimensional powder printed strontium-substituted magnesium phosphate scaffolds. Acta Biomater 31:401–411. https://doi.org/10.1016/j.actbio.2015.11.050
65. Demir AG, Previtali B, Colombo D et al (2012) Fiber laser micromachining of magnesium alloy tubes for biocompatible and biodegradable cardiovascular stents. p 823730
66. Wang L, Fang G, Qian L et al (2014) Forming of magnesium alloy microtubes in the fabrication of biodegradable stents. Prog Nat Sci: Mater Int 24:500–506. https://doi.org/10.1016/j.pnsc.2014.08.006
67. Al-Takhayneh O, Warren H, in het Panhuis M (2022) 3D printing of surgical staples. MRS Adv 7:489–494. https://doi.org/10.1557/s43580-022-00287-2
68. Toong DWY, Ng JCK, Huang Y et al (2020) Bioresorbable metals in cardiovascular stents: material insights and progress. Materialia (Oxf) 12:100727. https://doi.org/10.1016/j.mtla.2020.100727
69. Wang J, Zhou Y, Yang Z et al (2018) Processing and properties of magnesium alloy micro-tubes for biodegradable vascular stents. Mater Sci Eng, C 90:504–513. https://doi.org/10.1016/j.msec.2018.05.005
70. Demir AG, Previtali B (2023) Lasers in the manufacturing of cardiovascular metallic stents: subtractive and additive processes with a digital tool. Procedia Comput Sci 217:604–613. https://doi.org/10.1016/j.procs.2022.12.256
71. Zhang Z, Yu H, Wang S et al (2010) Microstructure and tensile properties of ZK60 alloy fabricated by simplified rapid solidification powder metallurgy (S-RS P/M) process. J Mater Sci Technol 26:151–155. https://doi.org/10.1016/S1005-0302(10)60025-4

Chapter 7
Clinical Translation, Regulatory Considerations, and Future Perspectives: Challenges in Magnesium-Based Biodegradable Materials for Biomedical Applications

Dayanidhi K. Pathak and Ajay Lamba

1 Introduction

A significant new area of medical study is tissue engineering, which attempts to create a viable tissue substitute to fix any tissue abnormalities. Structural scaffolds made of these materials deliver the intended tissue with adequate mechanical characteristics to encourage tissue revival, as well as effective transport of cell populations and medicinal medicines. However, issues like donor site disease and infection risk plague conventional tissue repair techniques like autografts. Consequently, the development of implantable allografts has been the main focus. At first, metals, ceramics, and polymers were the primary materials used to make implants [1–3]. Although these were effective in certain applications, there were major obstacles due to restrictions such as material fatigue and biocompatibility problems. MBMs have become a radically changing option among the many materials investigated, providing a special combination of properties that tackle the fundamental issues that conventional biomaterials encounter [4, 5]. The mechanical characteristics associated with bone are mirrored by magnesium (Mg), the eighth most prevalent constituent on the planet, which also has inherent biocompatibility and biodegradability [6, 7]. These applications offer a novel approach to medical treatments that may lower patient morbidity and healthcare expenses because of magnesium alloys' exceptional capacity to progressively deteriorate within the human body over time, doing away with the necessity for follow-up surgeries to remove implants. Because magnesium alloys have properties that are similar to those of real bone, they are especially well-suited for orthopaedic applications such as stents, scaffolds, and bone fixation devices. However, problems with

D. K. Pathak (✉) · A. Lamba
G.B. Pant DSEU Campus, New Delhi, Delhi, India
e-mail: dayanidhikrishana.pathak@dseu.ac.in

A. Dixit et al. (eds.), *Biodegradable Metallic Materials*, Biomedical Materials for Multi-functional Applications, https://doi.org/10.1007/978-981-95-2401-3_7

these materials' mechanical strength, corrosion behaviour, biocompatibility, and rate of deterioration have made clinical translation difficult. Furthermore, the regulatory environment for magnesium-based biodegradable products is complicated and must promote innovation in this new field while also guaranteeing patient safety. Due to its electrochemical characteristics, magnesium as a material is naturally prone to quick corrosion in aquatic settings. Magnesium hydroxide ($Mg(OH)_2$), which is produced when magnesium corrodes in a physiological setting, can raise the pH locally and might have negative tissue reactions [8]. Furthermore, the creation of gas pockets due to the release of hydrogen gas during degradation may result in local inflammation or impaired tissue healing [9, 10]. Because of Mg's remarkable biocompatibility and degradability, MBMs mark a significant advancement in implant engineering. Early in the twenty-first century, scientists worked to understand Mg's behaviour in vivo, especially its biocompatibility [11]. They examined how Mg and its alloys degrade in vivo and clarified the corresponding biological reactions. The basis for the later development of MBMs was laid during this research era. In biomedicine, MBMs were first used in the late nineteenth century. In 1878, ligatures made of Mg wires were used to stop bleeding. In order to guarantee safe bleeding cessation, absorbable Mg clips were soon adopted as a substitute for ligatures. In 1892–1898, pure Mg was utilised for the resorbability of MBM implants in a variety of medical fields, as well as for fixator pins, spheres, wires, plates, and nails. In 1906, some patients with fractures were successfully treated using Mg plates and steel screws, which led to the first use of both operations and crack treatment.

Researchers discovered in 1938 that MBM plates were inappropriate due to their quicker absorption when placed on the bone's surface. They demonstrated how to do rotation-resistant osteosynthesis by employing Mg–Mn as a thin screw and angled plate. In 1944, 34 cases of pseudoarthrosis were treated in biomedicine using Mg–Cd alloys. Serum Mg levels did not significantly rise in any of the 34 instances, and there were no overt signs of inflammatory responses to the implant. The majority of patients maintained their mechanical integrity for six to eight weeks. After 10–12 months, the implants were fully degraded, despite the researchers' failure to disclose the size of the implants. Similar to this, in 1945, Mg–10Al alloy was utilised to seal gunshot wounds; all patients experienced a fused bone fracture after six weeks, and the Mg plate and pins were undetectable after four and six weeks, respectively. For the treatment of fractures and bone grafts, Mg–Al–Mn alloys were later utilised to create implants. Consistent with earlier findings, there were no overt inflammatory responses to the implant and unsystematic reactions to the Mg–Al–Mn alloy application. The investigations confirmed the atoxic and biocompatible nature of MBMs. Implant applications account for the majority of the use of MBMs in these uses. As a novel cardiovascular implant technology, a degradable MBM alloy was initially used to make a stent in 2003. A bioabsorbable Mg stent that elutes drugs was later described. The use of MBMs in urology was investigated in 2013. Subsequent research examined pure Mg and Mg–6Zn alloys as the basis for bioresorbable implants intended for use in the urinary system. A Mg–3Y alloy was originally employed in 2014 as a possible extraluminal tracheal stent application candidate. The suitability of Mg alloy (AZ31) for the respiring system was also examined simultaneously. Orthopaedic clinical studies for Mg-based implants, such as screws and bone plates, have begun. Furthermore, Mg–5Ca–1Zn alloy screws were utilised by Lee et al. [12] in 2016 to treat distal radius fractures; within a year, new bone replaced the implants entirely. These implants showed improved biocompatibility, which reduced postoperative problems and accelerated fracture healing. MBMs have advanced quickly over the last 15 years and have seen early clinical use in a number of medical specialities, including dentistry, cardiovascular surgery, and orthopaedics [4, 13, 14]. Over the past 200 years, the achievements of MBMs in biomedical applications are depicted in Fig. 1 [15]. Simultaneously, the revolutionary effects of Mg-based materials on tissue engineering, emphasising their function in vascular repair, angiogenesis, bone regeneration, nerve remodelling, anti-inflammatory response, and antibacterial action, can be seen from Fig. 2 [16]. A comprehensive picture of MBMs and their various uses in medicine is shown in this image. It demonstrates how MBMs, which have anti-inflammatory and antibacterial qualities, aid in the development of bones, nerve remodelling, and vascularisation. The circular Fig. 2 highlights the therapeutic potential of MBMs in tissue engineering and infection prevention by connecting their physiological advantages with the cell functions and materials associated. In recent years, there have been few reviews of the literature that summarise these features, especially when it comes to its biological roles and related mechanisms.

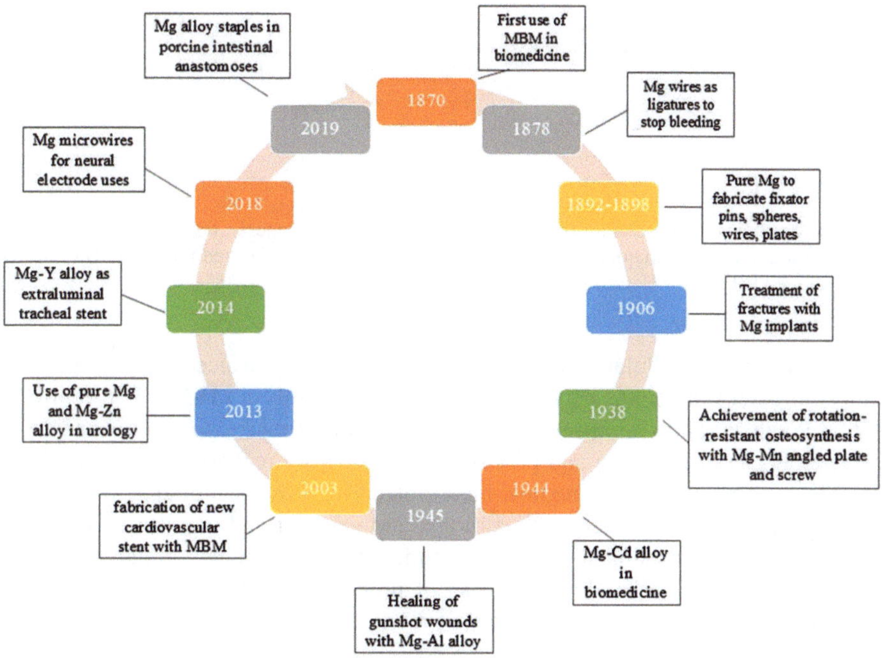

Fig. 1 Historical development of MBMs in biomedical uses

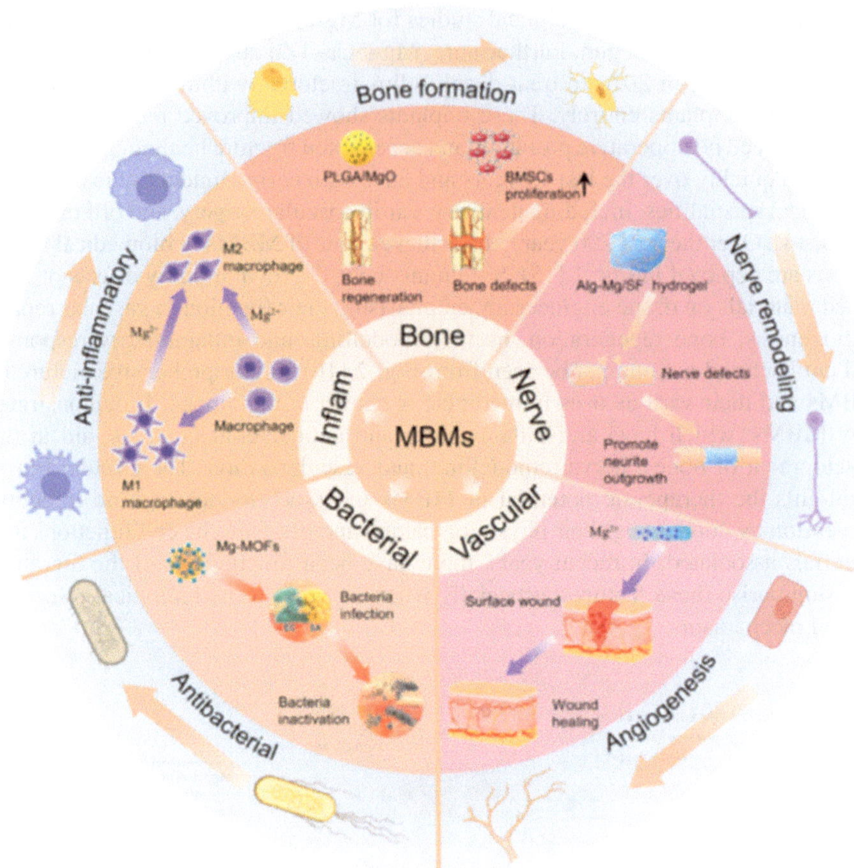

Fig. 2 Schematic representation of the role of MBMs in angiogenesis, bone regeneration, nerve remodelling, anti-inflammatory response, and antibacterial action. Reused under CC BY-NC-ND from source [16]

With improved cellular adhesion, growth, and spread, surface property optimisation strengthens the bonds between biomaterials and cells in MBMs. The control of cytokine release and immune cell activation promotes tissue regeneration and repair by reducing graft rejection and persistent inflammatory reactions. Because of this, MBMs, a cost-effective biomedical metal material, have a wide range of potential uses in orthopaedics, surgery, stomatology, and vascular and non-vascular lumens. They also have mechanical properties similar to living beings, notable biocompatibility and bioresorbability, and ease of processing and moulding. The use of biomaterials based on Mg in a variety of medical devices is shown in an anatomical image (Fig. 3). Their applications in dental and bone implants, tracheal, biliary, and cardiovascular stents, surgical clips, and devices that support gastrointestinal and urinary processes are highlighted. It highlights the significance of each biomaterial in

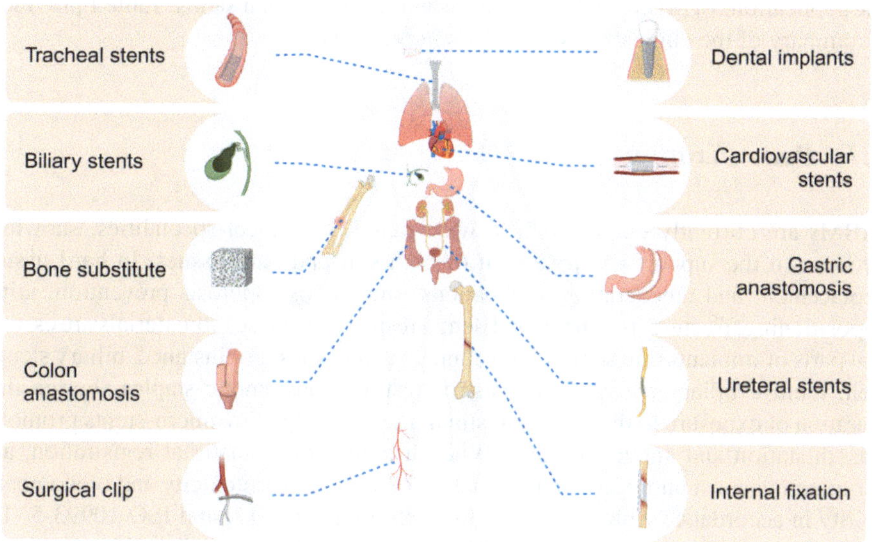

Fig. 3 Graphical representation of biomaterials based on magnesium in the human body. Reused under CC BY-NC-ND from source [16]

contemporary medical treatments and the development of biocompatible technology by graphically mapping its location and function within the human body.

An outline of the evolution and special benefits of MBMs in medicine has been given here, along with a synopsis of the signalling pathway that is pertinent to supporting tissue regeneration and repair. Furthermore, the recent clinical research that supports the potential applicability of MBMs in real-world settings has been incorporated. In this study, the obstacles to the clinical adoption of MBMs have been pointed out, and pertinent solutions have been offered. As a result, this chapter might offer recommendations for further study as well as for the application of MBMs in real-world clinical settings. In addition, given the limits of current research, this review has made several pertinent recommendations that could help with MBM design.

2 Clinical Applications of Mg-Based Materials in Biomedical Field

Although Mg alloys are a novel class of biomaterials, several technological and regulatory issues must be resolved before their use in medical applications. Material development, preclinical testing, regulatory assessment, and final clinical application are all included in the clinical translation process. The forthcoming sections discuss the applications of MBMs in bone fixation and cardiovascular stents. Table 1 provides a summary of the clinical uses of MBMs in recent years.

2.1 *Bone Fixation*

MBMs are currently employed in a wide range of medical specialities, showing promise in the support and treatment of both soft and hard tissues. In hard tissue replacement and regeneration applications, such as osteoporosis prevention, joint repositioning, fracture treatment, and dental restoration, these biomaterials are essential parts of implants that support and connect soft tissues. For instance, biliary stents help remove biliary obstructions, gastrointestinal anastomotic staples shorten the duration of exposure to the gastrointestinal tract, and Mg alloy micro-stents promote vasodilatation and anticoagulation. When it comes to biomaterial registration, all biomaterials must undergo extensive in vitro testing for cytotoxicity and biocompatibility in accordance with the standardised requirements [17] and ISO 10993-5/-12 [18]. Furthermore, in vivo clinical trials are required to guarantee the biosafety and effectiveness of new biomaterials. In 53 patients, Lee et al. [12] showed that there was very little hydrogen gas (H_2) surrounding the head of the Mg–5Ca–1Zn alloy screw that was inserted inside the distal radius fracture. Within two to four weeks, this H_2 was absorbed into the surrounding environment and disappeared, and within a year, new bone replaced the implants entirely (Fig. 4a). Ca–P-coated Mg–Nd–Zn–Zr alloy screws on outer layer were reported to be safe and effective when used to treat ankle fractures in grown females [19]. After 12.2 ± 4.9 months of surgical subsequent examination, a rupture site was found to have healed satisfactorily. A patient who experienced no pain and an implant that stayed remarkably stable throughout its recovery and maintained its in vivo functionality for 17 months before catastrophic degradation took place (Fig. 4b).

2.2 *Cardiovascular Stents*

Due to their potential for persistent inflammation and permanent foreign body status, vascular stents can cause extremely late thrombosis and restenosis. Patients who have compensatory vasodilatation or delayed endothelialisation, which results in the stent trabeculae becoming ectopic, may be at higher risk for late thrombosis. MBM stents are anticipated to be used in exterior vascular, neuro, and coronary procedures in the future [30]. Modest mechanical strength is needed for cardiovascular stents, but to guarantee procedure success, mechanical integrity must be maintained for a predetermined amount of time [31]. Only a handful of the five stents—Absorb BVS, DESolve, Fantom, Meres100, and Magmaris—are now undertaking clinical studies, despite the fact that they have all received the CE mark. The first to be approved for use in 2016 was the Magmaris stent, which was initially

Table 1 Recent clinical applications of MBMs

Implant material	Trial type	Implantation site	Surgery type	Healing effect/ conclusion	Degradation duration	Ref
MgYREZr (MAGNEZIX®CS)	Human	Toe hold	Hallux valgus	Compared to traditional titanium screws, magnesium screws showed superior results	19 months	[21]
Poly-L-lactide-coated laser-polished magnesium scaffold ([a]DREAMS)	Human (BIOSOLVE-I)	Coronary stent	Coronary stent implantation	No cardiac death or scaffold thrombosis	12 months	[22]
DREAMS-2G	Human (BIOSOLVE-II)	Coronary stent	Coronary stent implantation	Steady angiographic values from 6–12 months and a consistent, good safety profile for up to 12 months	12 months	[23]
DREAMS-3G	Human (BIOMAG-I)	Coronary stent	Coronary stent implantation	DREAMS-3G showed good strut fit, is very deliverable, and produces good safety and performance results	6 months	[24]
Mg–Nd–Zn–Zr alloy screws with Ca–P coating	Human	Malleolar fracture site	Medical malleolar fracture	Coated Mg implant screws showed an effective recovery and an alternative solution for bone repair	12 months	[20]

(continued)

Table 1 (continued)

Implant material	Trial type	Implantation site	Surgery type	Healing effect/ conclusion	Degradation duration	Ref
MAGNEZIX	Human	Left patella	Medical patellofemoral ligament reconstruction	There was no clinical indication of a systemic or local implant-related adverse response	18 months	[25]
MgYREZr screw	Human	Malleolar fracture site	Medical malleolar fracture	Radiographic strong union was attained. The follow-up revealed no issues. No individuals needed revision surgery or implant removal	12–24 months	[26]
UNITY-B	Human	Biliary	Post-cholecystectomy biliary strictures	The bilateral stents were found to be open, as shown by a CT scan 24 h later	8 weeks	[27]

(continued)

Table 1 (continued)

Implant material	Trial type	Implantation site	Surgery type	Healing effect/ conclusion	Degradation duration	Ref
Mg–0.45Zn–0.45Ca (ZX00) screws	Animal—sheep	Tibia	Fixation of osteotomy in the right tibia of the sheep	A degradation pattern similar to the control group's and produced total osteotomised bone consolidation	12 weeks	[28]
Mg–0.45Zn–0.45Ca (ZX00) pins and screws	Animal—rat	Tibia	Internal fixation	Degradation rates can be examined using typical comparisons with transcortical implantation of pin-shaped implants. Large animal models should be used for load-bearing efficiency and fracture stabilisation	6–24 weeks	[29]

[a] Drug-eluting absorbable metal scaffold

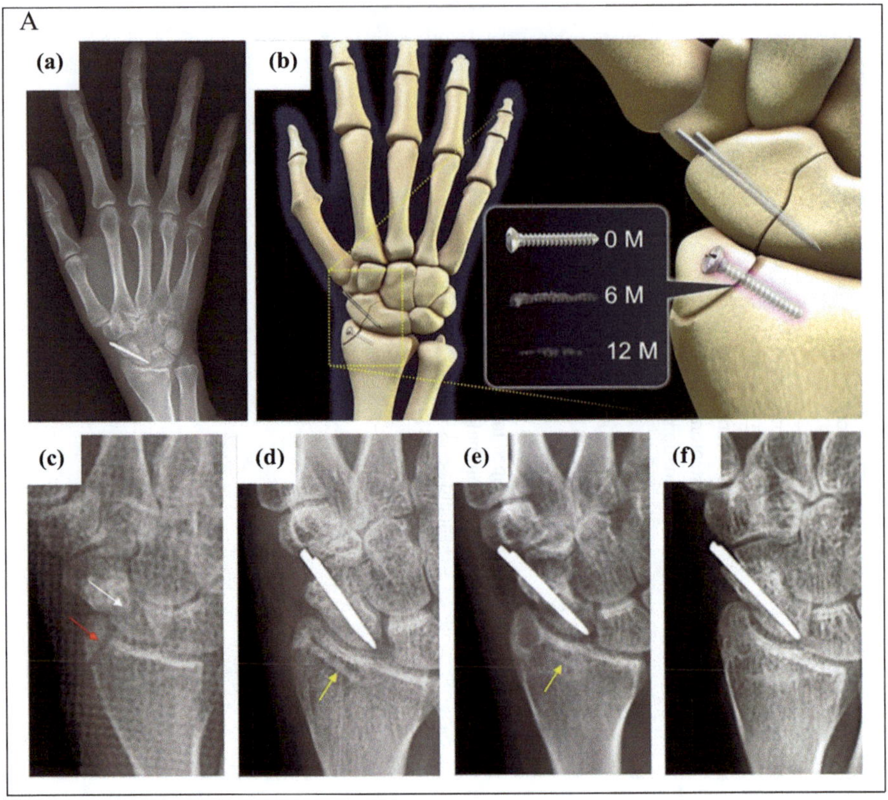

Fig. 4 Basic research and clinical use of MBMs in orthopaedics. **a** Clinical observation of the Mg alloy screw's total deterioration and bone healing over a one-year period. (a) A one-year subsequent X-ray of the patient who had a conventional SS implant for wrist fracture and a Mg alloy implant for distal radius fracture. (b) Diagrammatic representation of the implantation location and the evolution of the MBM at the time points of immediate, six months, and one year following implantation. X-rays of (c) before surgery, (d) after surgery, (e) six months, and (f) one year. Reused with permission [12]. **b** Radiographs taken before and after surgery of a young woman who had a trimalleolar fracture. Reused with permission [20]

made available in Hong Kong in 2018 and is made by Biotronik in Germany. When compared to bare resorbable metal stents, drug-eluting resorbable magnesium stents (DREAMS; Magmaris) demonstrated superior angiographic performance at 6 (mean 0–65 mm [SD 0.50]) and 12 (0.52 mm [0.39]) months, according to the 12-month results of the prospective multicentre first-in-human trial BIOSOLVE-I [22]. Late in-stent luminal loss after six months of use was lower than that at four months in the ongoing study utilising resorbable metal stents (1.08 mm [0–49]) [22, 32] (Fig. 5a). The BIOSOLVE-II research demonstrated the better performance of the second-generation DREAMS 2G with a target lesion failure (TLF) rate of 8% and no established or probable thrombosis at the 5-year post-surgery follow-up [33, 34] (Fig. 5b). Initial results from the first-in-human trial of BIOMAG-I, in conjunction

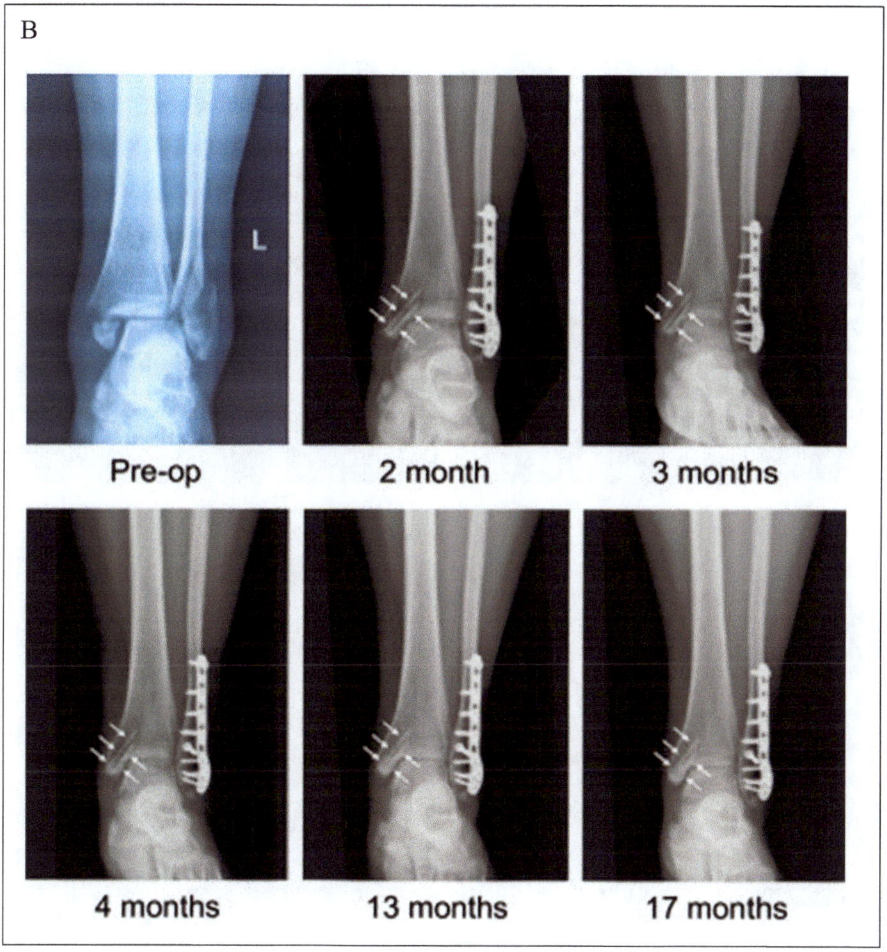

Fig. 4 (continued)

with almost entire imaging and clinical subsequent visit data, demonstrated that third generation DREAMS displayed favourable safety and performance outcomes, high transportability, and adequate scaffold bonding [24]. With a TLF rate of 0%, the DREAMS 3G demonstrated, like its predecessors, that its clinical execution was better than that of alternative degradable materials [24]. According to Erbel et al. [32], a biodegradable magnesium coronary stent (BIOTRONIK, Berlin, Germany) showed satisfactory apposition with the artery wall, and strong echo-reflectivity of the stent remains on post-surgery intravascular ultrasound (Fig. 5c). The echo-reflectivity significantly decreased throughout follow-up. Even without shadowing, tiny spherical patches that had a higher echo-reflectivity than the surrounding tissue could be seen in the initial position.

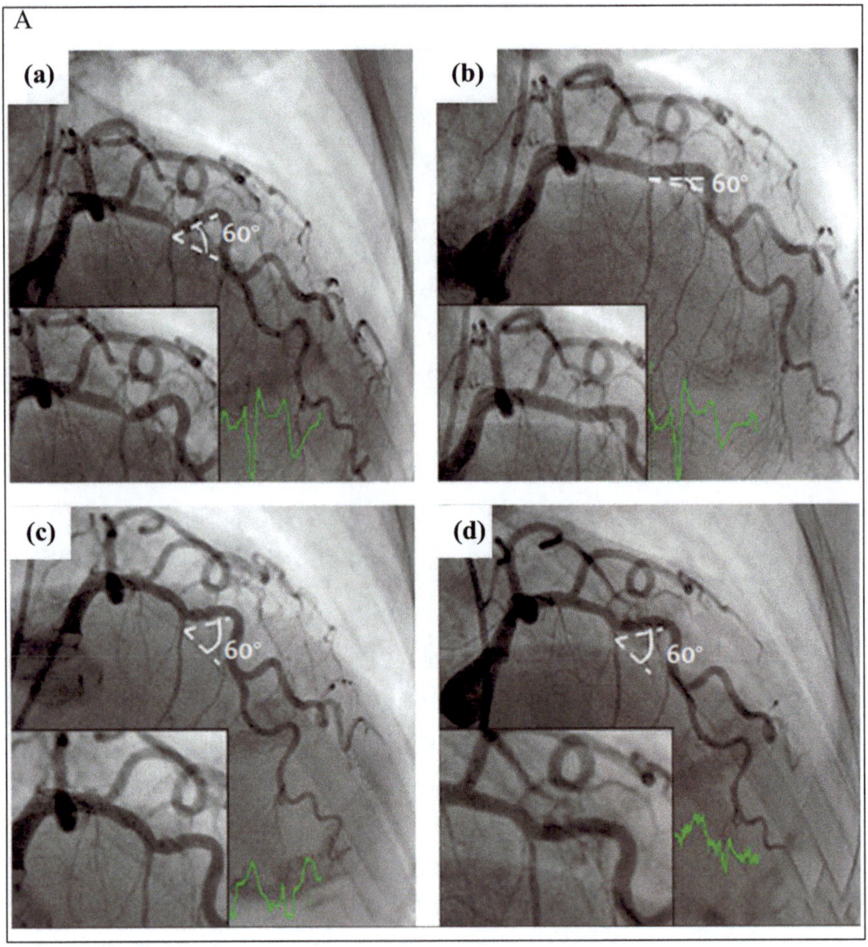

Fig. 5 Clinical uses of MBMs in heart surgery. **a** The scaffolded vascular section's angiographic appearance prior to the surgery (a), following scaffold implantation (b), and at the 6-month (c) and 12-month (d) subsequent visits. Reused with permission [22]. **b** Illustrations of intravascular imaging evaluations at the 6-month follow-up and during the procedure. Reused with permission [24]. **c** A Mg stent that is bioabsorbable: (a) following expansion, (b) prior to expansion, and (c) under the magnification of electron microscopy. Coronary angiography cross-sectional pictures and the associated intracoronary ultrasonography: (d) following surgery and (e) four months later. Reused with permission [32]

3 Regulatory Considerations in Mg-Based Biomaterials

Depending on their intended application, MBMs may have different regulatory mechanisms. For Mg-based biodegradable products to successfully enter clinical trials, the regulatory environment is essential. Tight regulatory procedures are in place to guarantee the safety, efficacy, and quality of biomaterials used in medical devices.

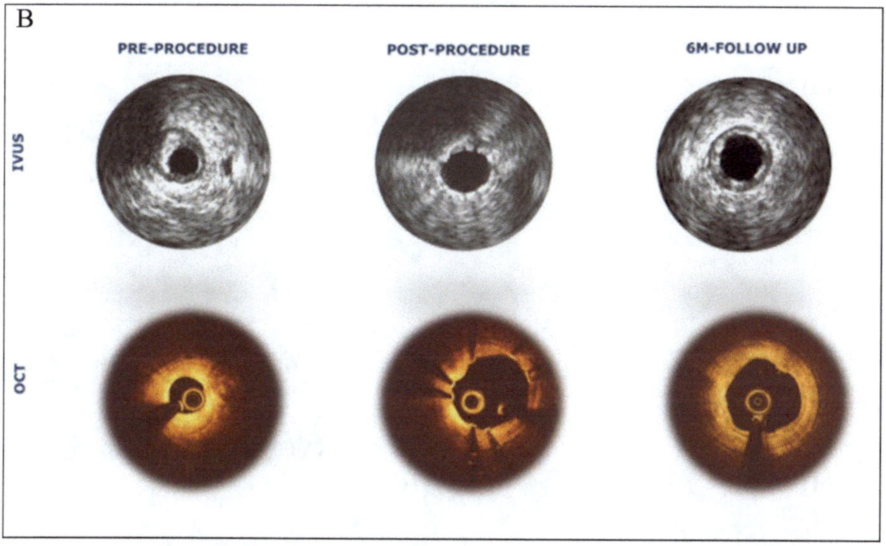

Fig. 5 (continued)

Common regulatory bodies, including the European Medicines Agency (EMA), China's National Medical Products Administration (NMPA), and the US Food and Drug Administration (FDA), have established approval channels for biomaterials and medical devices, albeit these procedures differ by location. Class I (low risk), Class II (moderate risk), and Class III (high risk) are the three categories into which the FDA divides medical devices in the United States [35]. Class II or III biomaterials, which include the majority of MBM implants, are subject to more stringent premarket approval procedures, including premarket notice (510(k)) or premarket approval (PMA). Depending on its intended use, the material may be governed in Europe by the In vitro Diagnostic Regulation (IVDR) or the Medical Devices Regulation (MDR). Before being approved, both areas need a great deal of proof and documentation.

3.1 Biocompatibility

Biocompatibility, or the ability to avoid adverse reactions when in touch with living tissues, is a prerequisite for the use of MBMs in medical devices. To assess these materials' cytotoxicity, genotoxicity, and immunogenicity, regulatory agencies like the European Medicines Agency (EMA) and the US Food and Drug Administration (FDA) demand thorough testing [36]. The rate at which Mg degrades must be investigated in addition to routine in vitro and in vivo research since its metabolites, including $Mg(OH)_2$, may affect nearby tissues and organs if discharged in excess.

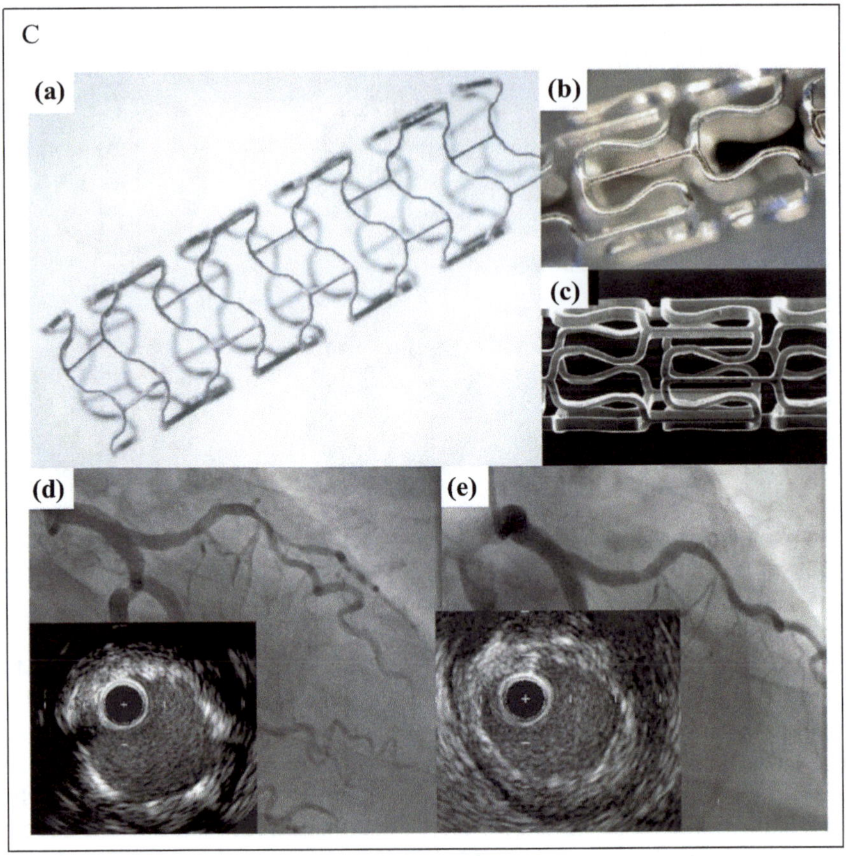

Fig. 5 (continued)

3.2 *Preclinical and Clinical Testing*

A Mg-based substance must pass a number of preclinical and clinical tests to determine its safety and effectiveness before it can be utilised in clinical settings. Both in vitro and in vivo investigations are commonly used in preclinical testing to evaluate the mechanical characteristics, corrosion behaviour, and biological response of the material. Key tests include cytotoxicity assays, in vivo animal model studies (e.g. rats, rabbits), and histological analysis to assess tissue compatibility and inflammation. In order to ensure that the material's corrosion rate is within allowable bounds for the particular application, it is assessed under physiologically simulated settings. According to Nachtsheim et al. [37], biocompatibility is also evaluated by looking at the material's interactions with living tissues, including any potential toxicity or immunological response. Animal research using bone repair models is necessary for orthopaedic applications to ascertain whether the substance can promote bone regeneration while deteriorating suitably. Clinical trials can begin once preclinical research shows the substance is safe and effective. Clinical trials typically take place in stages, beginning with small-scale research in patients or healthy volunteers to determine the material's safety profile. Larger trials are then conducted to determine the material's efficacy in the intended patient population.

3.3 Regulatory Approval Process

The device's categorisation and intended use determine the regulatory procedure for MBMs [36]. For example, the FDA normally classifies an orthopaedic implant or stent as a Class III device, which necessitates substantial preclinical and clinical data. The material's complexity, its intended application, and the particular regulatory procedure can all affect how long it takes for regulatory approval (FDA, 2021). Comprehensive information on the long-term degrading behaviour of MBMs and their possible effects on surrounding tissues is needed by regulatory bodies. It is important to carefully assess how Mg alloys affect the body's pH levels, immunological reactions, and electrolyte balance as they deteriorate over time. Manufacturers are required to submit information on their manufacturing processes, quality control methods, and post-market surveillance in addition to material-specific data. This guarantees that any possible problems with the material's performance, deterioration, or biocompatibility may be tracked and resolved after the device is approved.

3.4 Post-Market Surveillance

Regulatory bodies need continuous post-market surveillance of MBMs to keep an eye on their performance and safety after they are put on the market. Manufacturers may be asked to carry out more research or alter their product if safety concerns are raised, and they are expected to notify the proper authorities of any unfavourable incidents or device malfunctions. To safeguard patient health, MBMs' side effects, such as implant failure or delayed degradation, must also be carefully monitored.

4 Challenges of Mg-Based Biomaterials

The breakdown products of MBM implants might cause negative responses, such as infections or tissue damage, or they can cause oedema, pain, and inflammation at the insertion site. As a result, using biocompatible implant materials and making sure that the pre-procedural area is thoroughly sterilised are essential. Mg^{2+}, hydroxyl ions OH^-, and H_2 are usually produced during the breakdown of MBMs. It is imperative to keep such ions at concentrations that meet the very least demands for the protection and biological compatibility of the native tissue. Although MBMs typically show good biocompatibility, some magnesium alloys may have higher rates of corrosion. This increased corrosion may cause a significant discharge of associated byproducts, potentially making biocompatibility issues worse. Thus, this may have an impact on MBM's overall success [38].

4.1 Rapid Degradation Rate Under Physiological Environment

In an aqueous solution, Mg can be broken down into free Mg ions and H_2 gas under normal conditions. Its standard electrode potential is lower than that of H_2 (−2.372 V vs. 0 V). According to the electrochemical H_2 sensor's measurement results, the released H_2 in the subcutaneous tissue escaped considerably more quickly than that retained in the bone marrow cavity [39]. This could cause gas gaps to form surrounding the implants because of the gas buildup. Controlling the production of gases during the deterioration of MBM implants is essential due to the impaired cell migration, adhesion, and blood/bone ingrowth in the presence of gas voids at the interface between the implants and bone. Failure to do so may compromise the quality of tissue healing. In addition to H_2, the loose $Mg(OH)_2$ skin is formed when Mg ions are released under alkaline circumstances [40]. The matching responses are as follows:

$$\text{Anodic reaction: } Mg \rightarrow Mg^{2+} + 2e^- \tag{1}$$

$$\text{Cathodic reaction: } H_2O + 2e^- \rightarrow OH^- + H_2 \tag{2}$$

$$\text{Product formation: } Mg^{2+} + 2OH^- \rightarrow Mg(OH)_2 \tag{3}$$

When Mg-based interference screws degrade, Mg ions are released, hydrogen gas is produced, and alkaline chemicals are produced. This changes the peri-screw microenvironment [41], which in turn affects the quality of tendon graft-bone healing. The primary factors impacting the breakdown of Mg-based implants in the human body are inorganic ions (such as Ca^{2+}, Cl^-OH^-, $HPO_4{}^{2-}$, H_2PO^4, HCO^3, and $CO_3{}^{2-}$) [42] and organic molecules (such as glucose, amino acids, and protein) [43]. If the deterioration rate of the magnesium-based interference screws surpasses the expected gap, the ligament graft's fixation strength reduction after repair could lead to surgical failure because of knee instability. In addition to the mechanical integrity issues, one of the primary obstacles for clinical trials is the decrease in gas cavities brought on by the released hydrogen gas, which is directly tied to the rate at which the Mg-based interference screws degrade [44]. Therefore, in order to equal tissue healing without losing their mechanical integrity before curing, clinically viable Mg-based

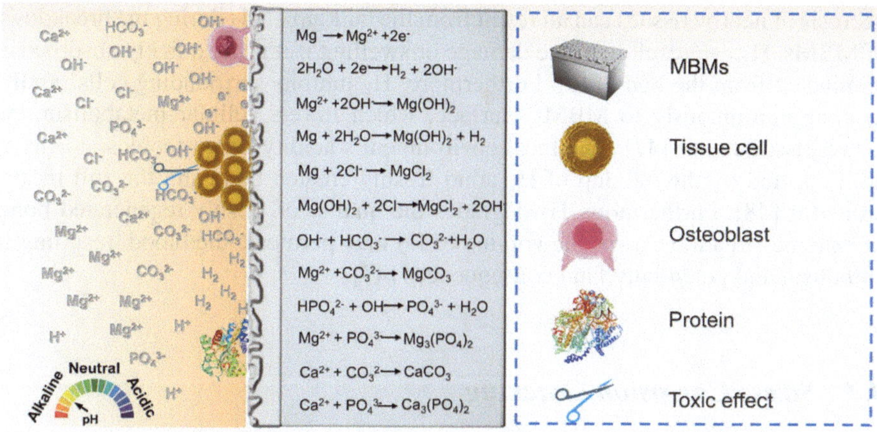

Fig. 6 Diagrammatic representation of MBMs' corrosion behaviour and potential chemical reactions in physiological settings. This picture shows the different chemical events that take place during the corrosion of MBMs, illuminating the intricate interactions of Mg^{2+} in the physiological milieu. In addition to physiological reactions like tissue cell activation, osteoblast function, and possible toxic effects on proteins, it emphasises the release of Mg^{2+} ions and their function in changing the local pH, highlighting the balance required for MBMs to be used safely and effectively in medical applications. Reused under CC BY-NC-ND from source [16]

interference screws should have the desired degrading behaviour. The human body's hydrodynamic state affects biological behaviour, corrosion, and local surface chemistry. For example, blood flow around MBMs can increase the pace of corrosion by moving corrosion products away from the implant surface and preventing their development. For example, because cartilage does not include blood vessels, Mg vascular stents may be more prone to quick corrosion than cartilage scaffolds composed of magnesium alloys when exposed to the blood flow in the arteries. The corrosion behaviour of MBMs can be influenced by the aforementioned characteristics in a variety of ways, along with several other unknown elements. This leads to a very complex corrosion scenario in the human body that is still not fully understood. Therefore, it is necessary to investigate a precise forecast of how well magnesium alloy implants will withstand corrosion in the human body. Some potential interactions with a corroding MBM in the body are shown in Fig. 6.

4.2 Rate of Hydrogen Gas Release

Mg and related alloys are widely used in medicine because of their intrinsic degradability. However, in a corrosive environment, they experience anodic polarisation, and water performs a cathodic reaction, producing H_2. For every 1 mg of Mg that is broken down, around 1 ml of H_2 is created [45]. Brittle fractures, gas pouches in nearby tissues, the splitting of degradation products from the Mg surface, and necrosis of nearby tissues can all result from the buildup of H_2 during the breakdown of MBMs. H_2 can result in tissue damage or swelling if it is not quickly absorbed or eliminated from the body [46]. Furthermore, H_2 inhibits surrounding cells' ability to cling continuously to MBMs' surface, which lowers cellular metabolism and delays tissue repair [47]. The local environment's acidity and osteoblastic activity are impacted by the buildup of H_2 amid freshly created bone and the soft tissues around it [48]. Furthermore, H_2 degrades the quality of freshly regenerated bone. In extreme situations, a sizable volume of H_2 may permeate the blood, resulting in embolism and potentially fatal consequences [49].

4.3 Stress Corrosion Cracking

Although magnesium alloys have many unique features that make their applications promising, the oxide film that forms on their surface is frequently slack and permeable, which does not adequately provide protection to the base. In addition, because of their active chemical properties, they are vulnerable to decomposition in the Cl^- containing milieu [50]. Furthermore, Mg alloys are susceptible to stress corrosion when utilised as structural elements because they are exposed to corrosive and stressful conditions when in use [51]. Given that the majority of magnesium alloy structural applications are thought to be subjected to corrosive conditions either permanently or temporarily, stress corrosion cracking (SCC) is essential to the components' integrity and utility. SCC commonly causes an abrupt brittle fracture, which is incredibly sneaky and random and can cause irreparable impairment [52]. Mg alloys are anticipated to be viable options for cardiovascular stents and orthopaedic implants in the biomedical industry [53]. Alloys must, however, withstand stresses while in use and deteriorate over time in the body [54]. Implants made of Mg alloy are particularly vulnerable to SCC when these external stressors and the intricate physiological milieu are combined. This causes early failure during usage, which significantly restricts the use of Mg alloys in the biomedical area and also impacts the therapeutic recovery effect [55]. In fact, even though Mg alloys have been employed in biomedical and industrial settings [56], the service life will be shortened due to the possibility of SCC in corrosive and stressful environments. For wider applications, it is crucial to comprehend the SCC behaviour of magnesium alloys and enhance their SCC resistivity. Few studies have emphasised SCC resistivity; most are limited to MBMs' in vitro degradation resistance and in vivo biodegradability. Consequently, further thorough investigation into the SCC behaviour of MBMs is needed. Furthermore, it is difficult to precisely reproduce the complex human regulatory system and in vivo milieu in in vitro research. Therefore, further in vivo animal research is needed to increase the accuracy of SCC tests and enhance comprehension of SCC's in vivo mechanisms for biomedical uses.

5 Future Perspective

With ongoing research aimed at resolving current issues, the future of MBMs in biomedical applications is bright. Innovation in this area will be fuelled by developments in manufacturing processes, surface modification, and material design. Additionally, novel hybrid biomaterials with improved qualities may result from the combination of Mg alloys with other materials like polymers, ceramics, and bioactive chemicals. The development of MBMs, which are distinguished by their distinct biocompatibility, biodegradability, and mechanical characteristics similar to those of real bone, shows a promising direction for tissue engineering innovation. The functional properties and design approaches of MBMs have been thoughtfully developed to address the concerns of mechanical integrity, degradation kinetics, and biological interactions while enhancing tissue repair, together with those of bone, arteries, and nerve tissues. Despite significant progress in the evolution and uses of MBMs, queries remain to be addressed before their full potential in clinical contexts can be realised. The advent of novel manufacturing processes like 3D printing has enabled the creation of complex, patient-specific implants with unique mechanical and degrading characteristics. A powerful method to improve tissue regeneration and provide a multimodal approach to complex tissue abnormalities is the integration of MBMs with bioactive chemicals, growth factors, and stem cell technologies. For the next generation of MBMs to be developed, interdisciplinary cooperation between materials researchers, biologists, engineers, and physicians is essential to conquering current obstacles. Such cooperative efforts will speed up the transition of MBMs from the bench to the bedside while simultaneously improving our understanding of their biological interactions. Additional research and improvement are needed in the areas of corrosion rate, possible cytotoxicity of corroded products, and host immune response modification. Additionally, metal compositions, surface treatments, and the addition of bioactive chemicals must all be adjusted to improve the interaction between cells and customise the absorption profile.

6 Conclusion

With its harmonic combination of mechanical adequacy, biocompatibility, and biodegradability, MBMs have a special place in the field of regenerative medicine. Because of their mechanical strength, biocompatibility, and capacity for natural bodily degradation, MBMs have great potential for the creation of novel biomedical devices. In many orthopaedic and vascular applications, innovative biodegradable MBMs are regarded as among the most capable alternatives to their titanium or polymer-based equivalents because of their superior biocompatibility, decreased stress shielding effects, and osteopromotive contributions. Nevertheless, issues with material design, degradation behaviour, biocompatibility, and regulatory approval impede their clinical translation. Overcoming these obstacles will mostly depend on the development of new technologies, legislative changes, and ongoing advances in material science. Magnesium alloys have the potential to transform the biodegradable biomaterials industry and provide fresh approaches for a variety of medicinal applications as research advances. Empowering creativity, encouraging teamwork, and being committed to the pursuit of excellence are essential as we map out the future. The development of MBMs from idea to medical realism is evidence of the individual spirit's unwavering capacity for healing and regeneration, showing the way to a day when sophisticated tissue regeneration is not only possible but a reality. This enlarged perspective not only captures the present situation of MBMs but also lays out a revolutionary course for the future, emphasising the critical role of creativity, versatile cooperation, and the unwavering chase of resolving current and upcoming obstacles.

References

1. Teo AJT, Mishra A, Park I, Kim YJ, Park WT, Yoon YJ (2016). Polym Biomater Med Implants Dev. https://doi.org/10.1021/acsbiomaterials.5b00429
2. Mishnaevsky L, Levashov E, Valiev RZ, Segurado J, Sabirov I, Enikeev N, Prokoshkin S, Solov'Yov AV, Korotitskiy A, Gutmanas E, Gotman I, Rabkin E, Psakh'E S, Dluhoš L, Seefeldt M, Smolin A (2014) Nanostructured titanium-based materials for medical implants: Modeling and development. https://doi.org/10.1016/j.mser.2014.04.002
3. Chevalier J, Gremillard L (2009) Ceramics for medical applications: a picture for the next 20 years. J Eur Ceram Soc 29. https://doi.org/10.1016/j.jeurceramsoc.2008.08.025
4. Amukarimi S, Mozafari M (2022) Biodegradable magnesium biomaterials—road to the clinic. https://doi.org/10.3390/bioengineering9030107
5. Das P, Sethy C, Pandey PM, Patel AK (2025) Evaluation of in vitro degradation behavior and biocompatibility assessment of Mg60Ti24.24Zn9.66Nb6.1 medium entropy alloy. J Alloys Compd 1010:178003. https://doi.org/10.1016/J.JALLCOM.2024.178003
6. Jahnen-Dechent W, Ketteler M (2012). Magnes Basics. https://doi.org/10.1093/ndtplus/sfr163
7. Das P, Pathak DK, Sharma P, Pandey PM (2024) A review on the mechanical and biocorrosion behaviour of iron and zinc-based biodegradable materials fabricated using powder metallurgy routes. Corros Rev. https://doi.org/10.1515/CORRREV-2023-0169/ASSET/GRAPHIC/J_CORRREV-2023-0169_CV_004.JPG
8. Xu L, Liu X, Sun K, Fu R, Wang G (2022). Corrosion behavior in magnesium-based alloys for biomedical applications. https://doi.org/10.3390/ma15072613
9. Chen J, Tan L, Yang K (2016). Recent advances on the development of biodegradable magnesium alloys: a review. https://doi.org/10.1080/10667857.2016.1212587
10. Das P, Kumar TSS, Sahu KK, Gollapudi S (2022) Corrosion, stress corrosion cracking and corrosion fatigue behavior of magnesium alloy bioimplants. https://doi.org/10.1515/corrrev-2021-0088
11. Gu XN, Zhou WR, Zheng YF, Cheng Y, Wei SC, Zhong SP, Xi TF, Chen LJ (2010) Corrosion fatigue behaviors of two biomedical Mg alloys—AZ91D and WE43—in simulated body fluid. Acta Biomater 6. https://doi.org/10.1016/j.actbio.2010.07.026
12. Lee JW, Han HS, Han KJ, Park J, Jeon H, Ok MR, Seok HK, Ahn JP, Lee KE, Lee DH, Yang SJ, Cho SY, Cha PR, Kwon H, Nam TH, Han JHL, Rho HJ, Lee KS, Kim YC, Mantovani D (2016) Long-term clinical study and multiscale analysis of in vivo biodegradation mechanism of Mg alloy. Proc Natl Acad Sci U S A 113. https://doi.org/10.1073/pnas.1518238113
13. Haghshenas M (2017). Mechanical characteristics of biodegradable magnesium matrix composites: a review. https://doi.org/10.1016/j.jma.2017.05.001

14. Xing F, Li S, Yin D, Xie J, Rommens PM, Xiang Z, Liu M, Ritz U (2022). Recent progress in Mg-based alloys as a novel bioabsorbable biomaterials for orthopedic applications. https://doi.org/10.1016/j.jma.2022.02.013
15. Akbarzadeh FZ, Sarraf M, Ghomi ER, Kumar VV, Salehi M, Ramakrishna S, Bae S (2024). A state-of-the-art review on recent advances in the fabrication and characteristics of magnesium-based alloys in biomedical applications. https://doi.org/10.1016/j.jma.2024.06.015
16. Chen Y, Zhang S, Bai J, Yang Y, Wang Y, Zhou Y, Jiang W, Wang J, Zhu J, Zhu C, Zhang X (2024) Magnesium-based biomaterials for coordinated tissue repair: a comprehensive overview of design strategies, advantages, and challenges. https://doi.org/10.1016/j.jma.2024.05.028
17. Oyane A, Onuma K, Ito A, Kim HM, Kokubo T, Nakamura T (2003) Formation and growth of clusters in conventional and new kinds of simulated body fluids. J Biomed Mater Res A 64. https://doi.org/10.1002/jbm.a.10426
18. Jung O, Smeets R, Hartjen P, Schnettler R, Feyerabend F, Klein M, Wegner N, Walther F, Stangier D, Henningsen A, Rendenbach C, Heiland M, Barbeck M, Kopp A (2019) Improved in vitro test procedure for full assessment of the cytocompatibility of degradable magnesium based on ISO 10993-5/-12. Int J Mol Sci 20. https://doi.org/10.3390/ijms20020255
19. Zhang D, Xu R, Chen S, Du H, Qian S, Peng F, Liu X (2023) Surface defect engineered-Mg-based implants enable the dual functions of superhydrophobic and synergetic photothermal/chemodynamic therapy. Bioact Mater 30. https://doi.org/10.1016/j.bioactmat.2023.07.012
20. Xie K, Wang L, Guo Y, Zhao S, Yang Y, Dong D, Ding W, Dai K, Gong W, Yuan G, Hao Y (2021) Effectiveness and safety of biodegradable Mg-Nd-Zn-Zr alloy screws for the treatment of medial malleolar fractures. J Orthop Translat 27. https://doi.org/10.1016/j.jot.2020.11.007
21. Baldini M, Coppa V, Falcioni D, Cusano G, Massetti D, Marinelli M, Gigante AP (2022) Resorbable magnesium screws for fixation of medial epicondyle avulsion fractures in skeletally immature patients: a comparison with Kirschner wires. J Child Orthop 16. https://doi.org/10.1177/18632521221136100
22. Haude M, Erbel R, Erne P, Verheye S, Degen H, Böse D, Vermeersch P, Wijnbergen I, Weissman N, Prati F, Waksman R, Koolen J (2013) Safety and performance of the drug-eluting absorbable metal scaffold (DREAMS) in patients with de-novo coronary lesions: 12 month results of the prospective, multicentre, first-in-man BIOSOLVE-I trial. Lancet 381. https://doi.org/10.1016/S0140-6736(12)61765-6
23. Haude M, Ince H, Abizaid A, Toelg R, Lemos PA, Von Birgelen C, Christiansen EH, Wijns W, Neumann FJ, Kaiser C, Eeckhout E, Lim ST, Escaned J, Onuma Y, Garcia-Garcia HM, Waksman R (2016) Sustained safety and performance of the second-generation drug-eluting absorbable metal scaffold in patients with de novo coronary lesions: 12-month clinical results and angiographic findings of the BIOSOLVE-II first-in-man trial. Eur Heart J 37. https://doi.org/10.1093/eurheartj/ehw196
24. Haude M, Wlodarczak A, van der Schaaf RJ, Torzewski J, Ferdinande B, Escaned J, Iglesias JF, Bennett J, Toth G, Joner M, Toelg R, Wiemer M, Olivecrona G, Vermeersch P, Garcia-Garcia HM, Waksman R (2023) Safety and performance of the third-generation drug-eluting resorbable coronary magnesium scaffold system in the treatment of subjects with de novo coronary artery lesions: 6-month results of the prospective, multicenter BIOMAG-I first-in-human study. EClinicalMedicine 59. https://doi.org/10.1016/j.eclinm.2023.101940
25. Baldini M, Coppa V, Falcioni D, Senigagliesi E, Marinelli M, Gigante AP (2021) Use of resorbable magnesium screws in children: systematic review of the literature and short-term follow-up from our series. J Child Orthop 15. https://doi.org/10.1302/1863-2548.15.210004
26. Kose O, Turan A, Unal M, Acar B, Guler F (2018) Fixation of medial malleolar fractures with magnesium bioabsorbable headless compression screws: short-term clinical and radiological outcomes in eleven patients. Arch Orthop Trauma Surg 138. https://doi.org/10.1007/s00402-018-2941-x
27. Lakhtakia S, Tevethia HV, Inavolu P, Shah S, Reddy DN (2021) Bilateral balloon expandable biodegradable stent (Y-stent) for postcholecystectomy perihilar biliary stricture. VideoGIE 6. https://doi.org/10.1016/j.vgie.2020.11.001

28. Holweg P, Berger L, Cihova M, Donohue N, Clement B, Schwarze U, Sommer NG, Hohenberger G, van den Beucken JJJP, Seibert F, Leithner A, Löffler JF, Weinberg AM (2020) A lean magnesium–zinc–calcium alloy ZX00 used for bone fracture stabilization in a large growing-animal model. Acta Biomater 113. https://doi.org/10.1016/j.actbio.2020.06.013
29. Grün NG, Holweg P, Tangl S, Eichler J, Berger L, van den Beucken JJJP, Löffler JF, Klestil T, Weinberg AM (2018) Comparison of a resorbable magnesium implant in small and large growing-animal models. Acta Biomater 78. https://doi.org/10.1016/j.actbio.2018.07.044
30. Chen X, Gao Y, Wang Y, Pan G (2021). Mussel-inspired peptide mimicking: an emerging strategy for surface bioengineering of medical implants. https://doi.org/10.1016/j.smaim.2020.10.005
31. Ali M, Hussein MA, Al-Aqeeli N (2019). Magnesium-based composites and alloys for medical applications: A review of mechanical and corrosion properties. https://doi.org/10.1016/j.jallcom.2019.04.080
32. Erbel R, Di Mario C, Bartunek J, Bonnier J, de Bruyne B, Eberli FR, Erne P, Haude M, Heublein B, Horrigan M, Ilsley C, Böse D, Koolen J, Lüscher TF, Weissman N, Waksman R (2007) Temporary scaffolding of coronary arteries with bioabsorbable magnesium stents: a prospective, non-randomised multicentre trial. Lancet 369. https://doi.org/10.1016/S0140-6736(07)60853-8
33. Haude M, Ince H, Abizaid A, Toelg R, Lemos PA, Von Birgelen C, Christiansen EH, Wijns W, Neumann FJ, Kaiser C, Eeckhout E, Lim ST, Escaned J, Garcia-Garcia HM, Waksman R (2016) Safety and performance of the second-generation drug-eluting absorbable metal scaffold in patients with de-novo coronary artery lesions (BIOSOLVE-II): 6 month results of a prospective, multicentre, non-randomised, first-in-man trial. Lancet 387. https://doi.org/10.1016/S0140-6736(15)00447-X
34. Haude M, Toelg R, Lemos PA, Christiansen EH, Abizaid A, von Birgelen C, Neumann FJ, Wijns W, Ince H, Kaiser C, Lim ST, Escaned J, Eeckhout E, Garcia-Garcia HM, Waksman R (2022) Sustained safety and performance of a second-generation Sirolimus-Eluting absorbable metal Scaffold: long-term data of the BIOSOLVE-II First-in-man trial at 5 years. Cardiovasc Revascularization Med 38. https://doi.org/10.1016/j.carrev.2021.07.017
35. Schuh JACL, Funk KA (2019) Compilation of international standards and regulatory guidance documents for evaluation of biomaterials, medical devices, and 3-D printed and regenerative medicine products. https://doi.org/10.1177/0192623318804121
36. Tian J, Song X, Wang Y, Cheng M, Lu S, Xu W, Gao G, Sun L, Tang Z, Wang M, Zhang X (2022). Regulatory perspectives of combination products. https://doi.org/10.1016/j.bioactmat.2021.09.002
37. Nachtsheim J, Burja J, Ma S, Markert B (2023) In vitro degradation behaviour of biodegradable magnesium alloys. PAMM 22. https://doi.org/10.1002/pamm.202200326
38. Gu X, Zheng Y, Cheng Y, Zhong S, Xi T (2009) In vitro corrosion and biocompatibility of binary magnesium alloys. Biomaterials 30. https://doi.org/10.1016/j.biomaterials.2008.10.021
39. Zhao D, Brown A, Wang T, Yoshizawa S, Sfeir C, Heineman WR (2018) In vivo quantification of hydrogen gas concentration in bone marrow surrounding magnesium fracture fixation hardware using an electrochemical hydrogen gas sensor. Acta Biomater 73. https://doi.org/10.1016/j.actbio.2018.04.032
40. Esmaily M, Svensson JE, Fajardo S, Birbilis N, Frankel GS, Virtanen S, Arrabal R, Thomas S, Johansson LG (2017). Fundamentals and advances in magnesium alloy corrosion. https://doi.org/10.1016/j.pmatsci.2017.04.011
41. Gonzalez J, Hou RQ, Nidadavolu EPS, Willumeit-Römer R, Feyerabend F (2018) Magnesium degradation under physiological conditions—best practice. https://doi.org/10.1016/j.bioactmat.2018.01.003
42. Zeng RC, Hu Y, Guan SK, Cui HZ, Han EH (2014) Corrosion of magnesium alloy AZ31: the influence of bicarbonate, sulphate, hydrogen phosphate and dihydrogen phosphate ions in saline solution. Corros Sci 86. https://doi.org/10.1016/j.corsci.2014.05.006
43. Gray-Munro JE, Strong M (2013) A study on the interfacial chemistry of magnesium hydroxide surfaces in aqueous phosphate solutions: Influence of Ca^{2+}, Cl^{-} and protein. J Colloid Interface Sci 393. https://doi.org/10.1016/j.jcis.2012.10.047

44. Thormann U, Alt V, Heimann L, Gasquere C, Heiss C, Szalay G, Franke J, Schnettler R, Lips KS (2015) The biocompatibility of degradable magnesium interference screws: an experimental study with sheep. Biomed Res Int. https://doi.org/10.1155/2015/943603
45. Noviana D, Paramitha D, Ulum MF, Hermawan H (2016) The effect of hydrogen gas evolution of magnesium implant on the postimplantation mortality of rats. J Orthop Translat 5. https://doi.org/10.1016/j.jot.2015.08.003
46. Zhao D, Wang T, Nahan K, Guo X, Zhang Z, Dong Z, Chen S, Chou DT, Hong D, Kumta PN, Heineman WR (2017) In vivo characterization of magnesium alloy biodegradation using electrochemical H2 monitoring, ICP-MS, and XPS. Acta Biomater 50. https://doi.org/10.1016/j.actbio.2017.01.024
47. Kim SM, Jo JH, Lee SM, Kang MH, Kim HE, Estrin Y, Lee JH, Lee JW, Koh YH (2014) Hydroxyapatite-coated magnesium implants with improved in vitro and in vivo biocorrosion, biocompatibility, and bone response. J Biomed Mater Res A 102. https://doi.org/10.1002/jbm.a.34718
48. Byun SH, Lim HK, Cheon KH, Lee SM, Kim HE, Lee JH (2020) Biodegradable magnesium alloy (WE43) in bone-fixation plate and screw. J Biomed Mater Res B Appl Biomater 108. https://doi.org/10.1002/jbm.b.34582
49. Razavi M, Huang Y (2019). Assessment of magnesium-based biomaterials: From bench to clinic. https://doi.org/10.1039/c9bm00289h
50. Miao H, Huang H, Shi Y, Zhang H, Pei J, Yuan G (2017) Effects of solution treatment before extrusion on the microstructure, mechanical properties and corrosion of Mg-Zn-Gd alloy in vitro. Corros Sci 122. https://doi.org/10.1016/j.corsci.2017.01.001
51. Chen L, Tseng CM, Qiu Y, Yang J, Chang CL, Wang X, Li W (2020) A layer-by-layer assembled coating for improved stress corrosion cracking on biomedical magnesium alloy in cell culture medium. Surf Coat Technol 403. https://doi.org/10.1016/j.surfcoat.2020.126427
52. Winzer N, Atrens A, Song G, Ghali E, Dietzel W, Kainer KU, Hort N, Blawert C (2005) A critical review of the stress corrosion cracking (SCC) of magnesium alloys. https://doi.org/10.1002/adem.200500071
53. Nasr Azadani M, Zahedi A, Bowoto OK, Oladapo BI (2022). A review of current challenges and prospects of magnesium and its alloy for bone implant applications. https://doi.org/10.1007/s40204-022-00182-x
54. Koo Y, Jang Y, Yun Y (2017) A study of long-term static load on degradation and mechanical integrity of Mg alloys-based biodegradable metals. Mater Sci Eng B 219. https://doi.org/10.1016/j.mseb.2017.02.009
55. Harandi SE, Singh Raman RK (2015) Appropriate mechanochemical conditions for corrosion-fatigue testing of magnesium alloys for temporary bioimplant applications. JOM 67. https://doi.org/10.1007/s11837-015-1387-7
56. Ding W (2016). Opportunities and challenges for the biodegradable magnesium alloys as next-generation biomaterials. https://doi.org/10.1093/RB/RBW003

Chapter 8
Composition and Design Strategies for Zinc-Based Biodegradable Alloys

Gaurav Tripathi, Shitanshu Arya, Arun Kumar, and Pulak Mohan Pandey

1 Introduction

Biodegradable materials have emerged as a transformative class of biomaterials for medical implants, offering significant advantages over permanent metallic implants, specifically in terms of stress shielding and the requirement of secondary or revision surgeries [1]. Traditional implant materials, such as stainless steel, cobalt–chromium alloys, and Ti alloys, exhibit excellent mechanical strength and corrosion resistance. However, their permanent presence in the body may cause complications such as stress shielding, chronic inflammation, and the necessity of revision surgeries. In response to these challenges, biodegradable metals, including magnesium, iron, and zinc alloys, have gained popularity because of their ability to gradually degrade or resorb in vivo during the healing process while promoting tissue regeneration, eliminating the requirement of secondary surgery, necessary in case of permanent implants [2]. Their effectiveness depends on using surface engineering techniques, controlling degradation rates, optimizing microstructure, and carefully choosing the base metals.

Among biodegradable metals, Zn is the most desirable due to its medium degradation rate and favorable mechanical properties in comparison to Mg- and Fe-based biodegradable alloys [3]. Zn possesses an electrochemical potential between that of

G. Tripathi · S. Arya · A. Kumar · P. M. Pandey (✉)
Department of Mechanical Engineering, Indian Institute of Technology Delhi, Hauz Khas, New Delhi, India
e-mail: pmpandey@mech.iitd.ac.in

G. Tripathi
e-mail: mez188588@mech.iitd.ac.in

S. Arya
e-mail: mez188599@mech.iitd.ac.in

A. Kumar
e-mail: mez188582@mech.iitd.ac.in

A. Dixit et al. (eds.), *Biodegradable Metallic Materials*, Biomedical Materials for Multi-functional Applications, https://doi.org/10.1007/978-981-95-2401-3_8

Mg and Fe, which allows for controlled degradation rates, mitigating the excessively rapid degradation seen in Mg-based alloys and the slow corrosion of Fe-based counterparts. Additionally, zinc has proved to be an important trace element contributing to numerous physiological processes, including enzymatic activity, immune response modulation, and bone metabolism, making it inherently biocompatible. Zn-based biodegradable alloys are primarily developed for biomedical applications such as temporary implants, vascular stents, and orthopedic devices. Despite the promising characteristics of pure Zn, its mechanical limitations—including low tensile strength and susceptibility to creep—necessitate alloying strategies to enhance its mechanical performance. Various alloying elements like magnesium, lithium, manganese, calcium, and silver have been explored to refine microstructure, improve mechanical integrity, and optimize degradation behavior. In addition, advanced fabrication techniques, including rapid prototyping and powder metallurgy, have been employed to tailor the characteristics of Zn-based alloys for specific medical applications. Their composition and design strategies must balance mechanical properties, biocompatibility, corrosion behavior, and degradation control [3].

As depicted in Fig. 1, Zn-based biodegradable alloys find their applications in various biomedical applications such as orthopedic scaffolds [4, 5], wound closure devices [6, 7], and cardiovascular stents [8, 9]. Preclinical investigations on biocompatible Zn alloys have been carried out by various researchers to investigate their therapeutic efficacy and biocompatibility in animal models. These studies provide important information about how shape design, elemental composition, and the choice of relevant animal models interact.

This chapter provides an overview of the composition and design strategies for zinc-based biodegradable alloys. It discusses the selection of alloying elements, their influence on mechanical and biological performance, and the processing methods that contribute to the development of clinically viable Zn-based implants. Understanding

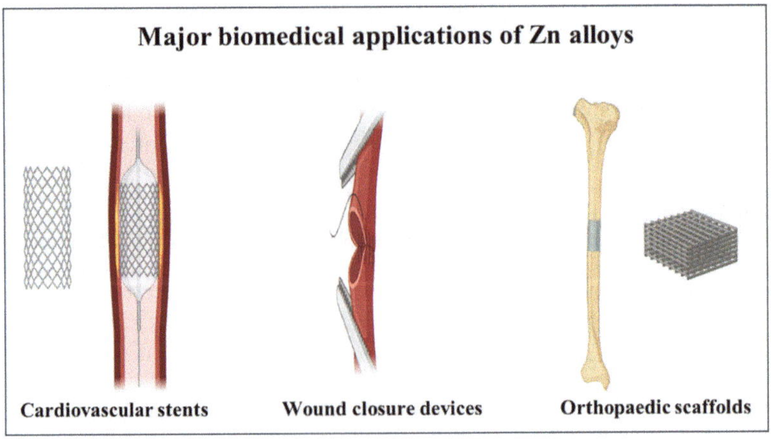

Fig. 1 Major applications of Zn alloys in the biomedical field

these factors is crucial in advancing Zn alloys as a reliable and effective class of biodegradable biomaterials for future medical applications.

2 Composition Strategies

There are several techniques to fabricate metals and alloys according to the requirements in different applications. However, major techniques used for Zn alloys are illustrated in Fig. 2. One of the most common techniques for producing zinc alloys is casting, as it makes alloy compositions easily customizable [10]. But casting is not good at creating the net forms needed for different fixation devices like plates, staples, or porous structures, which are crucial for promoting the right kind of tissue healing in fracture repair. Another technique that could be used is forging. Depending on the processing temperature in relation to the metal's recrystallization temperature, the forging process can be classified as either hot or cold. The most widely used wrought processes are drawing, extrusion, forging, and rolling [11].

Powder metallurgy is another technique that facilitates the fabrication of various metallic and non-metallic powders. It helps to create products with near-net shape and good surface quality. The method has been used by various manufacturers to fabricate Zn-based biodegradable alloys. Krystýnová et al. [12] used powder metallurgy techniques in conjunction with cold and hot pressing to process zinc powders with 7.5 and 150 μm particle sizes. Cold-pressed samples were further sintered. The study found that particle size and shape significantly influence deformability, with smaller $Zn_{7.5}$ particles showing minimal deformability and larger Zn_{150} particles showing significant deformability. Additionally, hot pressing significantly improved the bending properties of these Zn-based materials. Sotoudeh Bagha et al. [13] used uniaxial cold pressing, sintering, and mechanical alloying of matching powders to create dense Zn–4Mn discs. Zn–4Mn alloys' compressive yield strength was 290 MPa, and their corrosion rate was reduced to 0.72 mm/year.

Additive manufacturing (AM), with its state-of-the-art technologies, has largely contributed to the fabrication of metallic implants for biomedical applications [14].

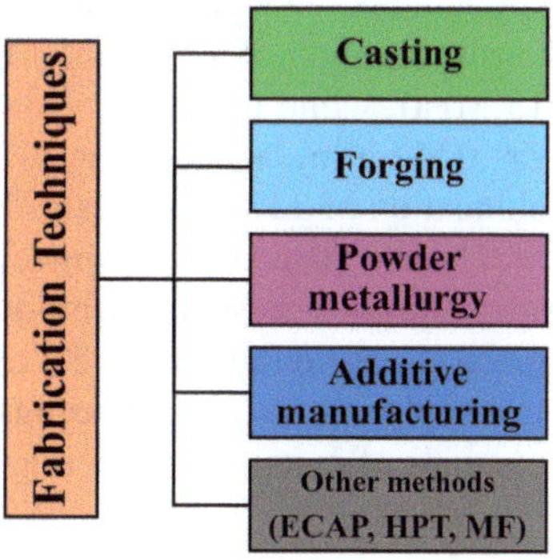

Fig. 2 Different fabrication techniques

Yang et al. [15] used selective laser sintering (SLM) to fabricate Zn-3 Mg alloy, which had good biocompatibility, an acceptable degradation rate, and good mechanical properties. Zhao et al. [16] used 3D printing to fabricate a Zn–Mg–Cu alloy helical scaffold with a yield strength of 41.15 $\pm$ 0.49 MPa and a refined grain size of about 1.21 μm. This alloy showed good antibacterial qualities and was effective in stimulating osteogenesis. Other than these methods, various plastic deformation processes include equal channel angular pressing (ECAP), torsion extrusion, high-pressure torsion (HPT), and multidirectional forging (MF) [17, 18].

Zinc metal has attracted much interest among degradable alloys due to its favorable mechanical qualities, suitable degradation rate, and biocompatibility. However, one of the major issues, such as the low mechanical strength of zinc alloys, limits its use in demanding situations. Nonetheless, Zn alloys' mechanical qualities can be successfully improved with the help of different techniques such as grain refinement, surface modification, alloying, and composite strategies.

2.1 Alloying Strategies

Adding elements like strontium (Sr), magnesium (Mg), lithium (Li), and copper (Cu) has been shown to be a practical way to improve biocompatibility. These components lessen the cytotoxic effects by limiting the discharge of Zn^{2+} ions from the alloy surface [19, 20]. The incorporation of Mg into Zn alloys leads to the formation of intermetallic compounds and grain refinement. Moreover, Zn–Mg alloy coatings containing less than 10% Mg have been reported to possess enhanced corrosion resistance in comparison to pure Zn coatings. Vojtech et al. [21] assessed the mechanical properties of biodegradable Zn alloys. They produced a Zn–1Mg alloy using the hot extrusion process and discovered that the alloy consists of primary Zn dendrites along with a Zn + Mg_2Zn_{11} eutectic mixture. This alloy exhibited enhanced corrosion resistance and excellent mechanical properties due to its relatively fine structure. Liu et al. [22] discovered that introducing Al into the casting process can refine the microstructure of pure Zn, thereby improving its mechanical properties. In fact, even small amounts of various elements have been shown to significantly influence tensile strength and elongation. Zhao et al. [23] developed a Zn–0.1Li binary alloy with a tensile strength of 274 $\pm$ 61 MPa, significantly exceeding that of pure Zn (116 $\pm$ 13 MPa), while maintaining a ductility of 17.7%. Dai et al. [24] reported that the Zn–0.5Li alloy achieved an ultimate tensile strength (UTS) of 364.9 MPa and an elongation of 22% in the extruded state.

The effort to optimize mechanical performance through grain refinement and specialized processing techniques marks a significant step toward addressing the mechanical limitations of Zn alloys, ultimately improving their potential for clinical applications. These findings highlight the significant influence that even minute amounts of metal elements may affect the mechanical properties of alloys.

2.2 *Surface Modification Strategies*

This method can modify the surface chemistry and morphology of Zn alloys, reducing their degradation rate and enhancing cell adhesion. Zhong et al. [25] performed carbon dioxide plasma surface modification of a Zn–1Mg alloy, followed by grafting choline phosphate chitosan to develop Zn–1Mg-PP-PCCs. The modified alloy demonstrated a substantially lower degradation rate of approximately 0.011 mm/year, which is half of the unmodified alloy. This reduction in degradation helps prevent cytotoxicity by virtue of excessive release of Zn^{2+}. Furthermore, the Zn–1Mg alloy surface modification increased the expression of osteogenic factors while suppressing the expression of inflammatory cytokines. Shomali et al. [26] modified the surface of Zn wires using 3-(trimethoxysilyl) propyl methacrylate (MPS) and applied a 1–12 μm film of poly L-lactic acid (PLLA) to create PLLA/MPS-coated Zn wires. Their study on the corrosion behavior revealed that, over the first 4–5 months, the corrosion rate of the PLLA/MPS-coated Zn wire was approximately half that of the unmodified Zn wire.

2.3 *Composite Material Strategies*

The exploration of polymer matrix composites reinforced with Zn particles or fibers has been conducted as a strategy to enhance biocompatibility. Wang et al. [27] applied a dopamine-modified strontium-doped hydroxyapatite (SrHA/PDA) coating to pure Zn, effectively reducing its corrosion rate while also providing antibacterial properties. Su et al. [28] obtained comparable results by developing a ceramic/PLA composite coating on pure Zn, which promoted cell viability in conjunction with the improvement of expression of osteogenic genes.

3 Characteristic Control Through Composition

The degradation of Zn-based alloys is a key factor that influences their performance, as it must match the rate of tissue healing and regeneration. Additionally, the mechanical strength of the material must be sufficient to support the body's load during the healing process. By adjusting the composition of Zn alloys, researchers can control these properties to improve the material's performance in specific clinical applications. Adding different alloying elements can increase biocompatibility, slow down or accelerate degradation rates, and improve mechanical qualities. Primary alloying elements in zinc-based alloys for industrial use are copper and aluminum [29, 30], which can improve the zinc's apparent mechanical characteristics. However, due to neurotoxicity [31], aluminum is not a good major alloying element for biomedical materials. The main alloying elements used in biodegradable zinc-based alloys are iron, lithium, magnesium, manganese, copper, silver, strontium, and calcium [32–34]. Zn-based alloys were reported to have an ideal degradation rate but possess poor mechanical properties for load-bearing requirements of bone implants [35, 36]. Various design strategies have been proposed to strengthen zinc-based biomaterials, which are discussed below.

An alloy design approach based on a solid solution strengthening first-principles analysis was proposed by Chen et al. [32]. Grain refinement, solid solution, and secondary phase are the three variables that regulate the alloying's strengthening effect. Fleischer's and Labusch's theories were used to assess the solid solution strengthening of six different elements (Li, Mg, Al, Mn, Cu, and Ag) for biodegradable Zn. The ranking of solutes' strengthening capabilities based on solubility was found to be Mn > Ag > Cu > Li > Al > Mg. The findings of the research are summarized in Fig. 3. A new ternary alloy system, Zn–0.8%Li–(Mg, Ag) alloys, was proposed by Zhang et al. [37] for directed bone regeneration employing casting and hot rolling of annealed alloys. According to reports, the primary secondary phase in Zn–0.8%Li–0.2%(Mg, Ag) alloys was the $LiZn_4$ nanoprecipitate. In the Zn–0.8%Li–0.2%Ag alloy, 0.2 weight percent Ag produced fine equiaxed grains (2.3 μm) with a finer microstructure, which promoted a uniform distribution of $LiZn_4$ nanoprecipitates with refined size and higher density. The Zn-0.8%Li-0.2%Ag alloy was found to have the best tensile strength, the best corrosion resistance, and good biocompatibility. Tang et al. [38] formulated the binary Zn–xCu ($x = 1$, 2, 3, and 4 wt%) alloy system as a biodegradable stent material via casting and extrusion techniques for cardiovascular applications. An increase in the amount of Cu was observed to result in a higher quantity of $CuZn_5$ phase precipitates. The $CuZn_5$ phase precipitates were fractured following the extrusion procedure. The extension of the Zn–4Cu was determined to be 50.6 ± 2.8%, which may be advantageous for the fabrication of micro-tubes in stent construction. Engineering stress–strain curves of the as-extruded pure Zn and Zn–xCu are shown in Fig. 4.

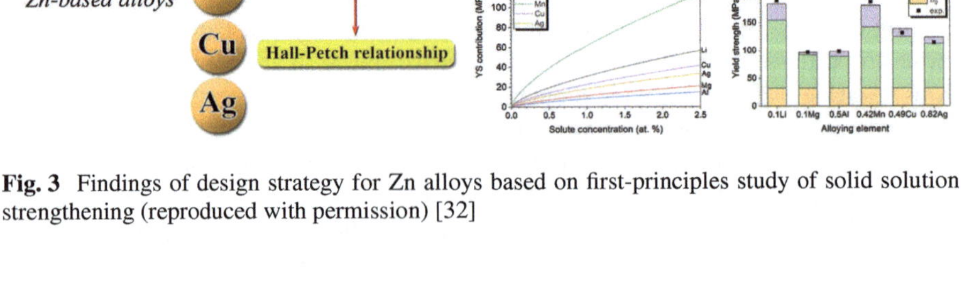

Fig. 3 Findings of design strategy for Zn alloys based on first-principles study of solid solution strengthening (reproduced with permission) [32]

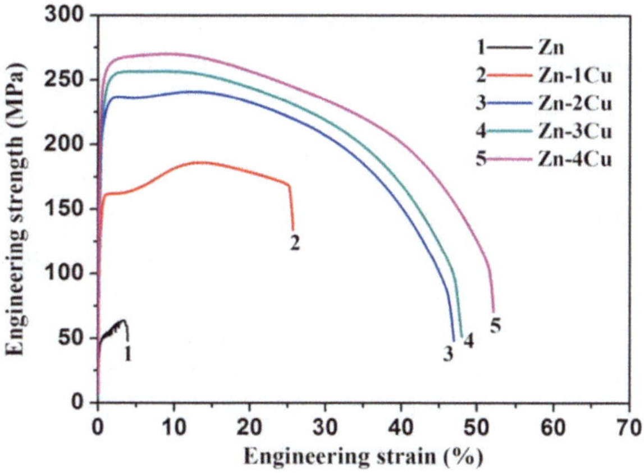

Fig. 4 Engineering stress–strain curves of the as-extruded pure Zn and Zn–xCu ($x = 1, 2, 3$, and 4 wt%) alloys (reproduced with permission) [38]

Bakhsheshi-rad et al. [39] fabricated binary Zn–0.5Al and Zn–0.5Al–xMg ($x = 0.1, 0.3, 0.5$ wt%) alloys by casting for biodegradable implant applications. The Zn–0.5Al and Zn–0.5Al–xMg alloys were determined to consist of α-Zn and $Mg_2(Zn, Al)_{11}$ phases featuring a fine lamellar structure. Ternary Zn–Al–Mg alloys exhibited superior tensile strength, micro-hardness, and corrosion resistance compared to binary Zn–Al alloys. Furthermore, the corrosion rates obtained from the immersion tests were discerned in the subsequent order. Zn–0.5Al–0.5Mg < Zn–0.5Al–0.3Mg < Zn–0.5Al–0.1Mg < Zn–0.5Al, with the Zn–0.5Al–0.5Mg alloy demonstrating superior survivability of MC3T3-E1 compared to the Zn–0.5Al alloy, indicating favorable biocompatibility. Yang et al. [15] incorporated hydroxyapatite (HAp) into the Zn matrix using spark plasma sintering to enhance biocompatibility (both in vitro and in vivo) and to achieve adjustable degradation rates. It was found that Zn–HAp composites have significantly enhanced cell viability of osteoblastic MC3T3-E1 cells, reduced thrombosis risk, and lower hemolysis rates, alongside adequate antibacterial properties, in comparison to pure Zn.

Furthermore, the Zn–5HAp composite demonstrated enhanced new bone production with prolonged implantation duration in comparison to pure Zn. Yu et al. [40] introduced a novel zinc–nanodiamond (Zn–ND) composite for biodegradable implant applications, utilizing the powder metallurgy (PM) method under varying sintering circumstances. It was observed that grain size was augmented with elevated sintering temperature and prolonged soaking duration. The maximum Young's modulus, at 10.95 and 27.32%, was achieved for Zn–1ND in comparison to pure Zn and Zn–5ND. The properties of various Zn-based biomaterials reported in different studies are shown in Table 1. Based on different studies, it could be deduced that different alloying elements affect the mechanical and biological characteristics of developed biomaterials to a great extent. Thus, depending on the specific medical application, different elements can be added to get the tailor-made characteristics of the biomaterial. Zn–Mg alloys offer faster degradation and enhanced mechanical

Table 1 Mechanical properties of Zn-based biomaterials reported in different studies

Material	Yield strength (MPa)	Ultimate tensile strength (MPa)	Elongation (%)	Hardness (HV)	Reference
Zn–Ca	119.12 ± 7.01	164.57 ± 13.92	2.10 ± 0.23	73.00 ± 7.43	[10]
Zn–Cu	227.0 ± 5.0	270.7 ± 0.5	50.6 ± 2.8	–	[38]
Zn–Fe	80 ± 5	136 ± 5	2.1 ± 1	56 ± 2	[41]
Zn–Mg	127.98 ± 10.72	184.84 ± 20.91	1.82 ± 0.23	78.26 ± 2.84	[10]
Zn–Mn	96.5–130	102–219	0.7–68	–	[42]
Zn–Li	178.2	238–330	17–81	–	[23, 37]
Zn–Sr	120.21 ± 6.08	171.40 ± 14.13	2.03 ± 0.22	61.88 ± 6.72	[10]
Zn–Mg–Ca	70–220	110–442	0.7–10	–	[43, 44]
Zn–Mg–Sr	80–225	120–260	1–9	87–101	[43, 45]
Zn–Ca–Sr	80–210	130–265	1–7.75	–	[43]
Zn–Mg–Mn	114.10	131.94	–	97.66	[46]
Zn–Mn–Ag	29–162	52–258	0–66	–	[42]
Zn–Mn–Ca	109–254	114–345	0.2–9.4	–	[42]
Zn–Li–Ag	253.7 ± 4.7	341.3 ± 4.8	30.6 ± 5.8	–	[37]
Zn–Li–Mg	196.2 ± 5.1	254.7 ± 3.9	97.9 ± 8.7	–	[37]

properties. Zn–Ca alloys, with their slower degradation rates and superior biocompatibility, are ideal for bone regeneration applications. Zn–Cu alloys offer enhanced mechanical strength and antimicrobial properties but require careful optimization to avoid cytotoxicity.

4 Degradation Rates, Biocompatibility, and Cytotoxicity Considerations

Zn-based biodegradable biomaterials offer promising potential for medical applications, particularly in bone implants and tissue engineering. The degradation rates of these materials can be controlled through alloying with other metals, surface coatings, and optimizing the environmental conditions. Zn-based materials demonstrate good biocompatibility, promoting cell proliferation and osteogenesis, which is beneficial for bone regeneration. However, to ensure their safety and efficacy, it is critical to control the degradation rates to avoid the release of toxic levels of Zn ions [47–49]. The standard electrode potentials for Zn, Mg, and Fe are − 0.762 V, − 2.372 V, and − 0.440 V, respectively. Considering bio-safety, zinc has been found to be one of the nutritionally vital components in the body, with an estimated per day need of 15 mg. Zinc is essential for various biological processes such as hormone mediation, DNA/RNA synthesis, and participation in over 300 enzymatic events [48]. Bowen et al. [50] investigated the mechanical properties and in vivo assessments of binary wrought Zn–xAl ($x = 1$, 3, and 5 wt%) alloys. In addition to meeting mechanical benchmarks, Zn–Al alloys demonstrated satisfactory biocompatibility with vascular tissues. Nonetheless, the Al-induced intergranular corrosion facilitated the oxidation process, which accelerated the expansion of volume, ultimately resulting in the fragmentation and implant fracture. Yu et al. [40] assessed the corrosion behavior of zinc–nanodiamond (Zn–ND) composite employing the powder metallurgy method. The Zn–5ND composite exhibited the lowest pH compared to all other Zn composite combinations over the whole immersion duration. The corrosion rates of the composites diminished with increased concentrations of ND, specifically 0.43 mg/day, 0.4 mg/day, and 0.44 mg/day after 30 days of immersion, and 0.36 mg/day, 0.26 mg/day, and 0.23 mg/day after 60 days of immersion for Zn–1ND, Zn–2.5ND, and Zn–5ND, respectively.

Yang et al. [15] examined the corrosion behavior of Zn–HAp composites by electrochemical assessments and immersion experiments. The electrochemical experiments indicated that the corrosion current density rose while the electrode potential significantly decreased with higher HAp concentration. The current density of Zn–10HAp was determined to be tenfold larger than that of pure Zn (51.04 ± 1.80 μA/cm^2 versus 4.90 ± 2.81 μA/cm^2). The immersion test revealed that localized corrosion predominated in Zn–HAp composites, with significant corrosion damage noted in the Zn–10HAp composite. Moreover, the Zn–5HAp and Zn–10HAp samples exhibited larger spherical precipitates and a greater quantity of corrosion products, with the corrosion rate of the Zn–10HAp composite being about five times that of pure Zn (24.88 ± 2.33 μmpy compared to 4.97 ± 1.32 μmpy). The findings of in vitro immersion study are depicted in Fig. 5. Xu et al. [50] reported that diluted extracts of the Zn–Cu–Fe alloy exhibited acceptable toxicity levels toward macrophages, endothelial cells, and MC3T3-E1 cells, all demonstrating great cell survival. Moreover, they employed a modified Chandler-loop model to illustrate that the alloy enhanced the coagulation and complement systems instead of adversely impacting blood cells. The in vitro tests demonstrated that Zn alloys exhibit adequate cytocompatibility.

Li et al. [50] conducted a study on the impact of Zn–4Ag on cell viability and osteogenic differentiation in a human periosteal cell line (TAg cells). The researchers found that Zn–4Ag decomposed at a rate four times greater than pure Zn under identical culture conditions, and the breakdown products did not negatively impact cell viability. The findings indicated that elevated levels of the Zn–4Ag extract decreased calcium accumulation in TAg cells. Nevertheless, low quantities facilitated it, indicating a dose-dependent impact of Zn^{2+}. This investigation further illustrated that extracellular Zn^{2+} can enhance calcium deposition and osteogenic differentiation in TAg cells, offering critical insight for the development of Zn-based alloys as bone implants.

Studies [51–53] assessing the biocompatibility of Zn-based biomaterials have shown that Zn and Zn alloys do not exhibit significant cytotoxicity at low concentrations of Zn^{2+} ions. Zn ions can stimulate cell proliferation and osteogenic differentiation, making these materials particularly suitable for bone regeneration applications. In vivo [1, 3, 54] studies on Zn-based alloys implanted in animal models have demonstrated good biocompatibility with minimal inflammatory responses. In

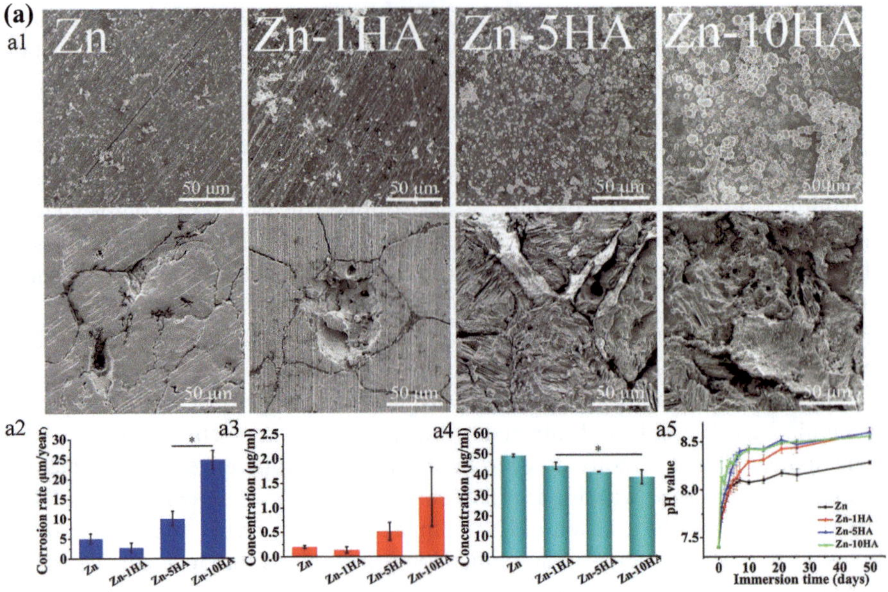

Fig. 5 **a** Degradation behavior in Hank's solution for 50 days: (a1) corrosion morphology of samples, (a2) calculated corrosion rates, (a3) Zn-ion and (a4) Ca-ion concentrations in Hank's solution, and (a5) change in pH values with immersion time (reproduced with permission) [15]

vivo and in vitro studies show that Zn-based alloys, when properly engineered, can effectively promote tissue regeneration without inducing significant cytotoxicity.

5 Conclusions and Future Directions

Zinc-based biodegradable alloys have emerged as promising materials for biomedical applications, particularly in temporary implants, orthopedic scaffolds, and cardiovascular stents. Their favorable degradation behavior, biocompatibility, and mechanical properties make them suitable candidates for medical use. This chapter explored the composition and design strategies of Zn-based alloys. Various alloying elements such as Mg, Li, Mn, Ca, and Ag have been shown to enhance mechanical properties, improve degradation rate, corrosion resistance, and optimize biocompatibility. It also discussed the different fabrication routes, such as powder metallurgy, hot pressing, and additive manufacturing. It also presented various studies discussing the effects of different surface modification strategies, such as plasma modification, grafting, and coating, on cell adhesion, cytotoxicity, and osteogenic factors. These composition, design, and surface modification strategies provide the opportunity to tailor the biomechanical properties of the Zn-based biomaterials. Despite these advancements, several challenges remain in the widespread clinical adoption of Zn-based biodegradable alloys. One major limitation is the relatively low mechanical strength, which restricts their use in load-bearing applications. Strategies such as grain refinement, secondary phase formation, and composite reinforcement have shown promise in addressing this issue. Additionally, the degradation rate of Zn alloys must be carefully controlled to prevent premature structural failure or excessive Zn-ion release, which could lead to cytotoxic effects.

Further research is needed to identify optimal compositions that balance mechanical strength, degradation rate, and biocompatibility. Multi-component alloying strategies and computational modeling could aid in predicting and designing new Zn-based materials with tailored properties. Use of advanced manufacturing techniques such as additive manufacturing can be explored to fabricate Zn-based implants with enhanced microstructure, mechanical properties, hierarchical structures, and porous architectures in order to improve osseointegration and tailorable degradation control. Additionally, further research needs to be carried out in novel surface modification techniques such as bioactive coatings, nanostructured layers, and plasma treatments to regulate degradation rates and enhance biological responses. More extensive in vitro and in vivo studies are required to fully understand the long-term biocompatibility, cytotoxicity, and immune response of Zn-based alloys. Standardized testing protocols should be established to facilitate regulatory approvals. Finally, collaborative efforts between researchers, medical practitioners, and industry stakeholders are essential to transition Zn-based biodegradable alloys from laboratory research to clinical applications. Regulatory guidelines must be addressed to ensure safety and efficacy in human trials. By addressing these challenges and exploring novel fabrication and modification techniques, Zn-based biodegradable alloys can become a viable alternative to conventional implant materials, ultimately improving patient outcomes and advancing the field of biomaterials science.

Conflict of Interest Statement The authors declare no conflict of interest with respect to the research, authorship, and/or publication of this article.

References

1. Yang H et al (2020) Alloying design of biodegradable zinc as promising bone implants for load-bearing applications. Nat Commun 11(1):1–16. https://doi.org/10.1038/s41467-019-14153-7
2. Hussain M, Ullah S, Raza MR, Abbas N, Ali A (2022) Recent developments in Zn-based biodegradable materials for biomedical applications. J Funct Biomater 14(1):1. https://doi.org/10.3390/jfb14010001
3. Rao J, Gao H, Sun J, Yu R, Zhao D, Ding Y (2024) A critical review of biodegradable zinc alloys toward clinical applications. ACS Biomater Sci Eng 10(9):5454–5473. https://doi.org/10.1021/ACSBIOMATERIALS.4C00210
4. Jia B et al (2020) In vitro and in vivo studies of Zn-Mn biodegradable metals designed for orthopedic applications. Acta Biomater 108:358–372. https://doi.org/10.1016/j.actbio.2020.03.009

5. Wang K et al (2021) Binary Zn–Ti alloys for orthopedic applications: corrosion and degradation behaviors, friction and wear performance, and cytotoxicity. J Mater Sci Technol 74:216–229. https://doi.org/10.1016/j.jmst.2020.10.031
6. Venezuela JJD, Johnston S, Dargusch MS (2019) The prospects for biodegradable zinc in wound closure applications. Adv Healthc Mater 8(16):1–15. https://doi.org/10.1002/adhm.201900408
7. Yang N, Venezuela J, Almathami S, Dargusch M (2022) Zinc-nutrient element based alloys for absorbable wound closure devices fabrication: current status, challenges, and future prospects. Biomaterials 280(July):121301. https://doi.org/10.1016/j.biomaterials.2021.121301
8. Bowen PK et al (2016) Biodegradable metals for cardiovascular stents: from clinical concerns to recent Zn-alloys. Adv Healthc Mater 5(10):1121–1140. https://doi.org/10.1002/adhm.201501019
9. Mostaed E, Sikora-Jasinska M, Drelich JW, Vedani M (2018) Zinc-based alloys for degradable vascular stent applications. Acta Biomater 71:1–23. https://doi.org/10.1016/j.actbio.2018.03.005
10. Li HF et al (2015) Development of biodegradable Zn-1X binary alloys with nutrient alloying elements Mg, Ca and Sr. Sci Rep 7:1–14. https://doi.org/10.1038/srep10719
11. Merson D et al (2020) The functional properties of Mg-Zn-X biodegradable magnesium alloys. Materials (Basel) 13(3):1–18. https://doi.org/10.3390/ma13030544
12. Krystýnová M, Doležal P, Fintová S, Březina M, Zapletal J, Wasserbauer J (2017) Preparation and characterization of zinc materials prepared by powder metallurgy. Metals (Basel) 7(10). https://doi.org/10.3390/met7100396
13. Sotoudeh Bagha P, Khaleghpanah S, Sheibani S, Khakbiz M, Zakeri A (2018) Characterization of nanostructured biodegradable Zn-Mn alloy synthesized by mechanical alloying. J Alloys Compd 735:1319–1327. https://doi.org/10.1016/j.jallcom.2017.11.155
14. Ni J et al (2019) Three-dimensional printing of metals for biomedical applications. Mater Today Bio 3(May). https://doi.org/10.1016/j.mtbio.2019.100024
15. Yang H et al (2018) In vitro and in vivo studies on zinc-hydroxyapatite composites as novel biodegradable metal matrix composite for orthopedic applications. Acta Biomater 71:200–214. https://doi.org/10.1016/J.ACTBIO.2018.03.007
16. Zhao D et al (2023) Material–structure–function integrated additive manufacturing of degradable metallic bone implants for load-bearing applications. Adv Funct Mater 33(16):1–16. https://doi.org/10.1002/adfm.202213128
17. Arab SM, Akbarzadeh A (2013) The effect of equal channel angular pressing process on the microstructure of AZ31 Mg alloy strip shaped specimens. J Magnes Alloy 1(2):145–149. https://doi.org/10.1016/j.jma.2013.07.001
18. Sitdikov O, Sakai T, Goloborodko A, Miura H, Kaibyshev R (2004) Effect of pass strain on grain refinement in 7475 Al alloy during hot multidirectional forging. Mater Trans 45(7):2232–2238. https://doi.org/10.2320/matertrans.45.2232
19. Zhu D et al (2019) Mechanical strength, biodegradation, and in vitro and in vivo biocompatibility of Zn biomaterials. ACS Appl Mater Interfaces 11(7):6809–6819. https://doi.org/10.1021/acsami.8b20634
20. García-Mintegui C et al (2021) Zn-Mg and Zn-Cu alloys for stenting applications: From nanoscale mechanical characterization to in vitro degradation and biocompatibility. Bioact Mater 6(12):4430–4446. https://doi.org/10.1016/j.bioactmat.2021.04.015
21. Vojtěch D, Kubásek J, Šerák J, Novák P (2011) Mechanical and corrosion properties of newly developed biodegradable Zn-based alloys for bone fixation. Acta Biomater 7(9):3515–3522. https://doi.org/10.1016/j.actbio.2011.05.008
22. Liu Z, Li R, Jiang R, Li X, Zhang M (2016) Effects of Al addition on the structure and mechanical properties of Zn alloys. J Alloys Compd 687:885–892. https://doi.org/10.1016/j.jallcom.2016.06.196
23. Zhao S et al (2017) Zn-Li alloy after extrusion and drawing: structural, mechanical characterization, and biodegradation in abdominal aorta of rat. Mater Sci Eng C 76:301–312. https://doi.org/10.1016/j.msec.2017.02.167

24. Dai Y et al (2019) Mechanical strengthening mechanism of Zn-Li alloy and its mini tube as potential absorbable stent material. Mater Lett 235:220–223. https://doi.org/10.1016/j.matlet.2018.10.001
25. Zhong C et al (2024) Biocompatibility and osteogenic potential of choline phosphate chitosan-coated biodegradable Zn1Mg. Acta Biomater 175:395–410. https://doi.org/10.1016/j.actbio.2023.12.014
26. Shomali AA, Guillory RJ, Seguin D, Goldman J, Drelich JW (2017) Effect of PLLA coating on corrosion and biocompatibility of zinc in vascular environment. Surf Innov 5(4):211–220. https://doi.org/10.1680/jsuin.17.00011
27. Wang B et al (2021) Electrodeposited dopamine/strontium-doped hydroxyapatite composite coating on pure zinc for anti-corrosion, antimicrobial and osteogenesis. Mater Sci Eng C 129(June):112387. https://doi.org/10.1016/j.msec.2021.112387
28. Su S, Tang Q, Qu D (2022) In vitro study of degradation and cytocompatibility of ceramics/PLA composite coating on pure zinc for orthopedic application. Front Bioeng Biotechnol 10(March):1–12. https://doi.org/10.3389/fbioe.2022.856986
29. Wang YH et al (2016) Corrosion behavior of Zn-Cu-Ti and Zn-Cu-Ti-Mg alloys in NaCl solution. Mater Corros 67(3):297–304. https://doi.org/10.1002/MACO.201508410
30. Abou El-Khair MT, Daoud A, Ismail A (2004) Effect of different Al contents on the microstructure, tensile and wear properties of Zn-based alloy. Mater Lett 58(11):1754–1760. https://doi.org/10.1016/J.MATLET.2003.10.058
31. Abd El-Rahman SS (2003) Neuropathology of aluminum toxicity in rats (glutamate and GABA impairment). Pharmacol Res 47(3):189–194. https://doi.org/10.1016/S1043-6618(02)00336-5
32. Chen C et al (2021) Alloying design strategy for biodegradable zinc alloys based on first-principles study of solid solution strengthening. Mater Des 204:109676. https://doi.org/10.1016/j.matdes.2021.109676
33. Tripathi G, Pandey PM (2024) Study of compaction-free fabrication of topologically ordered functionally graded iron-hydroxyapatite-zinc biodegradable composite implants. J Mater Sci 59(21):9333–9349. https://doi.org/10.1007/s10853-024-09786-z
34. Tripathi G, Pandey PM (2024) In vitro degradation behavior, cytocompatibility and hemocompatibility of topologically ordered functionally graded iron-hydroxyapatite-zinc composite biodegradable biomaterial fabricated using 3D printing and pressureless microwave sintering. Rapid Prototyp J (October). https://doi.org/10.1108/RPJ-07-2024-0286
35. Das P, Kumar TSS, Sahu KK, Gollapudi S (2022) Corrosion, stress corrosion cracking and corrosion fatigue behavior of magnesium alloy bioimplants. Corros Rev 40(4):289–333. https://doi.org/10.1515/CORRREV-2021-0088/ASSET/GRAPHIC/J_CORRREV-2021-0088_FIG_010.JPG
36. Pathak DK, Pandey PM (2021) Effect of 2 wt% addition of Fe on the compressive properties and corrosion behavior of Zn-HAp-Fe material. J Mater Eng Perform 30(5):3510–3523. https://doi.org/10.1007/s11665-021-05632-4
37. Zhang Y et al (2019) Investigation on the microstructure, mechanical properties, in vitro degradation behavior and biocompatibility of newly developed Zn-0.8%Li-(Mg, Ag) alloys for guided bone regeneration. Mater Sci Eng C 99(February):1021–1034. https://doi.org/10.1016/j.msec.2019.01.120
38. Tang Z et al (2017) Potential biodegradable Zn-Cu binary alloys developed for cardiovascular implant applications. J Mech Behav Biomed Mater 72(May):182–191. https://doi.org/10.1016/j.jmbbm.2017.05.013
39. Bakhsheshi-rad HR et al (2017) Fabrication of biodegradable Zn-Al-Mg alloy : Mechanical properties, corrosion behavior, cytotoxicity and antibacterial activities. Mater Sci Eng C 73:215–219. https://doi.org/10.1016/j.msec.2016.11.138
40. Yu M, George C, Cao Y, Wootton D, Zhou J (2014) Microstructure, corrosion, and mechanical properties of compression-molded zinc-nanodiamond composites. J Mater Sci 49(10):3629–3641. https://doi.org/10.1007/s10853-014-8066-x
41. Kafri A, Ovadia S, Goldman J, Drelich J, Aghion E (2018) The suitability of Zn–1.3%Fe alloy as a biodegradable implant material. Metals (Basel) 8(153):1–15. https://doi.org/10.3390/met8030153

42. Shi ZZ et al (2019) Effects of Ag, Cu or Ca addition on microstructure and comprehensive properties of biodegradable Zn-0.8Mn alloy. Mater Sci Eng C 99(February):969–978. https://doi.org/10.1016/j.msec.2019.02.044
43. Li H, Yang H, Zheng Y, Zhou F, Qiu K, Wang X (2015) Design and characterizations of novel biodegradable ternary Zn-based alloys with IIA nutrient alloying elements Mg, Ca and Sr. Mater Des 83:95–102. https://doi.org/10.1016/j.matdes.2015.05.089
44. Yang L et al (2019) Influence of Mg on the mechanical properties and degradation performance of as-extruded Zn-Mg-Ca alloys: In vitro and in vivo behaviour. J Mech Behav Biomed Mater 95(November):220–231. https://doi.org/10.1016/j.jmbbm.2019.04.029
45. Liu X et al (2016) Microstructure, mechanical properties, in vitro degradation behavior and hemocompatibility of novel Zn-Mg-Sr alloys as biodegradable metals. Mater Lett 162:242–245. https://doi.org/10.1016/j.matlet.2015.07.151
46. Liu X et al (2016) Micro-alloying with Mn in Zn-Mg alloy for future biodegradable metals application. Mater Des 94:95–104. https://doi.org/10.1016/j.matdes.2015.12.128
47. Tripathi G, Pandey PM (2023) Experimental investigations into pressureless microwave sintering of iron–hydroxyapatite composite for biodegradable implants. Proc Inst Mech Eng Part C J Mech Eng Sci 152. https://doi.org/10.1177/09544062231166488
48. Niu J et al (2016) Research on a Zn-Cu alloy as a biodegradable material for potential vascular stents application. Mater Sci Eng C 69:407–413. https://doi.org/10.1016/j.msec.2016.06.082
49. Tripathi G, Mohan P (2025) Dimensional analysis—based modelling of weight loss and degradation rate of topologically ordered functionally graded iron—hydroxyapatite—zinc composite biomaterial using Buckingham's π Theorem. Trans Indian Inst Met, pp 1–9. https://doi.org/10.1007/s12666-024-03518-2
50. Bowen PK et al (2018) Evaluation of wrought Zn-Al alloys (1, 3, and 5 wt% Al) through mechanical and in vivo testing for stent applications. J Biomed Mater Res B Appl Biomater 106(1):245–258. https://doi.org/10.1002/JBM.B.33850
51. Bahrani S, Hashemi SA, Mousavi SM, Azhdari R (2019) Zinc-based metal–organic frameworks as nontoxic and biodegradable platforms for biomedical applications: review study. Drug Metab Rev 51(3):356–377. https://doi.org/10.1080/03602532.2019.1632887
52. Li C et al (2020) Design of biodegradable, implantable devices towards clinical translation. Nat Rev Mater 5(1):61–81. https://doi.org/10.1038/s41578-019-0150-z
53. Liu Q et al (2023) Cytotoxicity of biodegradable zinc and its alloys: a systematic review. J Funct Biomater 14(4). https://doi.org/10.3390/jfb14040206
54. Kong L, Heydari Z, Lami GH, Saberi A, Baltatu MS, Vizureanu P (2023) A comprehensive review of the current research status of biodegradable zinc alloys and composites for biomedical applications. Materials (Basel) 16(13). https://doi.org/10.3390/ma16134797

Chapter 9
Fabrication Techniques and Characterisation for Zinc-Based Biodegradable Implants

Pooja Dwivedi, Sachin Maheshwari, and Arshad Noor Siddiquee

1 Introduction

In recent times, biomaterials (BMs) have been the attention seeker because of the new avenues created by technological integration of multidisciplinary branches such as manufacturing science, material science, medicine, and biology. BMs are natural or engineered substances that can support or replace the healing body tissues, which directly enhance the properties of live tissue, and they have long been employed as implant materials in the medical sciences [1]. Several implants need removal through additional surgical procedures after fulfilling their purpose. In bodily fluids, BMs can corrode over time and secrete corrosion products, which trigger the proper host reactions. These chemicals can then be absorbed, metabolised, or passed by cells or tissues. Interestingly, after serving to promote tissue healing, BMs completely dissolve and leave no implant remnant behind them [2].

In the medical field, metals have long been used as implant materials owing to their good mechanical properties, i.e. strength, wear resistance, toughness, ductility, and formability, which makes metals more attractive [3]. The initial generation of metallic materials utilised in implant applications had to be inactive in the various physiological environments in order to be considered biocompatible. These metals include pure Ti and its alloys (e.g. Ti–6Al–4V), CoCr alloys, and stainless steels (such as 316L). To ensure outstanding biocompatibility and avoid potential issues

P. Dwivedi (✉)
Department of Applied Science, Bharati Vidyapeeth's College of Engineering, New Delhi 110063, India
e-mail: pooja.dwivedi@bharatividyapeeth.edu

S. Maheshwari
Department of Mechanical Engineering, Netaji Subhas University of Technology, New Delhi, India

A. N. Siddiquee
Department of Mechanical Engineering, Jamia Millia Islamia, New Delhi, India

A. Dixit et al. (eds.), *Biodegradable Metallic Materials*, Biomedical Materials for Multi-functional Applications, https://doi.org/10.1007/978-981-95-2401-3_9

Table 1 Comparison of properties among three biodegradable alloys

Materials	Density (g/cm^3)	Tensile strength (MPa)	Young modulus of elasticity (GPa)	In vitro corrosion rate (mm/year)
Mg alloys	~ 2	100–400	50	0.10 ± 0.07
Zn alloys	~ 7	100–400	90	0.08
Fe alloys	~ 8	200–1400	200	0.012

brought on by the interaction of corrosion products with human cells, good corrosion resistance was required. Metals such as Mg, Fe, and Zn have recently gained popularity as implant materials because of their biodegradable and bioabsorbable nature. Table 1 shows the comparison of properties among these three materials [4, 5]. Fe is favourable in load-bearing applications because of its high strength and modulus; however, an intermediate corrosion rate makes Zn an ideal candidate for bone healing and vascular fields.

In recent years, the majority of literature on biodegradable materials and their application has been published. There has been a significant surge in scientific interest in Zn, as research indicates that it may be used to address several major problems associated with Mg and Fe [6].

1.1 Importance of Zn-Based Implants

Zn is the second most abundant trace metal after Fe in humans, wherein 85% of Zn is found in bone and muscles, 11% is in the liver and skin, and the remainder of Zn is found in other tissues [7]. Zn is required for various biological processes, as Zn-based biodegradable polymers have an excellent balance of degradability and biocompatibility and are being considered for orthopaedic applications. For enzymes to carry out their catalytic or regulatory functions, Zn is also an essential element [8, 9]. Zn has a major impact on an organism's growth and bone metabolism, and by promoting osteoblast and osteoclast differentiation, Zn supplementation improves bone production while simultaneously boosting bone strength, as bone health and weakness are linked to its deficiency [10, 11]. The immune system's growth and integrity are also impacted by Zn. The action of some crucial immunological mediators, which are made up of cytokines, thymic peptides, and enzymes, is significantly influenced by Zn. It plays a role in synaptogenesis, neurogenesis, neurotransmission, and neuronal development because it is released as a neuromodulator after being specifically stored in the presynaptic vesicles of particular neurons. Numerous clinical signs, such as developmental problems, birth deformities, and hypotension, can result from a Zn shortage. Zn deficiency is also linked to a number of illnesses, including sickle cell disease, kidney disease, and gastrointestinal disorders. However, too much zinc can potentially have negative effects. In uncoupled mitochondria, Zn^{+2} can prevent electron transport, and if too much Zn is consumed, it can be fatal for

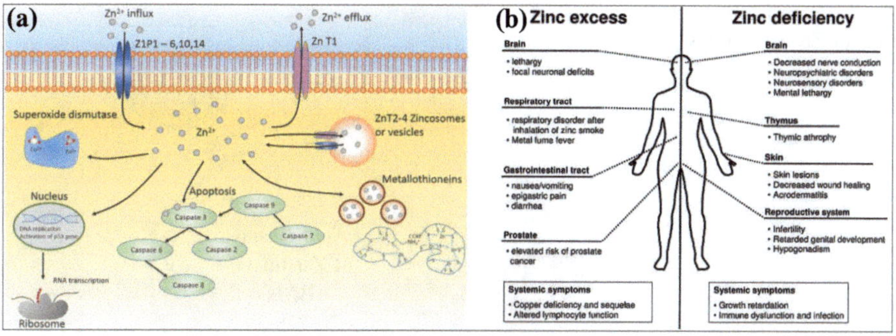

Fig. 1 Zn's biological roles and functions in the human body. **a** Zn's biological roles in the human body. **b** The effects of Zn excess and deficiency in the human body [13]

embryogenesis. According to recent reports, Zn^{+2} affects cell survival, adhesion, and proliferation in two phases. The cytocompatibility would be suppressed by a high Zn^{+2} concentration. The biological roles and functions of Zn in the human body are shown in Fig. 1 wherein the primary biological functions of Zn in cells and throughout the human body are depicted. The figure illustrates how Zn is tightly controlled by cell wall channels, and it has a variety of functions after it enters the cytoplasm, including controlling DNA replication, coordinating apoptosis, and regulating metal-based enzymes [12]. Figure 1b shows the consequences of excess and deficiency of Zn in human body. As excess Zn causes issues in the gastrointestinal tract and respiratory tract and increases the risk of prostate cancer, deficiency leads to delay in wound healing, mental lethargy, immune dysfunction, etc.

A lower rate of corrosion gives several benefits to Zn alloys. For instance, it is proposed that the low degradability of Zn for stent applications: (i) provides more flexibility for metallurgical manipulation and resulting strengthening, (ii) improves biocompatibility, and (iii) enables the engineering of lower profile stents, which is essential to reduce thrombosis and restenosis. The fact that Zn does not produce hydrogen gas during its corrosion process is another benefit over Mg [14, 15]. Pure Mg degrades naturally at a pace of hundreds of micrometres annually in an arterial environment, whereas pure Zn degrades at a rate of tens of micrometres annually [18]. Although this matter is still up for debate, it is possible that this corrosion byproduct could be harmful to the body, such as the development of gas pockets surrounding the disintegrating implant.

2 Applications of Zn-Based Biodegradable Implants

Zn-based BMs are attractive materials for a range of possible therapeutic applications, including bone implants, biodegradable batteries, cardiovascular stents, orthopaedic fixation devices, and wound closure devices, as shown in Fig. 2. They are stronger

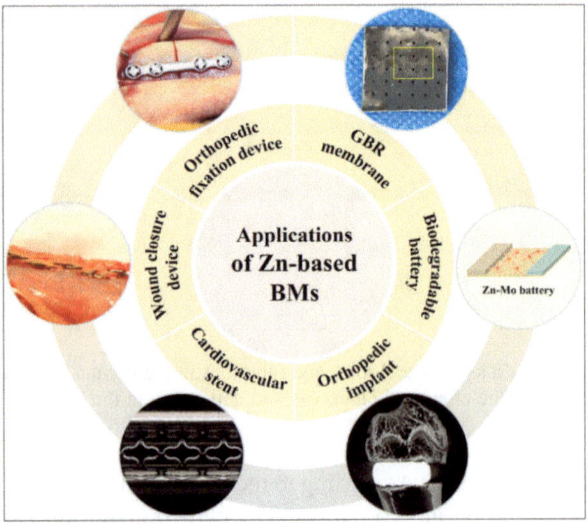

Fig. 2 Potential medical uses for Zn-based BMs [17]

and harder than polymers, as they offer a solution to wound closure. It is possible to make the staples thinner and less noticeable [16].

2.1 Orthopaedic Implants

In consequence of the fast ageing of the population, road accidents, sports injuries, and metabolic illnesses, bone fractures are getting more frequent [18, 19]. The yearly prevalence of fractures is 3.6%, while the lifetime frequency is approximately 40% [20]. The most typical fractures of the patella, tibia, fibula, or ankle in the lower leg are considered troublemakers [21]. In fracture curing, bone recreates and retrieves its biological and biomechanical characteristics [22]. In the internal fixation process, screws, pins, plates, etc. were used, which provide the mechanical stability to the fractured bone so that weight bearing, early limb use, and bone healing are possible. Furthermore, orthopaedic prosthesis is used to substitute or heal broken bones, and the severity of bone fractures determines the type of prosthesis that is used. Bone fractures are typically treated using metallic devices, but these devices have several detrimental effects on the human body, namely osteoporosis and delayed bone healing [23, 24]. Once the non-biodegradable device has fulfilled its intended purpose and the bone has healed, a second procedure is required to remove it. Biodegradable prostheses have been made to point out the problems with non-biodegradable devices. These devices resorb or decompose in the physiological environment throughout the recovery process [25, 26].

The two most appropriate biomaterials for creating biodegradable devices are based on Mg and Zn. The utilisation of Mg-based biomaterials in biomedical applications has been restricted due to their quick disintegration and excessive production of degradation products [27]. Biodegradable Zn alloys, in contrast to Mg-based alloys, have medium rates of decomposition and produce fully biodegradable byproducts that do not produce a lot of hydrogen gas [28].

2.2 *Other Biomedical Applications*

Notably, biomedical metallic implants are utilised extensively in clinical settings for bone plates, screws, hip and knee replacements, and dental implants. Biomedical metal implants are essential for treating serious illnesses. For example, in orthopaedics, medical metal implants can significantly improve patients' quality of life by efficiently treating issues including spinal abnormalities and joint diseases [29]. These implants are used in neurosurgery to support skull structures, stabilise intracranial blood vessels, and considerably lower the mortality and disability linked to cerebral haemorrhage, cerebral infarction, and other related conditions. Medical metal implants, such as artificial heart valves and coronary stents, are crucial for treating heart and circulatory disorders in the field of cardiology [30].

3 Zn-Based Biocompatible Materials: Opportunities and Challenges

The human body is characterised by its fluidic environment and exposure to mechanical forces, and these factors create the corrosive and mechanically ambience for biodegradable materials. Therefore, the implanted biodegradable material will be able to progressively break down at a pace that corresponds with the rate at which bone tissues recover. Figure 3 is the graphical and the visual presentation of the load sharing mechanism between bone and degradable implant as well as depicting the bone reformation process. Initially, biodegradable implant gives mechanical support to the fractured bone by forming the bone tissue around the fracture site, i.e. strengthening the bone during healing process. After regaining the mechanical integrity, the implant begins to degrade and disappear as bone takes over the full load.

3.1 *Biodegradability and Biocompatibility of Zn*

Various fixation implants, along with the quantity of degradable material, are needed for different kinds of fractures at different bone sites. Therefore, it is essential to

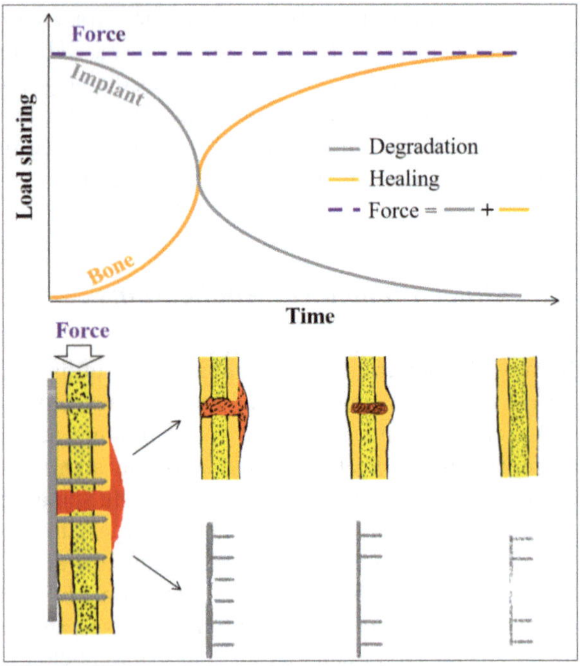

Fig. 3 Implant deterioration and bone repair processes in a perfect match scenario [13]

comprehend the degradation mechanisms and their rate and mechanical factors affecting the degradation of these materials when designing appropriate devices based on Zn with the right rate of deterioration to satisfy clinical fixation needs for different types of fractures. It is clear that the rate of degradation of pure Zn is comparatively low, and in order to adjust Zn's biodegradability, alloy elements or reinforcing materials are frequently added.

According to reports, bone implants should degrade at a pace of 0.2 to 0.5 mm y^{-1} in order to correspond with bone recovery [18]. Therefore, it is evident that pure Zn does not satisfy the specifications for biodegradable orthopaedic implants. One method to change the rate at which biodegradable metals corrode is to add more alloying elements. Accelerated degradation can result from alloying, and the elements that contributed most to corrosion acceleration were Fe, Ag, and Cu, followed by Li, Sr, Ca, and Mg [31]. In addition to alloying, another method of controlling the degradation rate of Zn is to incorporate reinforcement into Zn matrix composites. Spark plasma sintering (SPS) was used to create composites with hydroxyapatite (HA) as reinforcements and pure Zn as a matrix, wherein a wide range of degradation rates was achieved by altering the HA Concentration [32].

Biocompatibility is one of the most crucial parameters for internal bone fixation, in which the material can adjust the cell response, the living system, and the host reaction. Implants used in internal fixation directly discharge ions into the human body, which can further affect the blood, tissues, and surrounding cells and can have either positive or negative effects. Furthermore, one of the main reasons for implant failure in orthopaedic surgery is infections brought on by biomaterials. A postoperative wound infection may result in complications such as limb deformity and implant malfunction, as well as higher pain management costs. In light of its eventual implantation in the human body, it is crucial to investigate the biocompatibility of Zn and its alloys [33, 34]. Zn ions, among other metal ions, exhibit good antibacterial properties and can eradicate a variety of bacteria and fungi. With its inherent antimicrobial and osteoinductive properties, Zn is a necessary element. Zn complexes and ZnO nanoparticles are typically found in Zn-based antibacterial products. Complexes like Zn pyrithione and its derivatives are widely used in medicine and are well-known antifungal substances [35].

4 Fabrication Techniques for Zn-Based Implants

Traditional techniques like casting, sintering, foaming, and chemical vapour deposition have been employed to fabricate biodegradable metal porous scaffolds. These methods face limitations in application due to the inability to specifically regulate the pore size, shape, and quantity, as well as to create tailored geometries that ideally correspond to a patient's anatomical structure. Despite the unpredictable interior pore shape and imprecise alignment with the host bone defect, traditional processing procedures often fall short of the necessities for personalised customisation in clinical practice [36]. Furthermore, traditional procedures generate a very great material intake ratio of 10:1 to 20:1 when making complicated-shaped implant pieces since the majority of materials are withdrawn during conventional processing, resulting in significant material waste. Therefore, additive manufacturing comes into the picture.

4.1 Additive Manufacturing (3D Printing)

Additive manufacturing (AM) or 3D printing is a process of fabricating parts by depositing material layer by layer. AM is a promising technology for biofabrication because it minimises the technical difficulty and cost of production [37]. It has emerged as a transformative technology in the medical field, particularly for creating metallic biomaterials used in implants, prosthetics, stents, scaffolds, and drug delivery systems [38]. This process involves designing and fabricating three-dimensional structures, typically using computer-aided design (CAD) software. The designs are then produced through various advanced techniques, including selective laser sintering (SLS), selective laser melting (SLM), and fused deposition modelling (FDM) [39]. These methods enable the precise creation of complex geometries tailored to individual patient needs, enhancing the effectiveness and integration of medical devices. 3D-printed metallic biomaterials offer significant potential beyond implants, extending to applications in medical hardware, electronics, microfluidics, and tissue repair. In medical hardware and electronics, these materials enable the creation of customised, patient-specific devices, such as wearable and implantable bioelectronic devices, due to their high flexibility, conductivity, and biocompatibility. Furthermore, 3D printed materials show a vital part in dentistry surgical tools and orthopaedics.

The layer-by-layer deposition mode of additive manufacturing (AM) can offer the perfect foundation to fabricate intricate implant components. It can create porous scaffold structures and nearly entirely dense components in a variety of significant technical alloys straight from computer models. AM is a desirable substitute for the manufacturing of complicated-shaped medical implants because it has a number of benefits, including the ability to net-form complex forms, excellent material utilisation, and little machining. It has transformed a number of industries, including bioprinting, which has successfully produced printed livers, hearts, and vascularised alveoli [40]. Its medicinal uses in medical metal implants are depicted in Fig. 4.

Combining CAD and computed tomography (CT) technologies in orthopaedics helps streamline the process of designing bone implants that are tailored for certain patients or deformities [41]. The development of AM has also helped the dental industry by producing implants that closely mimic real teeth. This resemblance is especially noticeable in crucial regions like the dentin-enamel junctions and the buccal and lingual apices. Conventional processing techniques cannot achieve such exact implant forms. In case of skull implants, the space between the implant and the skull usually presents a serious danger of postoperative infection in the field of neurosurgery, wherein operations consisting of brain injuries, infections, and tumours commonly call for extensive cranial osteotomy.

Identifying and optimising materials for AM is one of the biggest obstacles. Biomedical metals require a careful balancing act between corrosion resistance, biocompatibility, and mechanical strength. However, anisotropy caused by columnar

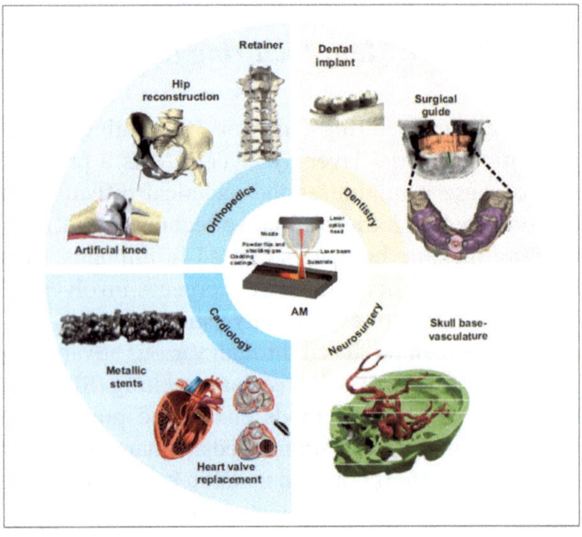

Fig. 4 Clinical use of additive manufacturing in medical metal implants [30]

grains, porosity, and crack formation are common issues in AM processes. These issues are typically closely related to production via layer-by-layer deposition, directional heat transfer, and material characterisation before printing. Researchers have investigated a range of innovative production techniques and post-processing procedures in an effort to overcome these obstacles [42]. The best methods for enhancing the microstructure and lowering the grain size are metal forming and severe plastic deformation, i.e. equal channel angular pressing (ECAP).

4.2 Casting Technique

Casting is the most widely used manufacturing technique for producing Zn-based alloys in large quantities. Casting is used in the mass manufacture of Zn-based alloys because it makes alloy composition tailoring simple. By fabricating intricate interior cavities in moulds, the casting process may also create complicated shapes. After casting, the material, especially Zn, is further processed and used in various medical devices, such as bone fixation screws, plates used in solid cylindrical rods, and flat plates [43]. Freeze casting is generally used for fabricating biomedical components. Various methods are involved in fabricating implanted scaffolds, such as solvent casting, freeze-drying, and gas foaming. However, some limitations like lack of uniformity in pore morphology, pinholes, some form of cracks, and shrinkage defects are present [44], and the dimensional accuracy of the casting is significantly lower than that of machining components. Biomaterials produced via freeze casting, commonly referred to as "ice-templating" because it utilises the principle of directional solidification of a liquid medium like water, are perfectly suited to support a methodical investigation that links structural characteristics to cellular responses [45].

4.3 Powder Metallurgy

This method is opted as an alternative technology over the conventional methods because of its precise control over porosity distribution for scaffolds, which alters the corrosion and mechanical properties of produced samples [46]. This fascinating technique is used for developing Mg–Zn series alloys and composites. According to Kun et al., a 10%β-$Ca_3(PO_4)_2$/Mg–6%Zn (mass%) composite made using powder metallurgy showed acceptable biocompatibility with experimental rabbit tissue. Yu et al. have fabricated the Zn–nanodiamond composite via powder metallurgy for enhancing the mechanical properties [47]. Notably, for research based on Mg–Zn series composites, it is crucial to examine the composition credentials and properties of sintered Mg–Zn series alloys [48]. The Mg–2.5%Zn alloy may be chosen as a good option for future biodegradable metals based on its mechanical characteristics and corrosion resistance. Additionally, the Mg–1%Zn alloy may be an excellent option for implants that do not require a lot of strength. The Mg–75%Zn and Mg–99%Zn alloys had the strongest corrosion resistance and high strengths, but they were too brittle to be employed as readily available implants [49].

5 Characterisation Techniques

5.1 Microstructural Analysis

In microstructural characteristics, secondary phases and their distribution within the matrix have a crucial role in the mechanical and corrosion behaviours of Zn-based alloys [50]. It is anticipated that enhanced characteristics of biodegradable Zn alloys will arise from refined microstructures and homogeneous dispersion of the second phase across the alloy.

In order to improve the microstructure and get better mechanical and degrading properties, Guo et al. conducted studies involving post-casting multi-pass extrusion used to fine-tune the grain size. Prior to extrusion, the as-cast alloy samples were heated. Following cooling, a multi-pass extrusion using the deformation method was carried out on the extruded alloy samples. The findings imply that plastic deformation has a significant impact on grain size [52]. Figure 5 depicts how corrosion affects the Zn-based implant and nearby areas with respect to time.

5.2 Mechanical Properties

Pure Zn exhibits incredibly low mechanical properties, such as yield strength (i.e. 28.9 MPa) and Young's modulus (i.e. 94.1 GPa) [16]. Therefore, it is challenging to satisfy the mechanical requirements for biodegradable metals with these property characteristics. By regulating the kind and composition of the second phase, a composite material was created with enhanced mechanical and biocompatibility. Zn–HA composites were fabricated in a study using the spark plasma sintering technique, wherein pure Zn served as the matrix phase and HA as a reinforcement. Furthermore, Zn-based material (i.e. Zn–5HAp–2Fe) reveals the promising role in bone fixation owing to its mechanical and degradation capabilities [53].

6 Concluding Summary

Regarded as the upcoming generation of metallic biomaterials that degrade naturally for application in medicine, such as wound closure devices, tissue regeneration, bone implants, and cardiovascular stents, Zn-based BMs have already attracted a lot

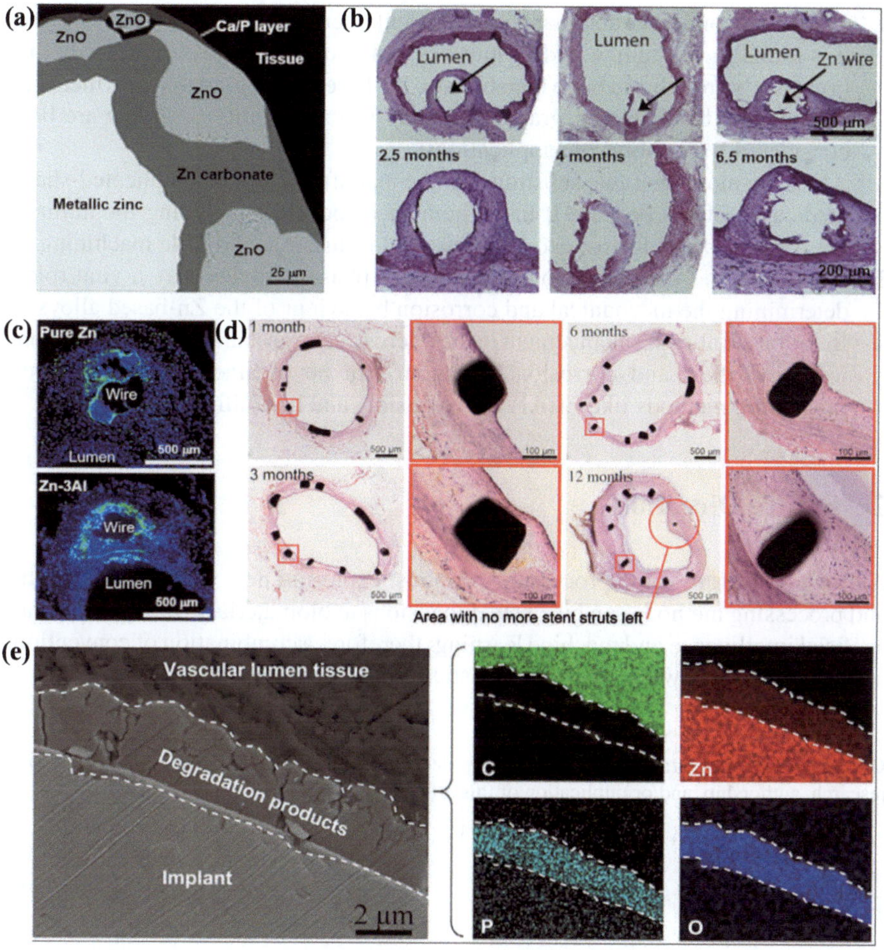

Fig. 5 Examples of Zn-based vascular implants in vivo [51]

of interest. Innovative clinical orthopaedic implants are becoming more and more in demand for bone problems associated with ageing. By adding alloying elements and using post-processing deformation techniques to refine the microstructure, materials based on Zn alloy can satisfy the necessary design requirements. Because of their desired mechanical and degrading qualities, these biodegradable materials may be significant orthopaedic implants to treat difficult bone ailments. The biological role of zinc, design standards for orthopaedic implant materials, and the functionality of these biodegradable alloys are all summed up in this chapter. The following conclusions were reached:

1. Zn is the second most abundant trace element after Fe in the human body, wherein it is found in the human body's muscles and bones, and is essential for both an organism's growth and bone metabolism. Zn-based biodegradable polymers have an excellent balance of degradability and biocompatibility, and they are being considered for orthopaedic applications.
2. AM is the most desirable substitute for the manufacturing of complicated-shaped medical implants because it has a number of benefits, including the ability to net-form complex forms, excellent material utilisation, and little machining.
3. Grain size and the dispersion of secondary phase particles play a vital role in determining the mechanical and corrosion behaviour of the Zn-based alloys.
4. In Zn-based alloys, enhancement in the microstructural characterisation, mechanical properties, and corrosion behaviour can be obtained via various post-processing methods like ECAP, hot extrusion, and hot rolling.

7 Future Direction

Although there are some processing methods, exploring new ways for preparation and processing methods can give new dimensions to biomaterials. Today is the era of artificial intelligence and machine learning; therefore, a combination of conventional and modern routes will give more realistic results in the field of research.

Conflict of Interest Statement The authors declare no conflict of interest with respect to the research, authorship, and/or publication of this article.

References

1. Toong DWY et al (2020) Bioresorbable metals in cardiovascular stents: Material insights and progress. Materialia 12:100727. https://doi.org/10.1016/j.mtla.2020.100727
2. Liu Y et al (2019) Fundamental theory of biodegradable metals—definition, criteria, and design. Adv Funct Mater 29(18):1805402
3. Manam NS et al (2017) Study of corrosion in biocompatible metals for implants: a review. J Alloys Compd 701:698–715
4. Vojtěch D, Kubásek J, Čapek J, Pospíšilová I (2014) Magnesium, zinc and iron alloys for medical applications in biodegradable implants. In: Proceedings of the 23rd international conference on metallurgy and materials, Brno, Czech Republic, pp 21–23
5. Limón I et al (2024) A Review of additive manufacturing of biodegradable Fe and Zn alloys for medical implants using laser powder bed fusion (LPBF). Materials (Basel) 17(24):6220
6. Zheng YF, Gu XN, Witte F (2014) Biodegradable metals. Mater Sci Eng R Reports 77:1–34
7. Hussain M, Ullah S, Raza MR, Abbas N, Ali A (2022) Recent developments in Zn-based biodegradable materials for biomedical applications. J Funct Biomater 14(1):1
8. Zhao C et al (2022) Conformational remodeling enhances activity of lanthipeptide zinc-metallopeptidases. Nat Chem Biol 18(7):724–732
9. Thompson MW (2022) Regulation of zinc-dependent enzymes by metal carrier proteins. Biometals 35(2):187–213

10. Zhu X, Shang X, Lin G, Li H, Feng X, Zhang H (2022) Effects of zinc glycinate on growth performance, serum biochemical indexes, and intestinal morphology of yellow feather broilers. Biol Trace Elem Res 200(9):4089–4097
11. Herrera-Quintana L, Vázquez-Lorente H, Molina-López J, Gamarra-Morales Y, Martín-López JI, Planells E (2022) Vitamin D status in critically ill patients with SIRS and its relationship with circulating zn and related parameters during ICU stay. Nutrients 14(17):3580
12. Bowen PK et al (2016) Biodegradable metals for cardiovascular stents: from clinical concerns to recent Zn-Alloys. Adv Healthc Mater 5(10):1121–1140
13. Liu Y, Du T, Qiao A, Mu Y, Yang H (2022) Zinc-based biodegradable materials for orthopaedic internal fixation. J Funct Biomater 13(4):164
14. Pathak DK, Pandey PM (2021) Evaluation of in vitro corrosion behavior of zinc–hydroxyapatite and zinc–hydroxyapatite–iron as biodegradable composites. J Biomed Mater Res Part B Appl Biomater 109(3):436–450
15. Venezuela J, Dargusch MS (2019) The influence of alloying and fabrication techniques on the mechanical properties, biodegradability and biocompatibility of zinc: A comprehensive review. Acta Biomater 87:1–40
16. Venezuela JJD, Johnston S, Dargusch MS (2019) The prospects for biodegradable zinc in wound closure applications. Adv Healthc Mater 8(16):1900408
17. Liu A et al (2024) Fabrication and performance of Zinc-based biodegradable metals: From conventional processes to laser powder bed fusion. Bioact Mater 41:312–335
18. Shuai C, Li S, Peng S, Feng P, Lai Y, Gao C (2019) Biodegradable metallic bone implants. Mater Chem Front 3(4):544–562
19. Fu R, Feng Y, Liu Y, Willie BM, Yang H (2022) The combined effects of dynamization time and degree on bone healing. J Orthop Res 40(3):634–643
20. Tucker A (2022) Management of common upper limb fractures in adults. Surg 40(3):184–191
21. Wu A-M et al (2021) Global, regional, and national burden of bone fractures in 204 countries and territories, 1990–2019: a systematic analysis from the Global Burden of Disease Study 2019. Lancet Heal Longev 2(9):e580–e592
22. Fu R, Feng Y, Liu Y, Yang H (2021) Mechanical regulation of bone regeneration during distraction osteogenesis. Med Nov Technol Dev 11:100077
23. Li F, Li S, Liu Y, Zhang Z, Li Z (2022) Current advances in the roles of doped bioactive metal in biodegradable polymer composite scaffolds for bone repair: a mini review. Adv Eng Mater 24(8):2101510
24. Xing F et al (2022) Recent progress in Mg-based alloys as a novel bioabsorbable biomaterials for orthopedic applications. J Magnes Alloy 10(6):1428–1456
25. Paiva JCC, Oliveira L, Vaz MF, Costa-de-Oliveira S (2022) Biodegradable bone implants as a new hope to reduce device-associated infections—a systematic review. Bioengineering 9(8):409
26. Pothupitiya JU, Zheng C, Saltzman WM (2022) Synthetic biodegradable polyesters for implantable controlled-release devices. Expert Opin Drug Deliv 19(10):1351–1364
27. Gu X, Li Y, Qi C, Cai K (2022) Biodegradable magnesium phosphates in biomedical applications. J Mater Chem B 10(13):2097–2112
28. Di T, Xu Y, Liu D, Sun X (2022) Microstructure, mechanical performance and anti-bacterial activity of degradable Zn-Cu-Ag alloy. Metals (Basel) 12(9):1444
29. Yao R et al (2023) Biodegradable porous Zn-1Mg-3βTCP scaffold for bone defect repair: In vitro and in vivo evaluation. J Mater Sci Technol 162:189–202
30. Li Y, Jiang D, Zhu R, Yang C, Wang L, Zhang L-C (2024) Revolutionizing medical implant fabrication: advances in additive manufacturing of biomedical metals. Int J Extrem Manuf 7(2):22002
31. Yang H et al (2020) Alloying design of biodegradable zinc as promising bone implants for load-bearing applications. Nat Commun 11(1):401
32. Yang H et al (2018) In vitro and in vivo studies on zinc-hydroxyapatite composites as novel biodegradable metal matrix composite for orthopedic applications. Acta Biomater 71:200–214

33. Yang H et al (2021) Zn-0.4 Li alloy shows great potential for the fixation and healing of bone fractures at load-bearing sites. Chem Eng J 417:129317
34. Li G et al (2019) Challenges in the use of zinc and its alloys as biodegradable metals: Perspective from biomechanical compatibility. Acta Biomater 97:23–45
35. Serrano-Aroca Á et al (2022) Scaffolds in the microbial resistant era: fabrication, materials, properties and tissue engineering applications. Mater Today Bio 16:100412
36. Zhang L-C, Wang J (2024) Stabilizing 3D-printed metal alloys. Science (80) 383(6683):586–587
37. Mobarak MH et al (2023) Recent advances of additive manufacturing in implant fabrication—a review. Appl Surf Sci Adv 18:100462. https://doi.org/10.1016/j.apsadv.2023.100462
38. Mirzaali MJ, Moosabeiki V, Rajaai SM, Zhou J, Zadpoor AA (2022) Additive manufacturing of biomaterials—design principles and their implementation. Materials (Basel) 15(15):5457
39. Nouri A, Shirvan AR, Li Y, Wen C (2021) Additive manufacturing of metallic and polymeric load-bearing biomaterials using laser powder bed fusion: a review. J Mater Sci Technol 94:196–215
40. Kanishka K, Acherjee B (2023) Revolutionizing manufacturing: a comprehensive overview of additive manufacturing processes, materials, developments, and challenges. J Manuf Process 107:574–619
41. Liang H et al (2022) Trabecular-like Ti–6Al–4V scaffold for bone repair: a diversified mechanical stimulation environment for bone regeneration. Compos Part B Eng 241:110057
42. Ji C, Ma A, Jiang J, Song D, Liu H, Guo S (2024) Research status and future prospects of biodegradable Zn-Mg alloys. J Alloys Compd 993:174669. https://doi.org/10.1016/j.jallcom.2024.174669
43. Eliaz N (2012) Degradation of implant materials. Springer Science & Business Media
44. Roshandel M, Dorkoosh F (2021) Cardiac tissue engineering, biomaterial scaffolds, and their fabrication techniques. Polym Adv Technol 32(6):2290–2305
45. Wegst UGK, Schecter M, Donius AE, Hunger PM (2010) Biomaterials by freeze casting. Philos Trans R Soc A Math Phys Eng Sci 368(1917):2099–2121
46. Das P, Pathak DK, Sharma P, Pandey PM (2024) A review on the mechanical and biocorrosion behaviour of iron and zinc-based biodegradable materials fabricated using powder metallurgy routes. Corros Rev 0
47. Yu M, George C, Cao Y, Wootton D, Zhou J (2014) Microstructure, corrosion, and mechanical properties of compression-molded zinc-nanodiamond composites. J Mater Sci 49:3629–3641
48. Yu K et al (2012) In vitro corrosion behavior and in vivo biodegradation of biomedical β-$Ca_3(PO_4)_2$/Mg–Zn composites. Acta Biomater 8(7):2845–2855
49. Yan Y et al (2018) Effects of the intermetallic phases on microstructure and properties of biodegradable magnesium matrix and zinc matrix prepared by powder metallurgy. Mater Trans 59(11):1837–1844
50. Dwivedi P, Maheshwari S, Siddiquee AN (2023) Fe-SiC-Sn-Mn reinforced surface composite via FSP: a comprehensive analysis. Mater Today Commun 35:105783
51. Hernández-Escobar D, Champagne S, Yilmazer H, Dikici B, Boehlert CJ, Hermawan H (2019) Current status and perspectives of zinc-based absorbable alloys for biomedical applications. Acta Biomater 97:1–22
52. Guo P et al (2021) Ultrafine-and uniform-grained biodegradable Zn-0.5 Mn alloy: Grain refinement mechanism, corrosion behavior, and biocompatibility in vivo. Mater Sci Eng C 118:111391
53. Pathak DK, Pandey PM (2021) Effect of 2 wt.% addition of Fe on the compressive properties and corrosion behavior of Zn-HAp-Fe material. J Mater Eng Perform 30(5):3510–3523

Chapter 10
Clinical Translation, Regulatory Considerations, and Future Perspectives: Challenges in Zinc-Based Biodegradable Materials for Biomedical Applications

Jasvinder Singh

1 Introduction

Biodegradable materials represent a revolutionary shift in biomedical science, offering a sustainable solution for temporary implants and devices. By naturally degrading within the human body, these materials eliminate the need for secondary surgeries, reducing patient risks and healthcare costs. In lieu of this, researchers have evolved several kinds of biodegradable materials in the categories of metals, polymers, their alloys, composites, and ceramics, which are highlighted in Fig. 1. Among metals with a biodegradable nature are Mg-, Fe-, Zn-, and Ti-based metals as well as alloys, which have found their applications in different regions and industries [1].

Within this domain, zinc-based biodegradable materials have gained significant traction due to their ideal balance of mechanical properties, biocompatibility, and controlled degradation rates. Their unique characteristics position them as potential candidates for applications such as orthopedic implants, cardiovascular stents, and wound-healing devices [2].

Historically, the use of biodegradable materials in medicine focused on polymers and magnesium-based alloys. While polymers were limited by their low mechanical strength, magnesium alloys suffered from rapid degradation and gas production in physiological environments, making them unsuitable for long-term applications [3]. Zinc, in contrast, offers an intermediate solution with favorable corrosion behavior, making it a promising alternative. As an essential trace element, zinc plays a critical role in cellular processes such as DNA synthesis and enzyme function, further enhancing its appeal in biomedical applications [4]. Orthopedic implants represent a

J. Singh (✉)
Department of Production and Industrial Engineering, Punjab Engineering College (Deemed to Be University), Chandigarh, India
e-mail: Jasvinder.singh.iitd@gmail.com

A. Dixit et al. (eds.), *Biodegradable Metallic Materials*, Biomedical Materials for Multi-functional Applications, https://doi.org/10.1007/978-981-95-2401-3_10

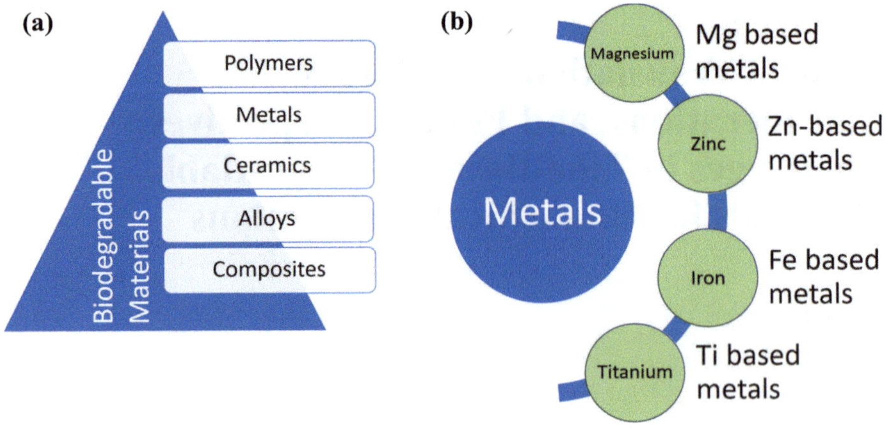

Fig. 1 **a** Different categories of biodegradable materials with **b** available varieties of metallic biomaterials

significant area where zinc-based materials have shown promise. Their strength and corrosion resistance provide a stable scaffold for bone regeneration while degrading at a rate compatible with tissue healing. Furthermore, zinc alloys can be tailored through alloying and surface modification techniques to meet specific clinical needs, such as improving osseointegration or controlling degradation rates [5]. In cardiovascular medicine, the use of zinc-based biodegradable stents offers the advantage of temporary vascular support, reducing the risks associated with permanent implants, such as late thrombosis or chronic inflammation. Studies on animal models have demonstrated their biocompatibility and favorable degradation profiles, paving the way for clinical trials [6]. Moreover, zinc's inherent antimicrobial properties and ability to promote epithelialization make it an attractive option for wound-healing devices, addressing the critical need for infection control in complex wounds [7]. Despite these advantages, several challenges must be addressed to realize the full clinical potential of zinc-based biodegradable materials. One major hurdle is the development of materials with predictable and consistent degradation rates across different physiological conditions. The interaction of zinc with bodily fluids and tissues can lead to variability in corrosion behavior, necessitating comprehensive preclinical evaluations [8]. Additionally, ensuring the biocompatibility of zinc alloys is critical, as excessive ion release or accumulation may lead to cytotoxicity or inflammatory responses [9]. Innovations in material engineering, such as alloying with biocompatible elements like magnesium or calcium, have shown promise in mitigating these effects [10].

The regulatory landscape presents another significant barrier. Biodegradable materials fall under stringent regulatory frameworks, such as those established by the US Food and Drug Administration (FDA) and the European Medicines Agency (EMA). These frameworks require extensive preclinical and clinical evaluations to ensure safety and efficacy. For zinc-based materials, specific standards related to their degradation behavior, ion release, and long-term biocompatibility are still under development, creating additional challenges for researchers and manufacturers [11]. From a manufacturing perspective, the scalability and reproducibility of zinc-based materials are critical. Advanced manufacturing techniques, such as additive manufacturing and powder metallurgy, are being explored to address these challenges. These methods not only enable precise control over material properties but also allow for the fabrication of complex geometries suitable for specific clinical applications [12]. Moreover, the cost-effectiveness of these manufacturing processes must be considered to ensure the widespread adoption of zinc-based devices in healthcare.

To overcome these challenges, a multidisciplinary approach is essential. Collaboration among material scientists, biomedical engineers, clinicians, and regulatory bodies can drive innovation and streamline the translation of zinc-based biodegradable materials from research to clinical practice. By leveraging advancements in material science and preclinical modeling, researchers can develop more robust and reliable zinc-based solutions. Furthermore, harmonizing global regulatory standards will play a pivotal role in facilitating the approval and commercialization of these materials [13]. In conclusion, zinc-based biodegradable materials hold immense potential to transform the landscape of biomedical implants and devices. Their favorable properties, coupled with ongoing research and development, make them strong contenders for clinical applications. However, addressing the challenges of degradation control, biocompatibility, manufacturing scalability, and regulatory approval is imperative. This review aims to provide a comprehensive overview of the current state of research, highlighting the key advances, challenges, and future directions in the field.

2 Zinc-Based Biodegradable Materials

2.1 Properties of Zinc and Its Alloys

Zinc exhibits moderate strength and ductility, making it suitable for load-bearing applications. When alloyed with elements like magnesium, calcium, or copper, the mechanical properties of zinc can be tailored to match the requirements of specific biomedical applications [14]. These alloys demonstrate tensile strength in the range of 100–150 MPa and elongation percentages of 10–20%, offering a balance between strength and flexibility [15]. Moreover, zinc alloys maintain mechanical stability during degradation, which is critical for applications like bone scaffolds and stents. Figure 2 shows the stress–strain curve comparison of zinc alloys with other biodegradable metals.

Zinc's degradation process is gradual and produces biocompatible by-products, minimizing the risks of rapid ion accumulation or toxicity. Compared to magnesium alloys, zinc corrodes at a slower rate in physiological environments, ensuring sustained mechanical support during the healing process [6]. Studies have shown

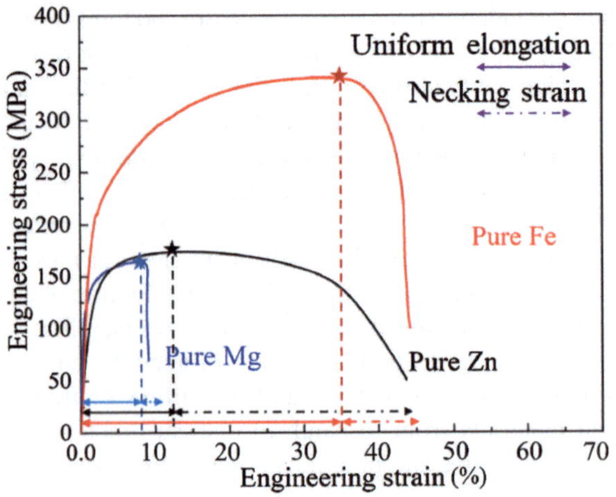

Fig. 2 Stress–strain curve comparison of zinc alloys with other biodegradable metals [16]

that zinc ions released during degradation play a role in cellular signaling pathways, promoting tissue regeneration and reducing inflammation. Figure 3 illustrates the in vitro degradation rates of pure zinc and zinc alloys in simulated body fluid (SBF).

Zinc corrodes by forming a protective layer of zinc oxide and zinc hydroxide on its surface. This layer modulates the degradation rate and prevents abrupt material loss. The corrosion behavior of zinc can be influenced by alloying elements, manufacturing processes, and surface treatments. Table 1 compares the corrosion rates of various zinc-based alloys with other biodegradable materials under similar conditions.

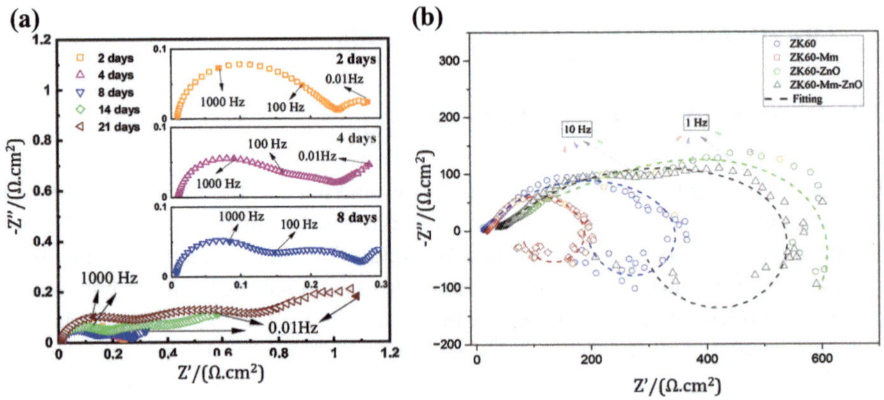

Fig. 3 In vitro degradation rates of pure zinc and zinc alloys in simulated body fluid (SBF) [17, 18]

Table 1 Corrosion rates and advantages of biodegradable materials [3]

Material	Corrosion rate (mm/year)	Key advantages
Pure zinc	0.02	Controlled degradation
Zn–Mg alloy	0.01	Enhanced mechanical properties
Zn–Cu alloy	0.015	Antimicrobial properties
Mg alloy	0.5	High strength
Polymers	Negligible	Biocompatibility

2.2 *Biomedical Applications*

Orthopedic applications require materials with high strength and controlled degradation. Zinc-based alloys, such as Zn–Mg and Zn–Cu, provide sufficient mechanical support for bone healing while degrading in tandem with the regeneration process. Preclinical studies have shown improved osseointegration with zinc alloys compared to traditional materials. The use of zinc implants in animal models has demonstrated significant improvements in fracture healing rates, as shown in Table 2.

Zinc-based stents offer temporary vascular support and gradually degrade to prevent complications like restenosis or thrombosis. Zinc's biocompatibility and controlled degradation ensure minimal inflammatory responses as reported by Zeng and team [9]. Recent advancements have focused on optimizing the microstructure of zinc stents to improve their mechanical strength and degradation behavior. Figure 4 illustrates the degradation profile of a zinc-based stent in an in vivo model.

Zinc's antimicrobial properties make it a promising candidate for wound dressings and tissue scaffolds. Studies have demonstrated that zinc ions promote epithelialization and angiogenesis, accelerating wound healing. Incorporating zinc nanoparticles into hydrogels and other biopolymer matrices has been shown to enhance wound closure rates. A recent study compared the efficacy of zinc-based dressings with conventional treatments, as summarized in Table 3.

Table 2 Comparative healing times and bone regeneration percentages for different implant materials in animal models [7]

Implant material	Healing time (weeks)	Bone regeneration (%)
Pure zinc	12	80
Zn–Mg alloy	10	85
Mg alloy	8	75
Titanium (control)	16	60

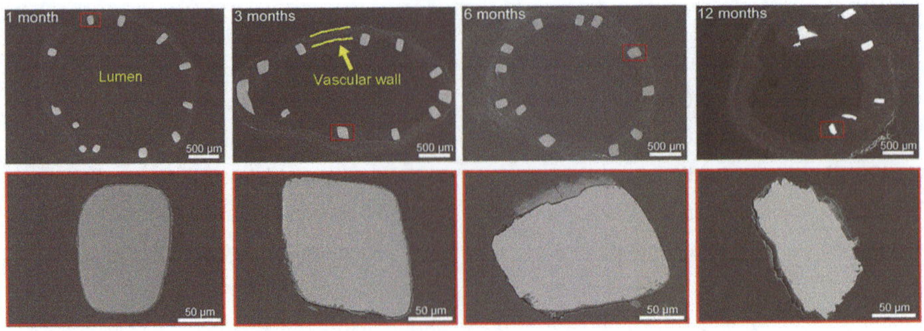

Fig. 4 Degradation profile of a zinc-based stent in an in vivo model [6]

Table 3 Healing times and infection rates for different dressing types [11]

Dressing type	Healing time (days)	Infection rate (%)
Zinc-based	7	5
Silver-based	10	8
Conventional	14	15

Table 4 Polarization values obtained for the ZK60, ZK60-Mm samples, and their ZnO coatings in SBF medium

Sample	E_{corr} (V × SCE)	I_{corr} (μA/cm^2)
ZK60	− 1.46 ± 0.02	18.08 ± 0.01
ZK60-Mm	− 1.49 ± 0.02	60.67 ± 0.01
ZK60-ZnO	− 1.43 ± 0.02	3.85 ± 0.01
ZK60-Mm-ZnO	− 1.47 ± 0.01	25.39 ± 0.01

2.3 *Challenges in Zinc-Based Materials*

2.3.1 Controlling Degradation Rates

The variability in zinc's degradation behavior under different physiological conditions poses a significant challenge. Researchers are exploring surface modifications, such as coating with biocompatible polymers, to achieve consistent degradation rates. Advanced alloying techniques, like incorporating rare earth elements, have also been investigated to fine-tune the degradation process. Table 4 highlights the impact of surface coating on the corrosion resistance of zinc implants.

2.3.2 Manufacturing and Scalability

Producing zinc-based devices with reproducible properties requires advanced manufacturing techniques. Additive manufacturing and powder metallurgy have emerged as viable solutions for fabricating complex structures with high precision. Challenges

Table 5 Comparative analysis of manufacturing methods for zinc-based materials [19]

Manufacturing method	Advantages	Limitations
Additive manufacturing	High precision	Expensive
Powder metallurgy	Scalability	Porosity issues
Casting	Cost-effective	Limited structural control

such as porosity control and surface finishing must be addressed to ensure the clinical reliability of zinc-based devices. Table 5 summarizes the pros and cons of different manufacturing methods for zinc-based materials.

2.4 Research Trends and Innovations

Alloying zinc with elements like lithium, calcium, or strontium enhances its mechanical and biological properties. Recent studies on Zn–Li alloys have shown improved corrosion resistance and biocompatibility. Zn–Cu alloys have demonstrated antimicrobial properties that further expand their potential applications in infection-prone environments [19]. Techniques like micro-arc oxidation, chemical vapor deposition, and laser texturing are being used to coat zinc surfaces. These surface treatments provide additional control over degradation and enhance cellular interactions. Researchers have also investigated biofunctional coatings incorporating growth factors to promote tissue regeneration [13]. Computational models are increasingly being used to predict the degradation behavior and mechanical performance of zinc-based materials. Finite element analysis and molecular dynamics simulations offer valuable insights into optimizing material properties and device designs for specific applications [4].

3 Clinical Translation

The clinical translation of zinc-based biodegradable materials involves the progression of promising experimental innovations from the laboratory into practical, clinical use. This phase is crucial in ensuring that the material is safe, effective, and suitable for human use. Biodegradable materials like zinc alloys have gained attention for applications such as bone scaffolds, vascular stents, and drug delivery systems due to their unique properties, such as controlled degradation, biocompatibility, and mechanical strength. Zinc-based materials can overcome limitations faced by traditional metals, such as permanent implants, offering a more favorable alternative for temporary medical devices [20].

3.1 Preclinical Studies and Bench Testing

Preclinical studies and bench testing serve as the first critical steps in the clinical translation of any material. For zinc-based biodegradable materials, these studies primarily involve in vitro testing for cytotoxicity, biocompatibility, and cell-material interactions, followed by in vivo testing for the long-term effects of material degradation and interaction with bodily tissues. These studies help understand the materials' biological safety and efficacy.

3.1.1 In Vitro Testing

In vitro studies involve testing zinc alloys with cultured cells to evaluate cellular responses, such as proliferation, migration, and differentiation. For instance, human osteoblasts or endothelial cells are often exposed to zinc alloy samples to observe how the materials influence cellular behavior, which is essential in determining their potential for bone tissue regeneration or vascular applications. Studies have shown that zinc-based materials support osteoblast activity and the formation of mineralized matrix, which is crucial for bone healing [21].

3.1.2 In Vivo Testing

In vivo testing, conducted in animal models, provides insight into how zinc alloys behave over time inside a living organism. The key parameters assessed during in vivo trials include the rate of material degradation, the inflammatory response, and the potential for infection or toxicity. Several animal models, including rats and rabbits, are used to test the biocompatibility and performance of zinc-based materials in bone defect repair or cardiovascular applications. In these models, the implants are typically observed for several months, during which time the degradation rate is closely monitored, and histological analysis is used to evaluate tissue integration and healing [3].

3.2 Regulatory Considerations for Clinical Translation

The clinical translation of zinc-based biodegradable materials requires rigorous regulatory approval before reaching the market. Regulatory agencies such as the US Food and Drug Administration (FDA) and the European Medicines Agency (EMA) play a pivotal role in ensuring that new materials are safe and effective for human use.

3.2.1 Regulatory Pathways

To obtain approval for clinical use, zinc-based biodegradable materials must undergo a series of tests and evaluations to meet the standards of biocompatibility, mechanical integrity, and degradation rate. These materials may need to be classified under specific regulatory pathways, depending on the intended use. For example, a material used as a bone scaffold might be categorized as a Class III medical device in the United States, which requires extensive clinical trials to demonstrate safety and efficacy [8, 22].

3.2.2 ISO Standards and Good Manufacturing Practices (GMP)

Compliance with ISO standards for biocompatibility (such as ISO 10993) is essential. These standards outline the testing procedures for materials, including cytotoxicity, irritation, and sensitization testing. Additionally, adherence to GMP ensures that the manufacturing process of zinc-based materials is consistent, reliable, and produces safe, high-quality products [7]. The different aspects of standards proposed by the United States and the European Union are described in Table 6.

3.3 Clinical Trials and Human Studies

After regulatory approval, zinc-based biodegradable materials enter the phase of human clinical trials, which are essential for validating their effectiveness and safety. Clinical trials are typically conducted in phases, starting with small-scale Phase I trials, followed by larger Phase II and Phase III trials. Each phase focuses on different aspects, such as safety, efficacy, and long-term outcomes.

3.3.1 Phase I Trials

Phase I trials primarily focus on assessing the safety of zinc-based materials in humans. These trials often involve a small number of participants and assess factors such as adverse reactions, local tissue responses, and the initial performance of the material in a clinical setting. In the case of bone regeneration, early clinical trials may involve zinc-based scaffolds or implants in patients with bone fractures or defects [21].

3.3.2 Phase II and III Trials

Phase II and III trials are designed to assess the material's efficacy and determine its long-term safety. These trials typically involve larger patient populations and

Table 6 Comparison of regulatory pathways for zinc-based biodegradable materials in the United States and the European Union [8, 23–28]

Aspect	United States	European Union
Regulatory authority	US Food and Drug Administration (FDA)	European Medicines Agency (EMA) or local regulatory bodies
Regulatory framework	FDA's Center for Devices and Radiological Health (CDRH) for medical devices; FDA's Center for Drug Evaluation and Research (CDER) for drugs	EU Medical Device Regulation (MDR) (Regulation (EU) 2017/745) and EU In Vitro Diagnostic Regulation (IVDR) (Regulation (EU) 2017/746) [3]
Approval process for medical devices	Premarket Notification (510(k)) or Premarket Approval (PMA) for devices	CE Marking (Conformity Assessment)
Approval process for drugs	Investigational New Drug (IND) application; New Drug Application (NDA)	Clinical trials (CT) approval, followed by market authorization (MA)
Biodegradable materials: specific regulations	FDA guidance on biocompatibility and degradation (e.g., ISO 10993 for medical devices)	EN ISO 10993 standards for biocompatibility under EU MDR
Safety and efficacy testing	Preclinical and clinical trials required for safety and efficacy evaluation	Preclinical studies and clinical trials required for the evaluation of safety and efficacy
Market authorization timeline	6 months to several years, depending on the classification (510(k) or PMA)	1–3 years, depending on the product type and risk class
Post-market surveillance	FDA's Medical Device Reporting (MDR) and post-market surveillance	Post-market clinical follow-up (PMCF) and vigilance reporting via Eudamed
Environmental considerations	Not always explicitly regulated but may be considered in environmental impact assessments for specific materials	Considerations under the EU REACH regulation and the Waste Framework Directive
Material-specific guidance	The FDA has specific guidance for materials in medical devices, including biodegradability	The EU has specific guidelines for medical devices made from biodegradable materials
Additional requirements	Biocompatibility testing (ISO 10993), degradation rate studies	Biocompatibility testing (ISO 10993), degradation studies for biodegradable medical devices

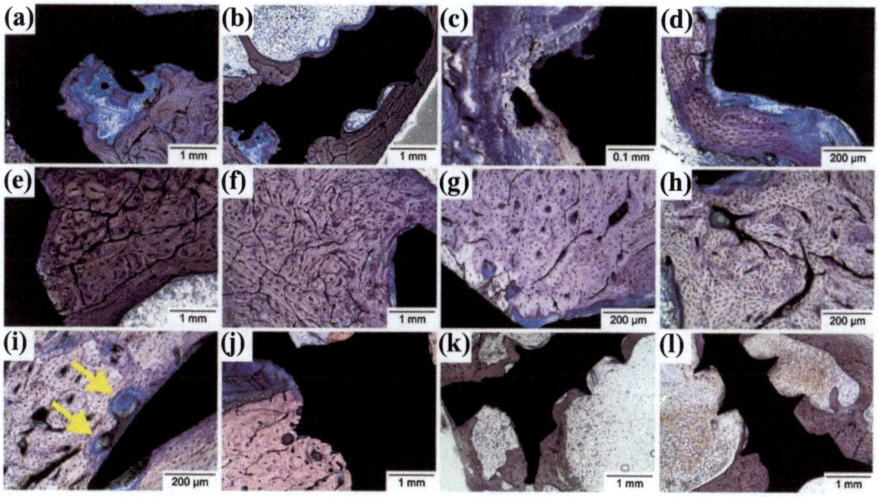

Fig. 5 Timeline of clinical trial phases for zinc-based implants [29]

more extended follow-up periods. During these trials, researchers gather data on the material's performance in specific clinical contexts, such as wound healing, bone repair, or vascular applications. Key outcome measures, including the rate of material degradation, tissue regeneration, and any adverse events, are represented in Fig. 5 [20].

Histological analysis of bone specimens: (a–f) 120 days post-implantation of the experimental screw; (g, h) 240 days post-implantation of the experimental Zn-based screw; (i, j) 360 days post-implantation of the experimental screw. Subtle alterations in the sharpness of the experimental screws indicate potential corrosion: (k, l) 360 days post-implantation of Ti-based screws. Osteon-like cavities are indicated by arrows. Figures 5a, c, and d are derived from our prior pilot work.

3.4 Challenges in Clinical Translation

The clinical translation of zinc-based biodegradable materials faces several challenges, including technical, regulatory, and biological hurdles.

One of the main challenges in clinical translation is achieving consistent and controlled degradation of the zinc material in vivo. The degradation rate of zinc can be influenced by factors such as pH, temperature, and the presence of ions in the surrounding tissue. If the degradation is too rapid, it may lead to premature loss of mechanical integrity, whereas slow degradation may impair the healing process [5].

Another challenge is the biological response to zinc alloys. While zinc is essential for various cellular functions, excessive zinc release can lead to cytotoxicity and an inflammatory response. Balancing the rate of zinc ion release with tissue healing

Table 7 Challenges in clinical translation of zinc-based biodegradable materials [30]

Category	Challenge
Material properties	Ensuring controlled degradation rates that match tissue healing timelines
Biocompatibility	Addressing local and systemic toxicity concerns due to zinc ion release
Corrosion behavior	Managing uniform and localized corrosion under physiological conditions
Surface modifications	Achieving consistent surface treatments to enhance biocompatibility and corrosion resistance
Clinical testing	Limited preclinical and clinical trials to validate performance and safety
Regulatory hurdles	Meeting stringent regulatory guidelines for medical device approval

requirements is critical for ensuring optimal outcomes [23]. Different kinds of challenges regarding the Zn implants observed in the clinical translation are listed in Table 7.

3.5 Future Perspectives and Opportunities

The future of zinc-based biodegradable materials in clinical settings is promising. Ongoing research aims to enhance the mechanical properties of these materials, improve their degradation rates, and optimize their biocompatibility. Emerging techniques such as nanotechnology and surface modifications can play a significant role in improving the performance of zinc alloys. By altering the surface properties of these materials, researchers can control the degradation rate, reduce adverse inflammatory responses, and enhance tissue integration. Nanoparticles, for example, can be used to enhance the osteogenic properties of zinc alloys, facilitating better bone regeneration [21].

Another promising avenue is the combination of zinc-based materials with other biomaterials, such as polymers or ceramics, to create composite materials with superior properties. These hybrid materials could offer improved mechanical strength, controlled degradation, and enhanced biological compatibility [4].

4 Regulatory Considerations

Regulatory considerations are a fundamental aspect of the clinical translation of zinc-based biodegradable materials for biomedical applications. This process involves ensuring that these materials meet the necessary standards for safety, performance, and quality before they can be introduced into clinical practice. The primary regulatory bodies involved in the approval of medical devices, such as the US Food and Drug Administration (FDA), the European Medicines Agency (EMA), and other global health organizations, play a critical role in ensuring the compliance of materials with legal and safety standards. The approval pathway varies depending on the intended application of the material—whether it is used for bone regeneration, drug delivery, or vascular implants.

4.1 Regulatory Pathways for Zinc-Based Biodegradable Materials

Zinc-based biodegradable materials can be classified under several regulatory categories depending on their intended use, characteristics, and risk profile. In this section, we examine the key regulatory pathways that such materials follow.

In the United States, the FDA regulates medical devices based on their level of risk to the patient. Zinc-based biodegradable materials are typically classified under Class II or Class III devices, depending on the complexity of the material and its intended use. For example, zinc-based bone scaffolds may be classified as Class II devices, requiring a 510(k) premarket notification, which involves demonstrating substantial equivalence to an already approved device. In contrast, devices such as zinc-based vascular stents may be classified as Class III, requiring a more extensive premarket approval (PMA) process, including clinical trials to establish safety and efficacy [24].

In the European Union, the regulatory approval process follows similar principles, though it is governed by the European Medicines Agency (EMA). Zinc-based biodegradable materials may be categorized as Class III medical devices, requiring conformity assessment through a Notified Body before receiving the CE mark, which is necessary for marketing in the European Union. The process includes risk assessment, biocompatibility testing, and extensive documentation of clinical data [25].

In addition to the FDA and EMA, other regulatory bodies such as Health Canada, the Japanese Pharmaceuticals and Medical Devices Agency (PMDA), and the Therapeutic Goods Administration (TGA) in Australia also regulate the approval of zinc-based biomedical materials. These agencies follow similar risk-based approaches but have their own specific requirements for clinical evidence, manufacturing quality, and post-market surveillance [31].

4.2 Biocompatibility and Toxicology Requirements

A critical component of the regulatory approval process for zinc-based biodegradable materials is biocompatibility testing. The materials must be evaluated to ensure that they do not elicit harmful effects when used in clinical applications, particularly in sensitive environments such as the human body.

ISO 10993 is a widely recognized international standard for the biological evaluation of medical devices. This standard includes various tests for cytotoxicity, genotoxicity, and sensitization, all of which are essential for ensuring the safety of zinc-based materials. The biocompatibility of these materials is particularly important as zinc ion release can have both beneficial and harmful effects depending on its concentration and the duration of exposure [32].

Excessive zinc release can lead to local toxicity or systemic toxicity, which could result in adverse inflammatory responses, organ dysfunction, or other negative health outcomes. Therefore, regulatory agencies require rigorous toxicology testing to assess the safety of the materials. Zinc concentrations, degradation rates, and the potential for excessive ion release must all be carefully evaluated in both in vitro and in vivo models [33].

Zinc-based biodegradable materials must also be tested for their local effects on surrounding tissues. For instance, zinc-based bone scaffolds should support osteoblast activity without triggering an inflammatory response. The systemic effects of zinc ions, particularly in relation to kidney and liver function, must also be evaluated to ensure that they do not exceed safe levels [34, 35].

4.3 Degradation and Mechanical Properties Testing

The degradation rate and mechanical properties of zinc-based biodegradable materials are crucial for their successful clinical application. These materials must degrade at a controlled rate that allows for the gradual replacement of the implant by natural tissue without compromising mechanical integrity.

The degradation rate of zinc alloys is influenced by factors such as pH, temperature, and the presence of electrolytes in the surrounding environment. Regulatory agencies require that materials undergo accelerated degradation testing to simulate the in vivo environment. The results of these tests must show that the material degrades in a predictable manner, ensuring that it does not lose its structural integrity too soon or too late during the healing process [36].

Mechanical testing, including tensile strength, fatigue resistance, and elongation at break, is essential to ensure that zinc-based biodegradable materials meet the required mechanical performance standards. These materials must support the mechanical load during their intended use (e.g., bone scaffold or stent) before they degrade. For instance, a zinc-based scaffold for bone regeneration must provide sufficient support to facilitate bone healing without fracturing prematurely [37].

4.4 Good Manufacturing Practices (GMP) and Quality Control

Adherence to Good Manufacturing Practices (GMP) is essential for ensuring the consistent quality of zinc-based biodegradable materials. GMP guidelines are intended to ensure that materials are produced in a controlled environment that minimizes the risk of contamination and variability.

Regulatory bodies such as the FDA and EMA require manufacturers to adhere to GMP standards during the production of medical devices, including zinc-based materials. GMP guidelines cover various aspects of the manufacturing process, such as raw material selection, equipment maintenance, staff training, and production processes. Compliance with these standards helps ensure that the final product meets the necessary quality and safety requirements [38].

Once zinc-based biodegradable materials are approved and marketed, post-market surveillance is critical for monitoring their long-term performance and detecting any unforeseen issues. Regulatory bodies mandate that manufacturers establish a robust system for tracking adverse events, collecting clinical data, and conducting periodic inspections to ensure continued compliance with safety standards [39].

4.5 Ethical and Regulatory Considerations in Clinical Trials

Clinical trials are a critical phase in the approval process for zinc-based biodegradable materials. However, ethical considerations, including patient consent and safety, must be carefully addressed (Fig. 6).

Regulatory agencies require that clinical trials be conducted in accordance with ethical guidelines, such as those outlined in the Declaration of Helsinki. These guidelines emphasize the importance of informed consent, patient confidentiality, and the protection of vulnerable populations. In clinical trials involving zinc-based materials, researchers must clearly communicate potential risks and benefits to participants, ensuring that they fully understand the nature of the trial before consenting to participate [40].

Clinical trials must be designed to meet regulatory standards and demonstrate the safety and efficacy of zinc-based materials. The trial design should include appropriate control groups, statistical analysis methods, and safety monitoring protocols. Continuous monitoring during trials ensures that any adverse effects are promptly identified and addressed, minimizing the risk to participants [41].

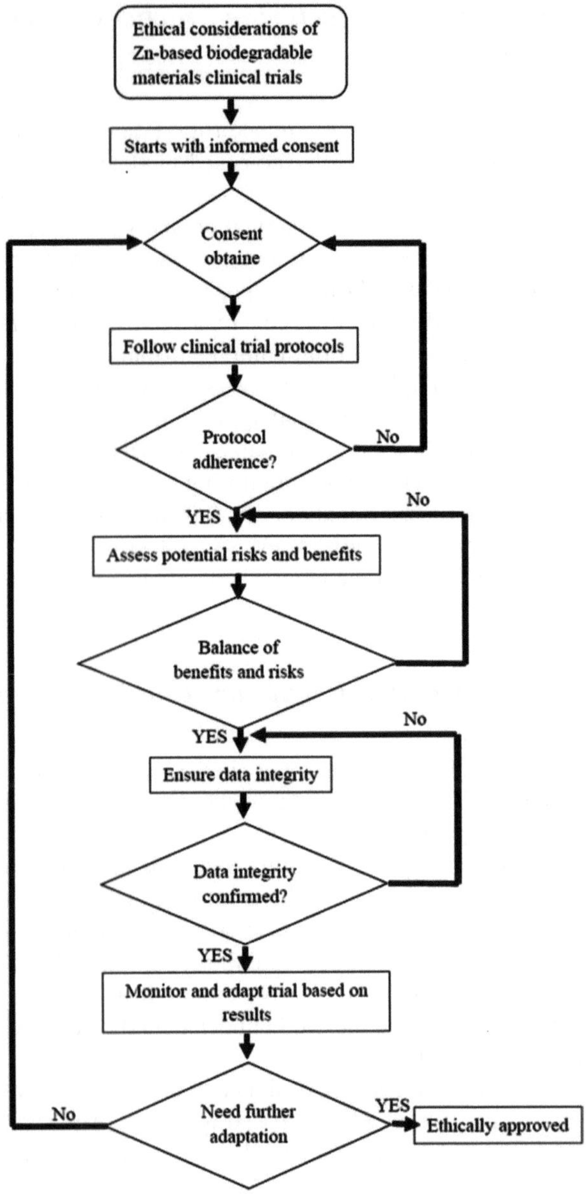

Fig. 6 Flowchart of ethical considerations in zinc-based biodegradable material clinical trials

5 Clinical Translation Stages

The clinical translation of zinc-based biodegradable materials into biomedical applications is a multifaceted process that involves rigorous regulatory considerations. These materials are used in a wide variety of applications, such as bone scaffolds, vascular stents, and drug delivery systems. However, their clinical implementation requires comprehensive testing and regulatory approval to ensure their safety, efficacy, and quality. Regulatory bodies such as the US Food and Drug Administration (FDA), European Medicines Agency (EMA), and other national regulatory authorities are responsible for establishing and enforcing standards for medical device approval. These standards address the biocompatibility, degradation, mechanical properties, and safety of the materials.

5.1 Regulatory Frameworks

Zinc-based biodegradable materials are subject to regulatory frameworks that vary across regions. Their approval depends on their classification, which is determined by their intended use, risks, and potential benefits. This section discusses the major regulatory pathways in the United States, Europe, and other regions.

In the United States, the FDA classifies medical devices based on their risk to patients and the intended use of the device. Zinc-based biodegradable materials, like stents and bone scaffolds, are typically classified into Class II or Class III devices depending on their complexity and risk profile.

- **Class II Devices**: These devices are moderate-risk and typically require a 510(k) premarket notification. A 510(k) application demonstrates that the device is substantially equivalent to a legally marketed device that is already approved by the FDA. Zinc-based devices, such as bone scaffolds, may fall into this category [24].
- **Class III Devices**: Devices with high risks, such as vascular stents, generally require premarket approval (PMA), which involves more rigorous clinical testing to establish safety and efficacy. PMA is a time-consuming process that includes clinical trials, long-term follow-up studies, and comprehensive documentation of all testing results [25].

The European Medicines Agency (EMA) regulates medical devices in the European Union through the Medical Device Regulation (MDR). Devices such as zinc-based biodegradable implants typically require conformity assessment through a Notified Body, which certifies that the device complies with EU regulations. Zinc-based devices often fall under Class III, requiring a more extensive review process compared to Class I or II devices [42].

- **CE Marking**: The Notified Body assesses whether the device complies with the European standards. The CE mark signifies that the product meets the safety and performance requirements under the MDR. The clinical evaluation for zinc-based materials includes biocompatibility, degradation rates, mechanical strength, and long-term safety studies.

Other global regulatory bodies such as the Japanese Pharmaceuticals and Medical Devices Agency (PMDA) and Health Canada also regulate zinc-based biodegradable materials. These agencies generally follow similar risk-based classification systems but may have differing requirements regarding the testing, documentation, and clinical trial processes. For example, Health Canada requires a risk assessment based on device type and intended use before classifying it as a Class I, II, or III device [43].

5.2 Toxicological Examination

For zinc-based biodegradable materials to be approved for clinical use, they must undergo comprehensive biocompatibility and toxicological testing. These tests ensure that the materials do not cause adverse effects, such as inflammation, cytotoxicity, or other systemic issues, when used in the human body.

ISO 10993, an internationally recognized standard, outlines the necessary biological evaluations for medical devices. Zinc-based materials are tested for cytotoxicity, sensitization, irritation, and systemic toxicity. These tests help regulatory bodies assess whether the material will interact negatively with tissues or organs upon implantation. For example, excessive zinc release may lead to toxicity, causing adverse tissue reactions [8].

- **Cytotoxicity Testing**: This test evaluates the potential for a material to cause cell death when in contact with cells. It is crucial to ensure that zinc-based materials do not damage surrounding tissue.
- **Genotoxicity and Sensitization Testing**: Zinc ion release must not lead to genotoxic effects or cause an allergic response in patients. Regulatory bodies require these tests to rule out any potential risks associated with prolonged exposure to zinc ions [33].

Toxicology studies must assess the amount of zinc released from the material and the potential for systemic effects. Zinc is an essential trace element, but excessive concentrations can lead to toxic effects, such as kidney damage or neurological impairments. Therefore, regulatory agencies require that these materials undergo extensive testing to ensure that zinc release remains within safe limits during degradation [44].

5.3 Sustainability Testing

Sustainability and working properties are critical aspects of zinc-based biodegradable materials, as these factors directly influence their performance and safety in clinical applications.

Zinc alloys must degrade at a rate that matches the healing process of the tissues they are meant to support. If the degradation occurs too rapidly, the implant may lose its mechanical integrity before the tissue has sufficiently healed. On the other hand, if the degradation rate is too slow, the material may remain in the body longer than necessary, potentially leading to chronic inflammation or infection.

To ensure controlled degradation, zinc alloys are subjected to accelerated degradation testing in both laboratory and animal models. Regulatory bodies require that manufacturers demonstrate the material's ability to degrade predictably under physiological conditions, including the presence of bodily fluids and varying pH levels [45].

The mechanical strength of zinc-based biodegradable materials must meet the requirements for their intended application. For example, a zinc-based scaffold for bone regeneration must have sufficient mechanical strength to support the bone during the healing process, but it must also degrade at a rate that allows for natural bone growth.

Mechanical testing involves evaluating parameters such as tensile strength, fatigue resistance, and modulus of elasticity. Regulatory bodies require that these tests be conducted in accordance with international standards to ensure that the material will perform as expected when used in clinical settings [46].

5.4 Manufacturing and Quality Control

Manufacturing zinc-based biodegradable materials at scale while maintaining consistent quality is another regulatory challenge. Regulatory bodies require that manufacturers adhere to Good Manufacturing Practices (GMP) to ensure product quality and safety.

GMP guidelines set out the standards for the manufacturing process of medical devices, ensuring that each product is consistent in quality and safe for use. The FDA and EMA have strict GMP requirements that cover all aspects of production, including raw material selection, cleanliness of the manufacturing environment, and testing of the final product [47].

Quality control ensures that the final product meets all required specifications, including biocompatibility, degradation rate, and mechanical properties. For zinc-based materials, the testing process must demonstrate that each batch is consistent with prior batches in terms of performance and safety.

Manufacturers must also validate their processes to ensure that the production methods are reliable and reproducible. This includes validating the degradation rate, mechanical strength, and other critical parameters before the material is approved for use in clinical settings [41].

5.5 *Clinical and Regulatory Challenges*

In addition to technical and manufacturing challenges, there are also clinical and regulatory hurdles that must be addressed to ensure the successful approval and adoption of zinc-based biodegradable materials.

The long-term safety of zinc-based materials remains a major concern. While these materials are biodegradable, their degradation products must be closely monitored to ensure that they do not cause long-term harm to patients. Regulatory agencies require extensive post-market surveillance to track the long-term effects of zinc-based implants on patients [48].

One of the challenges in the regulatory process is the lack of harmonization between global regulatory standards. Different countries may have different testing requirements, leading to delays in the approval process. Efforts are underway to harmonize regulatory standards across different regions, but this process is slow and may impact the global availability of zinc-based biodegradable materials [49].

In conclusion, the regulatory considerations for zinc-based biodegradable materials in biomedical applications are complex and multifaceted. These materials must meet rigorous safety, efficacy, and quality standards before they can be approved for clinical use. The regulatory pathways differ across regions, but all emphasize the importance of biocompatibility, mechanical properties, and controlled degradation. As the field continues to evolve, it is crucial to address the challenges in manufacturing, safety, and global regulatory harmonization to ensure the successful translation of zinc-based materials into clinical practice.

6 Challenges

Zinc-based biodegradable materials have gained significant attention in recent years due to their potential for use in biomedical applications such as orthopedic implants, cardiovascular stents, and tissue engineering scaffolds. Their unique properties, including moderate biodegradation rates and compatibility with physiological environments, make them promising candidates. However, several challenges need to be addressed for their successful translation from research to clinical applications. These challenges can be broadly categorized into material challenges, manufacturing challenges, and clinical and regulatory challenges.

6.1 *Material Challenges*

The primary advantage of zinc-based materials is their moderate biodegradation rate, which lies between magnesium's rapid degradation and titanium's lack of biodegradability. However, controlling this rate within a physiological environment remains challenging. Factors such as the local pH, presence of chloride ions, and enzymatic activity can lead to non-uniform degradation, causing localized corrosion or premature structural failure [50]. Alloying zinc with other elements such as magnesium, calcium, or strontium has been explored to refine its degradation behavior, but achieving consistent results is difficult due to the complex interplay between alloy composition and in vivo conditions [5]. Figure 7 represents the degradation behavior of Zn alloys when immersed in the different physiological solutions. Moreover, the composition of several physiological solutions is listed in Table 8.

Pure zinc has relatively low mechanical strength compared to other biodegradable metals like magnesium and iron. This limits its use in load-bearing applications. Although alloying and thermomechanical treatments can enhance the strength of zinc-based materials, balancing mechanical properties with biocompatibility and biodegradability remains a significant challenge (Table 9). The introduction of alloying elements may improve strength but could also adversely affect biocompatibility or lead to unpredictable degradation behavior [2].

While zinc is an essential trace element for human health, excessive accumulation of zinc ions due to rapid degradation can lead to cytotoxicity and adverse

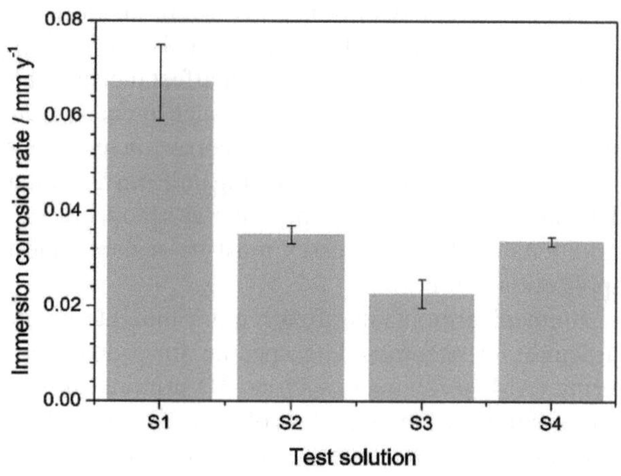

Fig. 7 Schematic representation of zinc degradation in different physiological environments [51]

Table 8 Chemical composition of blood plasma and the investigated physiological fluids [51]

Concentration (mmol L^{-1})	Na^+	K^+	Ca^{2+}	Mg^{2+}	Cl^-	HPO_4^{-2}	HCO_3^{-1}	SO_4^{-2}
Blood plasma	142	5.0	2.5	1.5	103.0	1.0	27.0	0.5
S1	103				103.0			
S2	103	2.0			103.0	1.0		
S3	130	2.0			103.0	1.0	27.0	
S4	131	2.0			103.0	1.0	27.0	0.5
Hanks' balanced salt solution	141	5.4	1.3	1.0	145	0.25	4.2	1.0
PBS	157	4.5			140	10		

Table 9 Comparison of mechanical properties of pure zinc, magnesium, and their respective alloys used in biomedical applications

Material	Ultimate tensile strength (MPa)	Yield strength (MPa)	Elongation (%)
Pure zinc	20–30	10–20	1–2
Magnesium alloys	150–250	80–200	5–20
Zinc alloys	50–150	30–100	1–5

tissue responses. Understanding the tolerable limits of zinc ion concentrations and ensuring that degradation by-products are non-toxic are critical. Furthermore, alloying elements introduced to improve mechanical or corrosion properties must not compromise the overall biocompatibility of the material [7].

6.2 *Manufacturing Challenges*

The production of zinc-based materials and their alloys requires careful control over melting, casting, and forming processes to ensure homogeneity and minimize defects. Zinc's low melting point (419.5 °C) makes it prone to oxidation during processing, which can introduce impurities that affect mechanical and corrosion properties [10].

Surface modification techniques such as coating, anodization, or plasma treatment are often employed to improve the corrosion resistance and biocompatibility of zinc-based materials. However, achieving uniform and durable surface coatings that do not delaminate or compromise the material's biodegradability is technically challenging. Additionally, these processes must be scalable and cost-effective for commercial applications [52].

Biomedical implants often require intricate geometries to mimic natural tissue structures or accommodate specific functional requirements. Additive manufacturing (AM) techniques, such as 3D printing, offer promising solutions for fabricating complex zinc-based implants. However, AM of zinc and its alloys is still in its infancy, with challenges including optimizing process parameters, achieving adequate mechanical properties, and ensuring material purity during fabrication [53] (Table 10).

Table 10 Summary of additive manufacturing challenges for zinc-based implants [53]

Challenge	Description
Process optimization	Ensuring consistent layer deposition and bonding
Mechanical properties	Achieving strength and ductility in printed parts
Material purity	Minimizing contamination during the printing process

6.3 *Clinical and Regulatory Challenges*

Preclinical studies on animal models have shown promising results for zinc-based materials, but translating these findings to human applications requires a deeper understanding of in vivo performance. Factors such as the variability in patient-specific physiological conditions and the interaction of zinc degradation products with biological systems need extensive investigation. Long-term studies are required to evaluate the safety and efficacy of these materials under clinically relevant conditions [54].

The lack of standardized protocols for testing and evaluating zinc-based biodegradable materials poses a barrier to their clinical adoption. Standards must be developed for material composition, degradation behavior, mechanical testing, and biocompatibility assessments. Consistent quality control during manufacturing is also critical to ensure the reproducibility of material properties [55].

The regulatory pathway for zinc-based biodegradable materials is complex and often requires substantial preclinical and clinical data to demonstrate safety and efficacy. Regulatory agencies such as the US Food and Drug Administration (FDA) and the European Medicines Agency (EMA) require rigorous testing to evaluate the biocompatibility, degradation behavior, and mechanical performance of these materials. Meeting these requirements is time-consuming and resource-intensive, posing a significant challenge for manufacturers [56].

Zinc-based biodegradable materials hold great promise for biomedical applications, offering a balance between biodegradability, biocompatibility, and mechanical performance. However, significant challenges related to material properties, manufacturing processes, and clinical translation must be overcome to unlock their full potential. Addressing these challenges requires interdisciplinary efforts involving materials scientists, biomedical engineers, and regulatory experts. Further research and innovation, combined with well-defined regulatory frameworks, will be key to advancing zinc-based biodegradable materials from bench to bedside.

7 Future Perspectives

Zinc-based biodegradable materials have emerged as a promising class of biomaterials for medical applications due to their unique combination of moderate corrosion rates, biocompatibility, and mechanical properties. Despite significant advancements in understanding their properties and applications, further development is necessary to realize their full potential. Future perspectives on zinc-based biodegradable materials encompass advancements in material design, innovations in manufacturing, exploration of clinical applications, and addressing regulatory challenges.

7.1 Advancements in Material Design

One major direction for future research is the design of zinc-based alloys with enhanced properties. Alloying zinc with elements such as magnesium, calcium, strontium, and manganese has shown promise in improving mechanical strength and corrosion behavior [50]. Research should focus on identifying optimal compositions that balance mechanical properties, degradation rates, and biocompatibility. Computational modeling techniques, such as density functional theory (DFT) and molecular dynamics (MD) simulations, can accelerate the discovery of novel zinc alloys [5].

Functional additives, including bioactive nanoparticles or growth factors, can be incorporated into zinc-based materials to enhance their biological performance. For example, hydroxyapatite nanoparticles can improve osseointegration, while vascular endothelial growth factors (VEGFs) can promote angiogenesis [2]. Future work should explore the synergistic effects of these additives on cellular responses and tissue regeneration.

Surface engineering techniques such as plasma treatments, anodization, and biopolymer coatings have the potential to improve the biocompatibility and corrosion resistance of zinc-based materials [7]. Innovations in this area include the development of stimuli-responsive coatings that adapt to changes in the physiological environment. For instance, pH-sensitive coatings can regulate zinc ion release to maintain a favorable biological response [57].

7.2 Innovations in Manufacturing

Multiple Zn-based BM raw bulks have been effectively generated through forming processes, including casting, extrusion, drawing, rolling, forging, powder metallurgy, and additive manufacturing [9]. Additive manufacturing (AM) is expected to play a significant role in the future development of zinc-based biomedical devices. AM enables the fabrication of patient-specific implants with complex geometries that cannot be achieved through conventional manufacturing techniques. Research should focus on optimizing AM parameters, such as laser power and scan speed, to ensure high-quality zinc parts [52].

Tailoring the microstructure of zinc materials through processes such as cryo-rolling or equal-channel angular pressing (ECAP) can significantly enhance their mechanical properties and corrosion behavior. Additionally, nanostructured zinc materials have shown improved bioactivity and mechanical strength, making them suitable for next-generation implants [58].

The integration of zinc with other biodegradable materials, such as magnesium or polymers, can lead to hybrid materials with superior properties. For instance, zinc-polymer composites can combine the strength of metals with the flexibility of polymers, offering a wider range of applications. Future research should explore the processing methods and interfacial properties of such composites [54].

7.3 Exploration of Clinical Applications

Zinc's moderate corrosion rate and biocompatibility make it an ideal candidate for cardiovascular stents. Future studies should focus on optimizing stent designs, including strut thickness and surface modifications, to improve endothelialization and reduce restenosis rates. Preclinical and clinical trials are essential to validate the safety and efficacy of zinc-based stents in vivo [20].

Zinc-based biodegradable materials are also promising for orthopedic implants such as screws, plates, and pins. These materials can eliminate the need for secondary surgeries to remove implants. Research should aim to enhance the mechanical strength and degradation behavior of zinc alloys for load-bearing applications while ensuring osseointegration and minimal inflammatory responses [56].

Zinc's role as an essential micronutrient makes it attractive for tissue engineering applications. Zinc-based scaffolds can provide structural support while releasing zinc ions that promote cell proliferation and differentiation. Future work should explore the fabrication of 3D-printed zinc scaffolds with controlled porosity and degradation rates tailored to specific tissue types [59].

7.4 Addressing Regulatory Challenges

The development of standardized testing protocols is critical for the clinical translation of zinc-based materials. Future efforts should focus on establishing guidelines for evaluating corrosion behavior, mechanical properties, and biocompatibility under clinically relevant conditions [60]. Comprehensive long-term studies are necessary to assess the safety of zinc-based materials in vivo. These studies should evaluate the accumulation of degradation products, potential cytotoxic effects, and the material's interaction with surrounding tissues. Such data are crucial for obtaining regulatory approval from agencies like the FDA or EMA [61].

Future research should also address the ethical and environmental implications of using zinc-based materials. This includes understanding the environmental impact of manufacturing processes and ensuring that materials sourced for medical applications are ethically obtained [3].

7.5 *Integration of Emerging Technologies*

Artificial intelligence (AI) and machine learning (ML) can revolutionize the development of zinc-based biodegradable materials by enabling predictive modeling of material behavior and optimizing manufacturing processes. For instance, AI algorithms can predict the corrosion behavior of novel alloys under various physiological conditions [62]. Zinc-based implants can be tailored to individual patient needs through advancements in personalized medicine. This includes designing implants with patient-specific geometries and degradation rates based on medical imaging data and computational modeling [63]. The integration of sensors into zinc-based implants can enable the development of smart implants that monitor degradation rates, mechanical stability, and local physiological conditions in real time. Such implants can provide valuable data for clinicians and improve patient outcomes [64].

The future of zinc-based biodegradable materials for biomedical applications is bright, with numerous opportunities for innovation in material design, manufacturing, and clinical applications. Addressing current challenges and leveraging emerging technologies will be key to unlocking their full potential. Collaborative efforts among researchers, clinicians, and regulatory bodies are essential to advance these materials from the laboratory to clinical practice. Continued investment in interdisciplinary research will ensure that zinc-based biodegradable materials play a pivotal role in the next generation of medical devices.

8 Conclusion

Zinc-based biodegradable materials represent a promising class of biomaterials for a variety of biomedical applications, offering a unique combination of moderate degradation rates, biocompatibility, and mechanical properties. These attributes make zinc and its alloys ideal candidates for applications such as orthopedic implants, cardiovascular stents, and wound-healing devices. By naturally degrading within the body, these materials eliminate the need for secondary surgeries, reduce long-term complications, and improve overall patient outcomes. However, the journey from laboratory research to clinical application is fraught with challenges that must be carefully addressed to unlock their full potential. One of the primary advantages of zinc-based materials lies in their moderate corrosion rates, which strike a balance between the rapid degradation of magnesium and the lack of biodegradability in titanium. This characteristic ensures that zinc-based devices provide sufficient mechanical support during tissue healing before safely degrading. Furthermore, zinc's role as an essential trace element enhances its biological appeal, as it actively participates in cellular functions such as DNA synthesis, enzyme activity, and tissue regeneration. Its inherent antimicrobial properties and ability to promote angiogenesis make it especially attractive for wound-healing applications, where infection control is critical.

Despite these advantages, controlling the in vivo degradation rates of zinc-based materials remains a significant challenge. Variations in physiological environments, such as pH, ion concentrations, and enzymatic activity, can lead to unpredictable corrosion behavior, potentially compromising the performance of these materials. To address this, researchers are exploring advanced alloying strategies by incorporating elements like magnesium, calcium, and copper to fine-tune the degradation process. Surface engineering techniques, including micro-arc oxidation, chemical vapor deposition, and polymer coatings, further enhance corrosion resistance, biocompatibility, and osseointegration. The scalability and reproducibility of manufacturing processes also present obstacles. Techniques like additive manufacturing (AM) and powder metallurgy are being employed to fabricate complex geometries tailored for specific applications, such as patient-specific implants. However, these methods face technical challenges, including porosity control, contamination risks, and high production costs. Overcoming these limitations will require continued innovation and collaboration between material scientists, engineers, and clinicians.

Another critical barrier to clinical translation is the stringent regulatory landscape. Regulatory agencies, such as the FDA and EMA, mandate comprehensive preclinical and clinical evaluations to ensure safety, efficacy, and long-term performance. For zinc-based materials, this includes biocompatibility testing, cytotoxicity studies, and degradation profiling under ISO standards. Harmonizing global regulatory standards and streamlining approval pathways will be essential to expedite the clinical adoption of these materials. Looking ahead, interdisciplinary collaboration will play a pivotal role in addressing these challenges. Advances in computational modeling, including finite element analysis and molecular simulations, offer valuable insights into optimizing material properties and predicting degradation behavior. Emerging technologies, such as nanostructuring and hybrid material development, promise to further enhance the mechanical and biological performance of zinc-based implants. Furthermore, integrating artificial intelligence into material design and manufacturing can accelerate innovation and reduce development timelines.

In conclusion, zinc-based biodegradable materials hold immense potential to transform the field of biomedical implants and devices. By addressing challenges related to degradation control, biocompatibility, manufacturing, and regulatory compliance, these materials can provide safer, more effective, and patient-specific solutions. Continued investment in research, regulatory harmonization, and technological innovation will ensure that zinc-based materials advance from the laboratory to clinical practice, reshaping the future of medical science. Furthermore, the coatings and surface improvement of Zn-based implants can ensure higher osteointegration properties for bone tissue engineering applications.

References

1. Modrák M, Trebuňová M, Balogová AF, Hudák R, Živčák J (2023) Biodegradable materials for tissue engineering: development, classification and current applications. J Funct Biomater 14(3). https://doi.org/10.3390/jfb14030159
2. Bowen PK, Drelich J, Goldman J (2013) Zinc exhibits ideal physiological corrosion behavior for bioabsorbable stents. Adv Mater 25(18):2577–2582. https://doi.org/10.1002/ADMA.201300226
3. Staiger MP, Pietak AM, Huadmai J, Dias G (2006) Magnesium and its alloys as orthopedic biomaterials: a review. Biomaterials 27(9):1728–1734. https://doi.org/10.1016/J.BIOMATERIALS.2005.10.003
4. Kabir H, Munir K, Wen C, Li Y (2021) Recent research and progress of biodegradable zinc alloys and composites for biomedical applications: biomechanical and biocorrosion perspectives. Bioact Mater 6(3):836–879. https://doi.org/10.1016/J.BIOACTMAT.2020.09.013
5. Mostaed E, Sikora-Jasinska M, Drelich JW, Vedani M (2018) Zinc-based alloys for degradable vascular stent applications. Acta Biomater 71:1–23. https://doi.org/10.1016/j.actbio.2018.03.005
6. Yang H et al (2017) Evolution of the degradation mechanism of pure zinc stent in the one-year study of rabbit abdominal aorta model. Biomaterials 145:92–105. https://doi.org/10.1016/J.BIOMATERIALS.2017.08.022
7. Zheng YF, Gu XN, Witte F (2014) Biodegradable metals. Mater Sci Eng R Reports 77:1–34. https://doi.org/10.1016/J.MSER.2014.01.001
8. "ISO 10993–1:2018(en), Biological evaluation of medical devices—Part 1: Evaluation and testing within a risk management process." Accessed: 21 Jan 2025. [Online]. Available: https://www.iso.org/obp/ui/en/#iso:std:iso:10993:-1:ed-5:v2:en
9. Zeng X et al (2024) The emerging role of zinc on alleviating inflammatory responses and enhancing immunity in grass carp (Ctenopharyngodon idella). Aquaculture 584:740645. https://doi.org/10.1016/J.AQUACULTURE.2024.740645
10. Hernández-Escobar D, Champagne S, Yilmazer H, Dikici B, Boehlert CJ, Hermawan H (2019) Current status and perspectives of zinc-based absorbable alloys for biomedical applications. Acta Biomater 97:1–22. https://doi.org/10.1016/J.ACTBIO.2019.07.034
11. Kokubo T, Takadama H (2006) How useful is SBF in predicting in vivo bone bioactivity? Biomaterials 27(15):2907–2915. https://doi.org/10.1016/J.BIOMATERIALS.2006.01.017
12. Mostaed E et al (2016) Novel Zn-based alloys for biodegradable stent applications: Design, development and in vitro degradation. J Mech Behav Biomed Mater 60:581–602. https://doi.org/10.1016/J.JMBBM.2016.03.018
13. "EMA adopts guideline on quality documentation for drug-device combination products | RAPS." Accessed: 21 Jan 2025. [Online]. Available: https://www.raps.org/news-and-articles/news-articles/2021/7/ema-adopts-guideline-on-quality-documentation-for
14. Jin H et al (2018) Novel high-strength, low-alloys Zn-Mg (< 0.1 wt% Mg) and their arterial biodegradation. Mater Sci Eng C 84:67–79. https://doi.org/10.1016/J.MSEC.2017.11.021
15. Rokkala U, Bontha S, Ramesh MR, Balla VK, Srinivasan A, Kailas SV (2021) Tailoring surface characteristics of bioabsorbable Mg-Zn-Dy alloy using friction stir processing for improved wettability and degradation behavior. J Mater Res Technol 12:1530–1542. https://doi.org/10.1016/J.JMRT.2021.03.057
16. Huang H et al (2022) Recent advances on the mechanical behavior of zinc based biodegradable metals focusing on the strain softening phenomenon. Acta Biomater 152:1–18. https://doi.org/10.1016/J.ACTBIO.2022.08.041
17. Huang S, Wu W, Su Y, Qiao L, Yan Y (2021) Insight into the corrosion behaviour and degradation mechanism of pure zinc in simulated body fluid. Corros Sci 178:109071. https://doi.org/10.1016/J.CORSCI.2020.109071

18. Santos MJA et al (2024) Zinc oxide coating impact on corrosion of ZK60 magnesium alloys in simulated body fluid. J Alloys Compd 1002:175339. https://doi.org/10.1016/J.JALLCOM.2024.175339
19. Jiang J, Huang H, Niu J, Zhu D, Yuan G (2022) Fabrication and characterization of biodegradable Zn-Cu-Mn alloy micro-tubes and vascular stents: microstructure, texture, mechanical properties and corrosion behavior. Acta Biomater 151:647–660. https://doi.org/10.1016/J.ACTBIO.2022.07.049
20. Li G et al (2019) Challenges in the use of zinc and its alloys as biodegradable metals: perspective from biomechanical compatibility. Acta Biomater 97:23–45. https://doi.org/10.1016/J.ACTBIO.2019.07.038
21. Jin X et al (2024) In vitro and in vivo studies on biodegradable Zn porous scaffolds with a drug-loaded coating for the treatment of infected bone defect. Mater Today Bio 24:100885. https://doi.org/10.1016/J.MTBIO.2023.100885
22. ISO 10993-5:2009, Biological evaluation of medical devices—Part 5: tests for in vitro cytotoxicity. Int Organ Stand 3:1–34
23. Reeve L, Baldrick P (2017) Biocompatibility assessments for medical devices—evolving regulatory considerations. Expert Rev Med Devices 14(2):161–167. https://doi.org/10.1080/17434440.2017.1280392
24. "Premarket Notification 510(k)|FDA." Accessed: 21 Jan 2025. [Online]. Available: https://www.fda.gov/medical-devices/premarket-submissions-selecting-and-preparing-correct-submission/premarket-notification-510k
25. Wilson DJ (1990) Regulation (EU) 2017/745 of the European Parliament and of The Council. Polio 2013(June):1–184. https://doi.org/10.1177/2165079915576935
26. The European Parliament and the Council of the European Union (2017) Regulation (EU) 2017/746 of the European parliament and of the council on in vitro diagnostic medical devices. Off J Eur Union 5(5):117–176
27. EMA (2008) Guideline M3(R2) on non-clinical safety studies for the coduct human clinical trails and marketing authorisation for pharmaceuticals [Online]. Available: www.ema.europa.eu
28. Reg.1906/2007 (2007) REACH concerning the registration, evaluation, authorisation and restriction of chemicals (REACH). Sustainaspeak, pp 219–219
29. Klíma K et al (2021) A complex evaluation of the in-vivo biocompatibility and degradation of an extruded ZnMgSr absorbable alloy implanted into rabbit bones for 360 days. Int J Mol Sci 22:13444. https://doi.org/10.3390/IJMS222413444
30. Guillory RJ, Oliver AA, Davis EK, Earley EJ, Drelich JW, Goldman J (2019) Preclinical in vivo evaluation and screening of zinc-based degradable metals for endovascular stents. JOM 71(4):1436–1446. https://doi.org/10.1007/S11837-019-03371-5/TABLES/1
31. Kramer DB, Xu S, Kesselheim AS (2012) How does medical device regulation perform in the United States and the European Union? A systematic review. PLOS Med 9(7):e1001276. https://doi.org/10.1371/JOURNAL.PMED.1001276
32. ISO 10993-4:2017, Biological evaluation of medical devices—Part 4: selection of tests for interactions with blood. Int Organ Stand. Accessed: 24 Apr 2019. [Online]. Available: https://www.iso.org/standard/63448.html
33. Liu Q et al (2023) Cytotoxicity of biodegradable zinc and its alloys: a systematic review. J Funct Biomater 14(4):206. https://doi.org/10.3390/JFB14040206/S1
34. Schoofs H, Schmit J, Rink L (2024) Zinc toxicity: understanding the limits. Mol 29(13):3130. https://doi.org/10.3390/MOLECULES29133130
35. Xie Y et al (2022) Benefits and risks of essential trace elements in chronic kidney disease: a narrative review. Ann Transl Med 10(24):1400–1400. https://doi.org/10.21037/ATM-22-5969
36. Farabi E et al (2021) Novel biodegradable Zn alloy with exceptional mechanical and in vitro corrosion properties for biomedical applications. ACS Biomater Sci Eng 7(12):5555–5572. https://doi.org/10.1021/ACSBIOMATERIALS.1C00763/ASSET/IMAGES/MEDIUM/AB1C00763_0019.GIF

37. Zhu D et al (2019) Mechanical strength, biodegradation, and in vitro and in vivo biocompatibility of Zn biomaterials. ACS Appl Mater Interfaces 11(7):6809–6819. https://doi.org/10.1021/ACSAMI.8B20634/SUPPL_FILE/AM8B20634_SI_001.PDF
38. Lietaert K, Vleugels J, Schrooten J, Mortensen A (2018) Design, manufacturing and validation of cellular Zn alloys as biodegradable implant materials. [Online]. Available: https://lirias.kuleuven.be/retrieve/521761
39. Gurman P, Rabinovitz-Harison O, Hunter TB (2012) Regulatory challenges in biomaterials: focus on medical devices. In: Biomaterials science. CRC Press, pp 235–282. https://doi.org/10.1201/B12303-14
40. World Medical Association (1974) WMA declaration of Helsinki: ethical principles for medical research involving human subjects, vol 353, no 1, pp 1418–1419. [Online]. Available: http://www.wma.net/en/30publications/10policies/b3/index.html
41. Rao J, Gao H, Sun J, Yu R, Zhao D, Ding Y (2024) A critical review of biodegradable zinc alloys toward clinical applications. ACS Biomater Sci Eng 10(9):5454–5473. https://doi.org/10.1021/ACSBIOMATERIALS.4C00210/ASSET/IMAGES/MEDIUM/AB4C00210_0007.GIF
42. Fink M, Akra B (2023) Comparison of the international regulations for medical devices–USA versus Europe. Injury 54:110908. https://doi.org/10.1016/J.INJURY.2023.110908
43. “Guidance on Medical Device Establishment Licensing (GUI-0016): Summary—Canada.ca.” Accessed: 21 Jan 2025. [Online]. Available: https://www.canada.ca/en/health-canada/services/drugs-health-products/compliance-enforcement/establishment-licences/directives-guidance-documents-policies/guidance-medical-device-establishment-licensing-0016.html
44. Latka K, Kolodziej W, Domisiewicz K, Lasowy P, Latka D (2024) Medical implant heavy metal contents and effect on patients. Biomed Biotechnol Res J 8(3):267–273. https://doi.org/10.4103/BBRJ.BBRJ_148_24
45. Fernandes MT, Martins Da Silva L, Nilo Mendes PS, Vergílio De Queiroz A, Adilson De Castro J, Dos Santos C (2021) A degradation kinetics model of Mg–Zn–Mn–Ca alloys in Kokubo solution. J Mater Res Technol 11:887–895. https://doi.org/10.1016/J.JMRT.2021.01.063
46. Venezuela J, Dargusch MS (2019) The influence of alloying and fabrication techniques on the mechanical properties, biodegradability and biocompatibility of zinc: a comprehensive review. Acta Biomater 87:1–40. https://doi.org/10.1016/J.ACTBIO.2019.01.035
47. Quality System (QS) Regulation/Medical Device Current Good Manufacturing Practices (CGMP)|FDA. Accessed: 21 Jan 2025. [Online]. Available: https://www.fda.gov/medical-devices/postmarket-requirements-devices/quality-system-qs-regulationmedical-device-current-good-manufacturing-practices-cgmp
48. Drelich AJ, Zhao S, Guillory RJ, Drelich JW, Goldman J (2017) Long-term surveillance of zinc implant in murine artery: surprisingly steady biocorrosion rate. Acta Biomater 58:539–549. https://doi.org/10.1016/J.ACTBIO.2017.05.045
49. Pawar H, Patel M (2024) Harmonization of regulatory frameworks for medical devices in BRICS countries: a path to enhanced trade and investment. Ann Pharm Françaises. https://doi.org/10.1016/J.PHARMA.2024.08.007
50. Li H, Zheng Y, Qin L (2014) Progress of biodegradable metals. Prog Nat Sci Mater Int 24(5):414–422. https://doi.org/10.1016/J.PNSC.2014.08.014
51. Almathami S, Venezuela J, Yang N, Wang Y, Mardina Z, Dargusch M (2023) Exploring the influence of biologically relevant ions on the corrosion behavior of biodegradable zinc in physiological fluids. ACS Biomater Sci Eng 9(5):2301–2316. https://doi.org/10.1021/ACSBIOMATERIALS.2C00901/ASSET/IMAGES/LARGE/AB2C00901_0011.JPEG
52. Li N, Zheng Y (2013) Novel magnesium alloys developed for biomedical application: a review. J Mater Sci Technol 29(6):489–502. https://doi.org/10.1016/J.JMST.2013.02.005
53. Liang W et al (2023) Recent advances in 3D printing of biodegradable metals for orthopaedic applications. J Biol Eng 17(1):1–24. https://doi.org/10.1186/S13036-023-00371-7/FIGURES/7
54. Bowen PK et al (2016) Biodegradable metals for cardiovascular stents: from clinical concerns to recent Zn-alloys. Adv Healthc Mater 5(10):1121. https://doi.org/10.1002/ADHM.201501019

55. He J et al (2020) A study of degradation behaviour and biocompatibility of Zn—Fe alloy prepared by electrodeposition. Mater Sci Eng C 117:111295. https://doi.org/10.1016/J.MSEC.2020.111295
56. Han HS et al (2019) Current status and outlook on the clinical translation of biodegradable metals. Mater Today 23:57–71. https://doi.org/10.1016/J.MATTOD.2018.05.018
57. Kubásek J, Vojtěch D (2012) Zn-based alloys as an alternative biodegradable materials. Met 2012 Conf Proceedings, 21st Int. Conf. Metall. Mater., pp 1355–1361
58. Jain T, Jain JK, Saxena KK (2022) Design and comprehensive study of biodegradable zinc–based implants for bio–medical applications. Adv Mater Process Technol 8(sup2):519–536. https://doi.org/10.1080/2374068X.2021.1939555
59. Qin Y et al (2019) Additive manufacturing of biodegradable metals: current research status and future perspectives. Acta Biomater 98:3–22. https://doi.org/10.1016/J.ACTBIO.2019.04.046
60. Roesner M, Zankovic S, Kovacs A, Benner M, Barkhoff R, Seidenstuecker M (2023) Biocompatibility assessment of zinc alloys as a new potential material for bioabsorbable implants for osteosynthesis. Mater 16(15):5224. https://doi.org/10.3390/MA16155224
61. Shi Y et al (2023) Surface modification on biodegradable zinc alloys. J Mater Res Technol 25:3670–3687. https://doi.org/10.1016/J.JMRT.2023.06.149
62. Mozafari M (2025) How artificial intelligence shapes the future of biomaterials? Next Mater 7:100381. https://doi.org/10.1016/J.NXMATE.2024.100381
63. Ferraris S, Spriano S (2016) Antibacterial titanium surfaces for medical implants. Mater Sci Eng C 61:965–978. https://doi.org/10.1016/J.MSEC.2015.12.062
64. Kowalski PS, Bhattacharya C, Afewerki S, Langer R (2018) Smart biomaterials: recent advances and future directions. ACS Biomater Sci Eng 4(11):3809–3817. https://doi.org/10.1021/ACSBIOMATERIALS.8B00889/ASSET/IMAGES/MEDIUM/AB-2018-008896_0002.GIF

Chapter 11
Biodegradable Metallic Materials: Recent Advances, Futuristic Opportunities, and Challenges

Anil Kumar, Mukesh Kumar, Anurag Dixit, and Dayanidhi K. Pathak

1 Introduction

The emergence of biodegradable metallic materials has revolutionized the field of biomedical materials, especially in areas like tissue engineering, orthopaedics, and cardiovascular treatments. Due to their progressive degradation within the body, these materials—mainly magnesium (Mg), iron (Fe), and zinc (Zn) alloys—offer a possible substitute for conventional permanent implants [1–3]. This eliminates the need for surgical removal and reduces long-term consequences. This introduction explores the latest developments, new prospects, and current issues with biodegradable metallic materials. The development of biodegradable alloys with specialized characteristics to satisfy the requirements of biological applications has advanced significantly. Because of their advantageous mechanical characteristics and biocompatibility, magnesium alloys like WE43 and AZ91 have been the subject of much research [4, 5]. However, their quick rates of deterioration frequently result in an early loss of structural integrity. Researchers have concentrated on alloying elements, including calcium (Ca), zinc (Zn), and rare earth elements, to improve mechanical strength and modify corrosion rates to lessen this [6]. Because of their slower rates of degradation, iron-based alloys—particularly pure iron and its composites—are appropriate for load-bearing applications [7]. The addition of elements like copper and phosphorus has been investigated recently to enhance mechanical qualities and further regulate deterioration. Because of their advantageous mechanical properties and intermediate rates of deterioration, zinc alloys have drawn interest as a compromise between iron and magnesium [8].

A. Kumar (✉) · M. Kumar · A. Dixit · D. K. Pathak
Department of Mechanical and Automation Engineering, G. B. Pant DSEU Campus, New Delhi, India
e-mail: anil.kumar1@dseu.ac.in

A. Dixit et al. (eds.), *Biodegradable Metallic Materials*, Biomedical Materials for Multi-functional Applications, https://doi.org/10.1007/978-981-95-2401-3_11

Research is being done to create Zn-based alloys with ideal compositions and microstructures [9]. Biodegradable metallic implants have been made using conventional manufacturing techniques, including forging and casting. Nevertheless, these methods frequently fall short in terms of the accuracy needed for intricate geometries and patient-specific designs. The creation of biodegradable metallic implants has undergone a radical change with the introduction of new manufacturing processes, especially additive manufacturing (3D printing) [10, 11]. This method makes it possible to create complex structures that nearly resemble the architecture of genuine bone with regulated porosity and mechanical characteristics. Furthermore, new possibilities for implant design have been made possible by the creation of biodegradable metallic foams. These foams, which are made using techniques like sintering and powder metallurgy, provide high surface area and lightweight structures that encourage tissue integration and cell adhesion. Recent studies have focused on the clinical application of biodegradable metallic materials. Vascular stents and bone fixation devices are only two of the many uses for magnesium-based alloys. The first-generation absorbable metal stent (AMS-1), made from a WE43 alloy, demonstrated the feasibility of biodegradable stents [12]. However, problems including restenosis and quick degradation prompted the creation of better variants, such as the DREAMS-2G stent and AMS-2.1, which include changes to increase performance [13, 14]. Because iron-based implants degrade more slowly, they have shown potential in load-bearing applications. Their potential for bone healing has been shown in clinical trials, with research showing positive results in terms of mechanical strength and biocompatibility. Although they are still being studied, zinc-based alloys have shown promise in applications that call for modest rates of deterioration. They are still being developed, with research concentrating on improving alloy compositions and production methods to boost efficiency. Personalized treatment is possible with the combination of biodegradable metallic materials, 3D printing, and sophisticated imaging methods. The effectiveness and results of medical procedures can be enhanced by creating implants that are tailored to each patient's unique anatomical and physiological needs [15].

In tissue engineering applications, biodegradable metallic materials can act as scaffolds, offering structural support while progressively breaking down as natural tissue grows again. Combining biodegradable metals with growth hormones and stem cells may result in more sophisticated restorative treatments. Controlling the pace of deterioration and the products that are emitted into the surrounding environment, which coincide with the healing process, is one of the main issues in the development of biodegradable metallic materials (Fig. 1a) [15]. These products may be poisonous and cause additional harm to the body. While slow deterioration may result in long-lasting inflammation and problems, rapid degradation may cause the loss of mechanical integrity before tissue repair is finished. The healing process in cardiovascular applications consists of three stages: inflammation, granulation, and remodelling (Fig. 1b) [16]. Fractured bones heal through three stages: inflammation, repair, and remodelling. During degradation, magnesium alloys in particular are susceptible to the development of hydrogen gas, which can build up at the implant site and result in local tissue injury. Creating surface coatings to prevent corrosion and alloying with elements that produce less hydrogen gas are two ways to lessen this problem. It is difficult to strike a compromise between the application's needs and the mechanical characteristics of biodegradable metallic materials. Although magnesium alloys have outstanding strength-to-weight ratios, more research and development are necessary because their mechanical qualities might not always be comparable to those of natural bone. There are many obstacles in the way of moving from laboratory research to clinical use, such as strict regulatory regulations and the demand for lengthy clinical studies to prove safety and efficacy. The practical use of biodegradable metallic implants is made more difficult by the intricacy of long-term degrading behaviours and the variation in patient reactions.

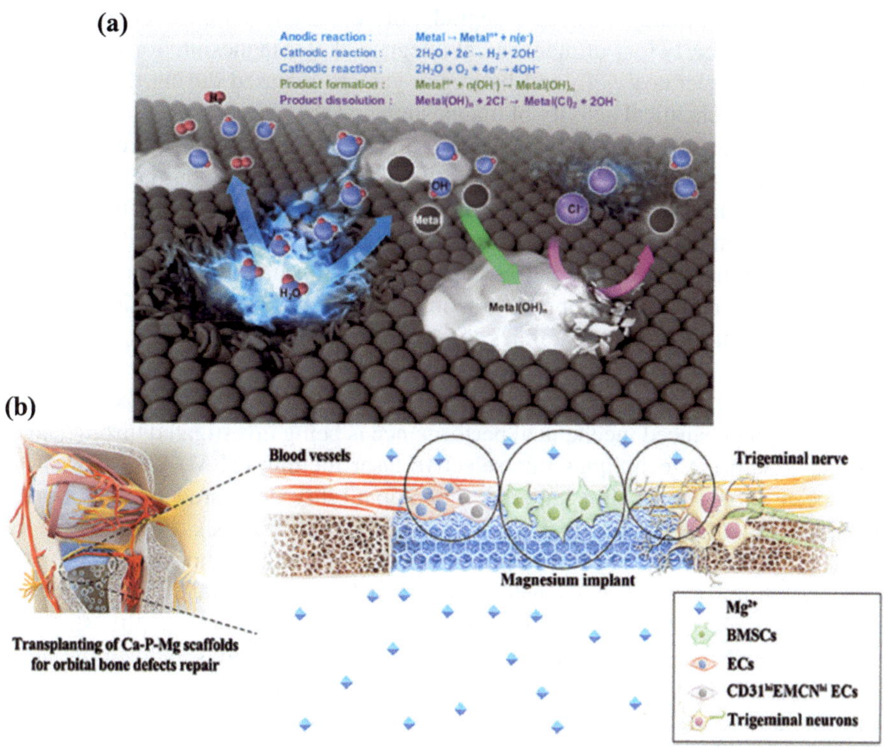

Fig. 1 **a** Corrosion mechanism of biometals [15]. Reused after permission from Elsevier. **b** Ca–P-coated Mg–Zn–Gd scaffolds are used to implant trigeminal neurones in an orbital bone defect model. Trpv1 triggers the release of the neuropeptide calcitonin gene-related peptide (CGRP), which promotes angiogenesis and osteogenesis to hasten the healing process of bone defects [16]. Reused after permission from ACS

2 Magnesium-Based Biodegradable Materials

Mg shows the important correlation between Young's modulus and the anisotropic characteristics of human cortical bone [17]. Moreover, Mg shows a significant osteopromotive effect and biocompatibility during degradation in physiological media [18]. Mg has been found to be extensively used in medical applications due to its increased availability. The initial reports of its use date back to 1878, when Edward C. H. successfully employed Mg wires as ligatures for blood vessels [19, 20]. Magnesium and its alloys have high biocompatibility density near bone (which lessens stress shielding), are a vital component for human metabolism, and have easy access. However, the hydrogen gas evolution during degradation may result in gas pockets surrounding the implant site, a rapid degradation rate in physiological environments, which causes an early loss of mechanical integrity [17]. Magnesium ions, which the body normally tolerates well, and hydrogen gas are released during this reaction. The high electrochemical activity of magnesium and the presence of chloride ions in the physiological environment, which speed up the corrosion process, are frequently blamed for the rapid corrosion. The most recent development is the alloying of magnesium with rare earth elements (e.g. calcium (Ca), strontium (Sr), zinc (Zn), yttrium, neodymium, and gadolinium), which have all been shown to be useful in slowing down the rate of degradation and enhancing mechanical behaviour [19, 21]. To prevent degradation and improve biocompatibility, surface modification methods like coating with polymers, ceramics, or bioactive agents are also frequently used [22].

The development of complex magnesium alloy implants with customized degradation rates and enhanced mechanical performance is being investigated through additive manufacturing techniques like selective laser melting (SLM) [23]. Mg alloys have been shown to exhibit superior biocompatibility compared to biopolymers such as poly L-lactic acid (PLLA), poly-ether ketone (PEEK), and polyglycolic acid (PGA). It also outshines bio-ceramics such as tricalcium phosphate (TCP) and hydroxyapatite (Hap), as these materials possess low mechanical strength for bone implants [24]. Mg alloys are easily sterilized due to their surface quality, stiffness, and plasticity. Moreover, the degraded by-products of Mg alloys and Mg implants, particularly Mg ions, are nontoxic and do not present any health risks. In fact, they demonstrate unique osteopromotive traits [25]. Degraded Mg ions, aided by transporter proteins, increase Osterix levels and serve to effectively promote the process of bone regeneration. Troitskii and Tsitrin documented cases in which fractures were effectively treated with the use of screws and plates containing an Mg–Cd alloy [26]. The formation of robust calluses during fracture repair was found to be promoted by these Mg–Cd implants. Nevertheless, in some instances, failures arose from the formation of gas cysts or infections. It is significant that none of the patients exhibited acute inflammatory reactions or abnormal serum magnesium level changes [21]. Mg-based studies have advanced (see Fig. 2), indicating that prior research concentrated on managing the degradation rate and corrosion of Mg alloys using ceramic coatings such as hydroxyapatite (HA) and titanium dioxide (TiO_2), as well as through the

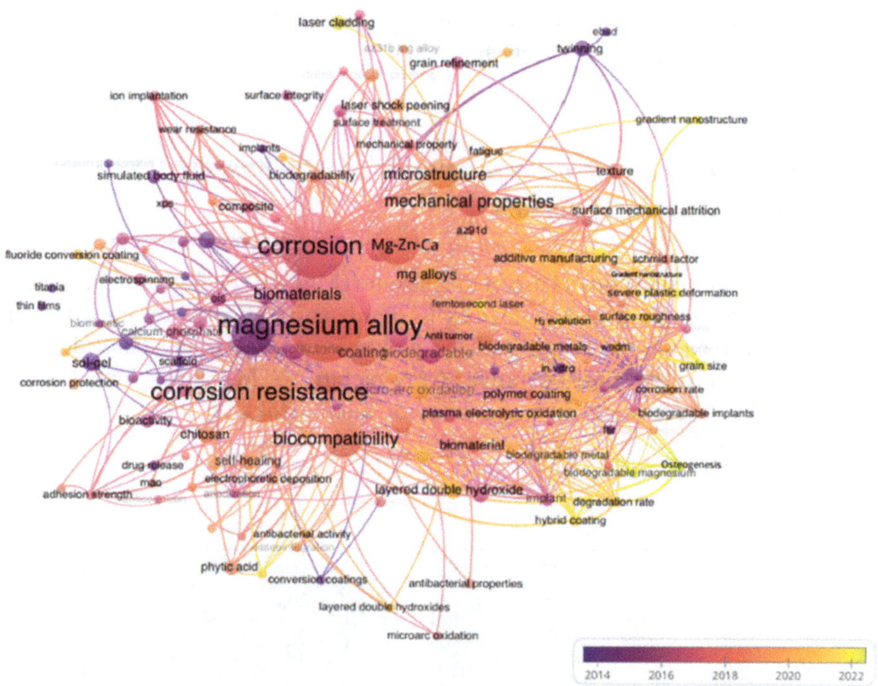

Fig. 2 A neural analysis of advancements in Mg studies to elicit biomedical applications [24]. Reused after permission from Elsevier

use of inert elements in alloying [27, 28]. Recent trends indicate that, in addition to corrosion resistance, the application of coatings is now being explored to provide various therapeutic responses. Besides chemical coatings, a number of thermal (such as laser cladding and laser shock peening) and mechanical methods (like SMAT and FSP) are becoming increasingly popular for customizing the grain size and texture of Mg alloys [28].

2.1 *Degradation Behaviour of Mg-Based Biodegradable Materials*

Since the dissolution of magnesium alloys must be regulated to guarantee implants continue to function for the necessary amount of time, this degradation process is essential in biomedical applications. Implant stability is impacted by the development of corrosion products such as $Mg(OH)_2$, $MgCO_2$, and $Ca_5(PO_4)_3(OH)$. Designing magnesium alloys for medical applications with enhanced corrosion resistance is made easier with an understanding of these phases [17]. For bioimplants to perform safely in vivo, corrosion resistance, cytocompatibility, and hemocompatibility are essential. Orthopaedic implants made of Mg and Mg-based materials exhibit a corrosion rate that exceeds biosafety limits [29].

2.2 In Vivo and in Vitro Performance of Mg-Based Biodegradable Materials

The main obstacle to the clinical use of magnesium (Mg) is its degradation behaviour. The in vivo performance of Mg devices is significantly affected by the chemical processes involved in corrosion, such as hydrogen evolution and chloride solubility. Figure 3 displays post-operative X-ray pictures of Mg alloys for 12 weeks and 24 weeks. There were no indications of instrumentation loosening or cage displacement [30].

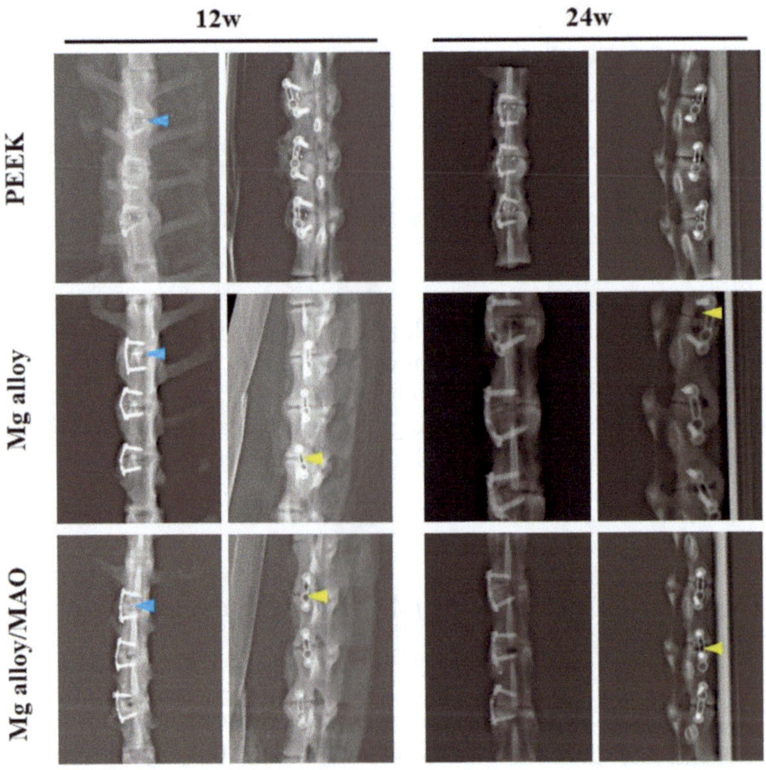

Fig. 3 Three groups of sheep's lumbar radiographs with implantations were taken 12 and 24 weeks following surgery. Yellow arrow, gas cavity; blue arrow, cage [30]. Reused under CC BY-NC-ND License of the journal

On the other hand, when Mg alloys degrade at a slower rate (under 0.5 mm/year), this promotes more effective bone healing. Kirkland et al. [31] stated that localized corrosion in most Mg alloys is caused by galvanic coupling, which creates micro-galvanic cells and accelerates the degradation process. Mg alloys comprise elements such as Cu, Ni, and Al, which are less electronegative than Mg. Consequently, they can cause damage to the protective layer $Mg(OH)_2$. Rods were prepared using Mg alloys (WE43, AZ31, LAE442, and AZ29) by Witte [32]. The alloy composition was the primary factor influencing the degradation. Of all the alloys, LAE442 exhibited the lowest corrosion rate, while WE43 Mg displayed Ca- and P-rich apatite mineralization after extended immersion [33]. In addition to alloying, various surface modification techniques have been developed to enhance the mechanical properties and corrosion resistance of Mg and Mg-based implants.

Gu et al. [34] performed an extensive investigation into the in vivo and in vitro behaviour of Mg–Sr alloy, showcasing its exceptional qualities such as the highest ultimate tensile strength (203 MPa), a finer microstructure (23 μm), enhanced osteogenic properties, and a reduced degradation rate (0.37 mm/year). Furthermore, it has been demonstrated that adding zinc (Zn) and calcium (Ca) can improve the biocompatibility of Mg alloys [35]. It is worth mentioning that the generation of calcium phosphate (Ca–P) during in vivo degradation aids in enhancing bone mineralization. Zn supplementation aids in gene expression, enhances nervous system function, supports protein synthesis, and shows anti-atherosclerosis effects. Schulz et al. carried out an exhaustive examination of the corrosion behaviour of pure Mg, AZ91D, and AZ31 across different in vivo and in vitro settings. They assessed biocompatibility and in vivo weight loss by implanting Mg/Mg alloys into mice [1, 36]. It is remarkable that they noted a reduced biocorrosion rate in the in vivo subcutaneous environment as opposed to the in vitro simulated conditions. Their study's histological findings suggested that Mg/Mg alloys are biocompatible with the heart, liver, skin, lung, and kidney tissues of mice.

3 Zinc-Based Biodegradable Materials

Zn-based materials have also been of interest due to their natural antibacterial activity and their critical role in cellular metabolism. However, the problems are substantial, as the mechanical strength is lower and there is less variability of degradation behaviour of PFs relative to the physiological demands of the bone [37, 38]. For biological applications, antibacterial qualities are crucial. Ag can efficiently destroy a variety of pathogens, including bacteria, yeast, viruses, and fungi, whereas zinc can prevent the growth of bacteria and the development of biofilm. In terms of silver's antibacterial capabilities, it can engage with the peptidoglycan layer of the cell wall, bacterial DNA, the plasma membrane, and various bacterial proteins, ultimately exerting a bactericidal action (Fig. 4a). As expected, Zn–Ag-based alloys exhibited broad-spectrum antibacterial activity against both Gram-negative bacteria, such as *Escherichia coli* (*E. coli*), and Gram-positive bacteria, including *Streptococcus goeth and Staphylococcus aureus (S. aureus)* [39]. Such useful properties of Zn-based materials make them appealing as they are biocompatible and mechanically compatible with the bone tissue.

Nevertheless, the underlying degradation mechanisms, especially in vivo, are poorly understood. Future work will be needed to maximize mechanical strength, refine corrosion resistance, and control and predict degradation for safe use in patients. Zinc (Zn) is one of the biodegradable metals, and it is well known for its high biocompatibility and physiological relevance, as it is the second most abundant trace element in the human body, most of which are found in bones and muscles [40]. In terms of mechanical properties, it is well matched to human bone and is more compatible with human bone than stainless steel and titanium. More than 600 enzymes are Zn^{2+} dependent in terms of Ze catalytic and structural integrity. Although Zn has biological significance, it has not been widely utilized as a primary bioresorbable medical material essentially due to the limited amount of literature, especially in the cardiovascular and orthopaedic fields. Zn alloys have been limited for biomedical use due to potential toxicity [41].

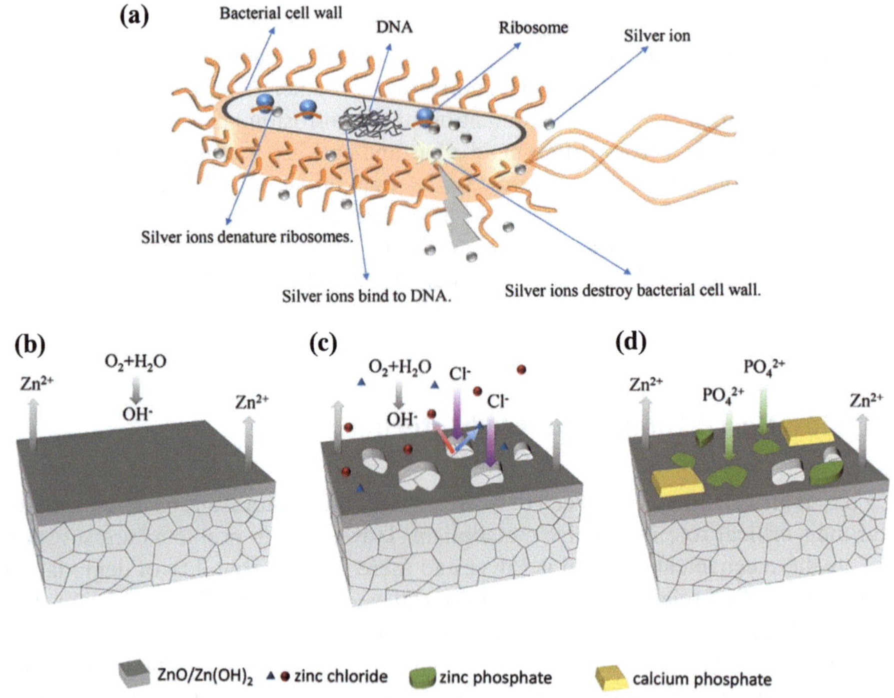

Fig. 4 **a** Mechanisms behind silver ions' antimicrobial action. Diagrammatic representation of Zn–Ag alloy deterioration: The first stage of degradation results in **b** $ZnO/Zn(OH)_2$ precipitation; **c** $ZnO/Zn(OH)_2$ is broken down by chloride ions to yield soluble zinc chloride salts; and **d** the presence of phosphate groups causes the formation of zinc phosphate and calcium phosphate precipitates [49]. Reused with permission from Sage

Zinc plays a crucial role in maintaining protein function and stability, participating in over 300 enzymatic processes and supporting the structural integrity of protein subdomains, such as DNA-binding regions within RNA polymerases and transcription factors [42]. At present, Zn alloys, especially Zn–Mg alloys, are under investigation as potential alternatives for bioresorbable metallic stents and orthopaedic implants due to their excellent tissue compatibility and improved mechanical characteristics. The corrosion of Zn is of anodic dissolution and cathodic oxygen reduction nature and is affected by the environmental pH. At a pH of 7.3, ZnO and $ZnCl_2$ form as part of the corrosion process [43]. Instead of magnesium, Zn corrodes more slowly without hydrogen gas evolution. In rat arteries, pure Zn wires used as stents for up to 6 months did not cause significant inflammation or thrombosis, and performed well with the tissue [43]. These cells should be studied further in vivo to evaluate the influence of surface oxide layers on biodegradation processes [36, 44].

3.1 Degradation Behaviour of Zn-Based Biodegradable Materials

A good biodegradable orthopaedic implant will provide enough mechanical strength within the bone healing process, and then it should break down as time goes on. Degradation of this nature should coincide with tissue regeneration to prevent the loss of support and implant failure when premature [45]. However, the host and implant-related factors influence the degradation behaviour [46–48]. For Zn-based materials, the major cause of degradation is electrochemical corrosion in the presence of body fluids as electrolytes. At the anodic Zn^{2+} ion production reaction, the metallic Zn is converted via anodic reaction, whereas, at the same time, during the cathodic oxygen reduction, the electrons are used, which increase local pH and accelerate corrosion [20]. Figure 4b–d illustrates the Zn–Ag alloy deterioration process. One notable benefit of Zn–Ag alloys over Mg-based alloys may be that no gas is produced during the breakdown process. ZnO and $Zn(OH)_2$ are the primary precipitates generated throughout the degradation process. Different immersion solutions result in different degradation products [49]. Surface cavities and pits may be formed, and the process may lead to this localized corrosion [50]. Zn ions released from the electrode interact with environmental ions to generate corrosion products such as phosphates, hydroxide, and carbonates, serving as a protective layer to temporarily reduce further degradation [51–53]. Furthermore, cathodic reactions may result in hydrogen gas by-product [54].

Local alkalinity is also increased by the degradation process, which also promotes the absorption of organic molecules and the formation of surface hydroxides. The formation of calcium phosphate layers also promotes the degradation dynamics [55, 56]. In vivo degradation is more complex than in vitro studies that take into account such factors as surrounding tissue interactions, variable pH, fluctuating ion concentrations, and protein adsorption [51, 53]. The body may be subject to different types of corrosion. The premature corrosion of the less noble material is due to galvanic corrosion when Zn alloys contact more noble metals. Among them, the localized electrochemical reactions, material heterogeneity, and mechanical wear also cause pitting, crevice, intergranular, and fretting corrosion [57–59]. Other causes of deterioration are corrosion fatigue, resulting from applications of cyclic mechanical loads in corrosive environments.

Bones regenerated and mineralized in the presence of Zn^{2+} ions [60]. OH^- ion release may create an alkaline environment that could enhance the activities of osteoblasts and support osteogenesis. Corrosion behaviour and biocompatibility of Zn alloy implants placed in bone depend on local factors, including the water content, vascularization, and cellular matrix interactions, and these all affect degradation [61, 62]. Degradation is usually determined in vitro by parameters such as weight loss, hydrogen evolution, pH change, and residual mechanical strength. Corrosion susceptibility is also estimated using electrochemical methods, such as open circuit potential and current density. The degradation and biocompatibility are evaluated in vivo in animal models. Challenges of the rapid degradation of Mg alloys are often associated with hydrogen gas evolution. Nevertheless, Zn–5Ge, Zn–0.8Mn-0.4Cu, Zn–1.5Mg-0.5Ca, Zn–3Cu–1Fe, and Zn–0.8Mn–0.4Ag were found to exhibit good corrosion resistance and can be promising for biomedical use [63–66].

Overall, it is important to understand and control the degradation behaviour of Zn- and Mg-based implants to optimize their performance. The degradation products of such materials can support bone healing, and the local pH changes can influence osteogenesis favourably, thus making these materials good candidates as next-generation orthopaedic devices.

3.2 *In Vivo and In Vitro Performance of Zn-Based Biodegradable Materials*

With numerous studies being conducted on zinc (Zn) as a promising material for bioresorbable metallic stents, its bioactivity, compatibility with cells, and degradation behaviour have been increasingly recognized. Zn is an essential trace element and is responsible for important biological processes [67]. Zn has been shown to have excellent hemocompatibility in vivo experiments and has not been noted to be antiatherogenic or to harm endothelial cells [23, 68]. In Hank's solution, its corrosion rate lies between that of Mn and Fe [32]. According to the research by Torné et al. [69], Zn implants in plasma and whole blood form passivating surface films of inorganic substances and biomolecules, rather than with PBS and Ringer's solution. Electromechanical studies confirmed further that the same protective layer forms in blood [69]. In addition, the high concentration of protein in the human serum (approximately 60–80 g/L protein) is also a factor that influences the corrosion behaviour of Zn implants, as proteins affect how Zn-based alloys degrade [70].

Cytotoxic effects on fibroblasts have been related to high concentrations of Zn ions released during degradation [71, 72]. In fact, pure Zn is reported to affect the MC3T3-E1 [68] and MG-63 cell [73] differentiation and growth negatively. Krus et al. [70], however, mention that using metals such as magnesium (e.g. Zn–3 Mg) alloying Zn can reduce genotoxic risks. In support, Kubasek et al. [74] have demonstrated that Zn–0.8Mg alloys are not toxic to U–2OS cells. Finally, Zn–0.05Mg and Zn–0.5Al–xMg alloys were also shown to increase cell proliferation. Degradation of Zn–Mg alloys also favours osseointegration with the formation of osseointegrated bone-like material such as hydroxyapatite. Ca and Sr have also been found to be elements that enhance the cytocompatibility of Zn-based implants [75].

Yang et al. [76] implanted Zn stents into the abdominal aortas of rabbits, which did not develop significant inflammation, thrombosis, or platelet buildup or intimal thickening of the stents and maintained structural integrity. Zhou et al. [77] had similarly discovered that Zn–0.8Cu alloy stents on the coronary artery for two years did not cause thrombosis or inflammation. Herelin et al. [78] used Zn–3Ag alloy stents to line the iliofemoral arteries of juvenile pigs, arguing that they were healed and free of vascular blockage or thrombosis in the other study. Implantation of Zn–xMg alloys into the rat abdominal aorta induced only slight inflammation and intimal activation, in contrast to what was observed by Jin et al. [79]. Zhao et al. [80] also found mild inflammatory responses on Zn–0.1Li alloy implants in the same model. Rat abdominal aortas were loaded with Zn–xAl alloy strips, and inflammation was observed at only a moderate level post-implantation [34].

3.3 Surface Modification of Zn-Based Biodegradable Materials

To enhance the mechanical performance of Zn-based biomaterials without perturbing the basic matrix tendency, efforts have been made in surface modification techniques [81]. These methods for improving surface properties are well suited for orthopaedic applications as they enhance the surface properties. But research in this area is still limited, and several strategies appear promising. For example, chrome-based coatings were once commonly utilized to gain corrosion resistance, wear resistance, and adhesion [82]. Despite their toxicity and carcinogenic risk, their use has been globally prohibited [83].

The Zn–P chemical conversion coating is a widely accepted method that greatly enhances the surface wear resistance, corrosion protection, and adhesion. Zinc is important for numerous biological processes, particularly in enzyme synthesis, and, among other things, catalyses gene regulation, apoptosis, DNA metabolism, signal transduction, and energy production. Corrosion resistance of the implant surface can be improved by coating with zinc-based coatings [84, 85]. Surface modification strategies have been applied to Zn–Cu stents that were balloon-crimped and subsequently experienced pressure balloon rupture following two years of coronary implantation [86]. In contrast, electrophoretic deposition has been employed as a coating method for biodegradable Zn alloy implants [87].

4 Iron-Based Biodegradable Materials

In non-load-bearing applications, biodegradable metallic biomaterials with modified mechanical and biodegradation characteristics may be the future perfect bone substitutes. The manufacturing, design, performance, and biodegradation behaviour of those promising metallic biomaterials can all be improved. Because of its specific surface and composition, iron may easily overcome magnesium and zinc. Compared to bone, iron is more rigid and degrades more slowly, but it is also highly biocompatible [88]. When a metallic sample encounters bodily fluids, electrochemical dissolution takes place, leading to iron deterioration through a corrosion mechanism. Several ways to get around some of iron's drawbacks will be examined in more detail. The abundance of recent articles in the Science Direct database is a clear indication of the significance of iron for biomedical applications (Fig. 5). These frequent articles further highlight how significant and real this topic is. It draws attention to the difficulty of creating intricate and useful architecture, such as the structure of bones, which provides a vital cellular microenvironment. In fact, the medical and societal demands of temporary medical implants drive research on iron, with specific requirements and innovative approaches for biomedical applications. Different strategies have been considered to solve these important challenges. Some of the most recent developments in the creation of biodegradable porous iron biomaterials for regenerative medicine will be highlighted here. In addition to the processes and biodegradation profile of porous iron biomaterials in various compositions.

Since iron-based biodegradable materials may have mechanical characteristics more akin to those of human bone [89], they are seen as potential substitutes for permanent metallic implants in non-load-bearing applications [90]. The outermost layer of the structure is cortical bone, also referred to as compact bone, which has a porosity of only 3–5%. On the other hand, cancellous or trabecular bone is a network of pores that resembles a sponge and has a porosity of 50–90%. It is packed with marrow and blood vessels [91]. The biodegradability of the components determines the suitability of a biodegradable iron alloy implant for medical use. The elements alloyed with iron should have the proper mechanical strength, elastic modulus, high cytocompatibility, zero or low toxicity, and accelerated rate of disintegration [88].

In general, alloying elements like Pd, Pt, W, C, S, Si, and Ga accelerate the pace at which iron-based alloys deteriorate [92]. The production of antiferromagnetic Fe alloys that are compatible with magnetic resonance imaging (MRI) is another goal of alloying elements. Through solid solution and second phase strengthening, it has been discovered that the addition of gold, silver, tungsten, platinum, and palladium improves the mechanical characteristics and speeds up the rate of iron alloy degradation [93]. Co, W, C, or S are the best elements to use when combining cytocompatibility, hemocompatibility, degradation rate, and mechanical qualities. When Ca and

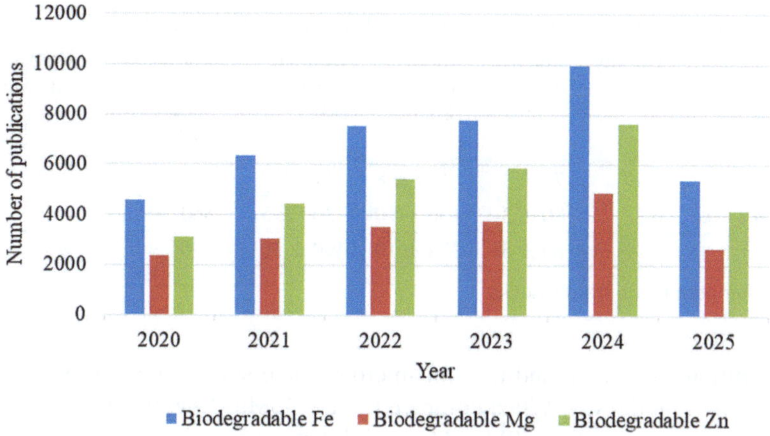

Fig. 5 The number of articles from 2020 to 2025 about biodegradable metals, iron, and porous iron, based on data from the Science Direct database (as of April 23, 2025). Growing interest in the subject of biodegradable iron biomaterials is indicated by a consistent rise in the number of publications

Mg are added to FeMnSi alloys, fine precipitates are formed, which accelerate the rate of corrosion [94]. Using a composite material, which is made possible by adding calcium phosphates, carbon nanotubes, or Fe_2O_3 to an Fe matrix, is another way to adjust the mechanical and degrading properties. Implants frequently alter the pH of bodily fluids in vivo, ranging from 7.4 to 4 (acidic) or 9 (alkaline), which results in electrochemical deviations from the human body's equilibrium state and increases the risk of pathogenic infections. Furthermore, it is well known that elements with significant potential differences, such as Ag (+ 0.7966 V) and Au (+ 1.83 V) compared to Fe (− 0.44 V), can accelerate iron corrosion. The anode (Fe) is corroded by the Ag or Au ions, which function as a cathode element and enhance the microgalvanic cells made of Ag/Fe or Au/Fe [95]. In addition, Ag is a biomaterial that has antiseptic and antibacterial qualities, making it a potential treatment for inflammatory and antibacterial bone disorders. Therefore, it may be essential to combine Ag and Fe for medicinal applications to greatly boost iron corrosion and decrease bacterial infection-related disorders [96].

The cytotoxicity, antibacterial activity, corrosion rate, and mechanical properties of nano-structured Fe–Mn–Ag alloys with varying concentrations of Ag were investigated by Sotoudehbagha et al. [97]. Given the significance of assessing iron corrosion, an electrochemical test was conducted at 25 °C with potentiodynamic polarization using a Hanks' Balanced Salt Solution (HBSS). Table 1 shows the mechanical characteristics and corrosion behaviour. Sotoudehbagha et al. [97] claim that the inclusion of Ag raises the alloys' density, ultimate shear stress, and microhardness. Additionally, when the amount of Ag in the austenite matrix increased, the scientists saw a more uniform dispersion of Ag.

Table 1 Mechanical characteristics and corrosion behaviour of Fe–Mn alloys [97]

Sample	Density (g/cm^3)	Relative density (%)	Micro hardness (HV)	Ultimate shear stress (MPa)	Corrosion current density (μA/cm^2)	Corrosion rate (mmpy)
Fe–30Mn	5.49	71 ± 3	119 ± 8	172 ± 7	800	2.61
Fe–30Mn–1Ag	6.2	80 ± 3	156 ± 10	360 ± 5	860	2.49
Fe–30Mn–3Ag	6.92	89 ± 5	174 ± 10	490 ± 10	890	2.31

Reused after permission from Elsevier

In addition, Ag was found to offer microgalvanic sites, which raise the alloys' corrosion current density. When compared to Fe–30Mn–1Ag, the iron alloy with a larger percentage of Ag (Fe–30Mn–3Ag) exhibited poorer cellular metabolic activity towards human umbilical vein endothelial cells (HUVEC), despite having a better antibacterial rate against *S. aureus* and *E. coli* bacteria (Fig. 6). These factors led to the Fe–30Mn–1Ag being regarded as the finest biodegradable antibacterial alloy [97].

In a different investigation, Mandal et al. [98] created a new Fe–Mn–Cu alloy using powder metallurgy and copper (Cu). Because copper has antibacterial qualities and is known to precipitate in the iron matrix beyond its solid solubility limit, improving mechanical properties and speeding up degradation via local galvanic cells, it was chosen for this work [98]. Even with the highest proportion of copper utilized in this investigation, only the γ-austenite phase (FCC) was found, and the addition of Cu did not significantly alter the size of the iron grains or cause any precipitation.

As anticipated, the percentage of copper added determines how quickly iron alloys corrode. Because copper and iron create a solid solution, the corrosion rate is

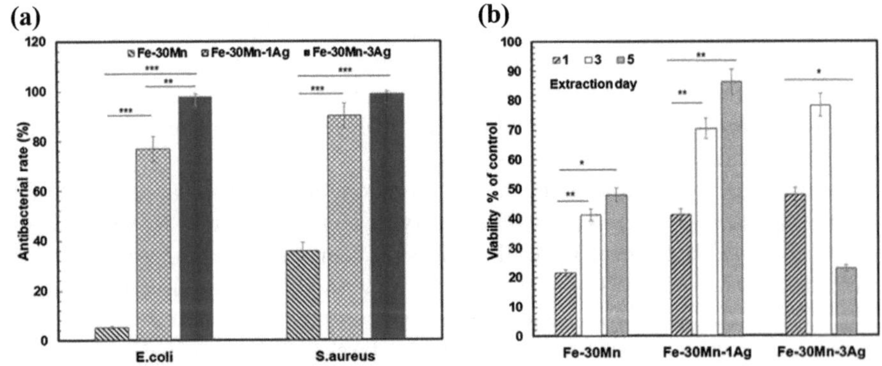

Fig. 6 **a** Antibacterial activity of the alloys against *S. aureus* and *E. coli*, and **b** HUVEC cell viability following a 24-h exposure to the alloys' extract after 1, 3, and 5 days of incubation (***, **, and * indicate $p < 0.001$, $p < 0.01$, and $p < 0.05$, respectively) [97]. Reused after permission from Elsevier

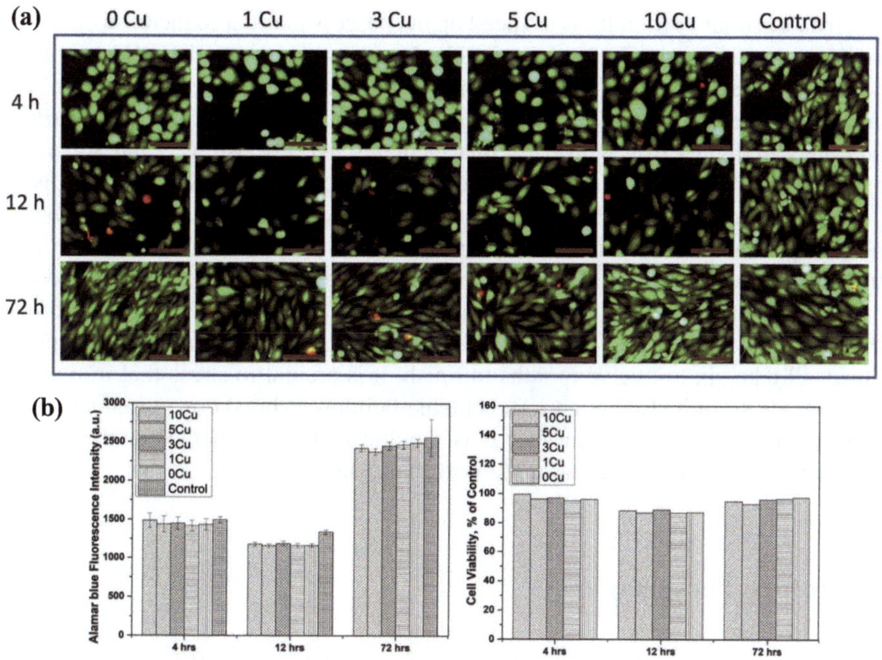

Fig. 7 Fe–Mn–Cu alloy: **a** live/dead imaging and **b** MG63 cell Alamar blue assay at 4, 12, and 72 h [98]. Reused after permission from Elsevier

decreased up to 3 weight percent of copper in the iron matrix [98]. Copper precipitates in the Fe matrix when its concentration is above the limit of its solubility in iron. This phenomenon causes the creation of local microgalvanic cells, which accelerate the pace of alloy corrosion [98]. To demonstrate the bactericidal action of the copper in the alloys, antimicrobial tests were conducted with *E. coli* bacteria present. It is well known that iron is a necessary cofactor for the metabolic pathway of pathogenic microorganisms, and that adding copper to the iron matrix might decrease bacterial adherence [98]. Furthermore, the cytocompatibility of Fe–Mn–Cu alloys was assessed in osteoblastic MG63 cells. As shown in Fig. 7 [98], the addition of copper to Fe–Mn alloys did not significantly alter cell viability, and the results showed no cytotoxicity of the alloys examined.

4.1 *Corrosion Behaviour and Biocompatibility of Fe-Based Materials*

Cytotoxicity results from the body's ion content levels rising due to the breakdown of metallic implants. Controlling the rate of breakdown and ensuring that absorption happens at the same pace as tissue regeneration are necessary to avoid cytotoxicity

[2]. Iron's primary difficulty is to speed up the process in order to increase the rate of deterioration. Finer microstructures and high-energy grain boundaries in iron alloys contribute to an increased rate of corrosion [99]. Additional techniques have been employed to regulate the pace of degradation, such as the creation of porous scaffolds and the alloying of iron with particular elements like Pd, Mn, Ca, and Ag, which enhance mechanical qualities and encourage iron corrosion [99]. Diffusion is the primary mechanism by which the porous iron scaffold in SBF degrades [100]. The harsh aqueous environment found in bodily fluids causes metals to corrode electrochemically. By striking a balance between the pace at which corrosion products are released and the body's capacity to absorb and eliminate them, one can reduce the cytotoxicity caused by metals [101]. It is anticipated that the remodelling procedure will take six to twelve months to finish. It is crucial to stress that iron should not degrade so quickly that an unacceptable buildup of breakdown products forms surrounding the implantation site. It is thought to be reasonable for the stent to disintegrate completely 12–24 months after implantation [35].

5 Future Perspective

The present review provides a thorough description of the traits, attributes, and actions of biomaterials. The fabrication and medical usage of materials like magnesium, zinc, and iron have been the main topics of research. Although there is a lot of potential for biomaterials in biomedical applications, medical research lately has concentrated on finding novel biodegradable metals that work better as biomedical implants. The complexities and restrictions exhibited by current fabrication techniques must be addressed. When choosing biometals, corrosion behaviour, nontoxicity, and biocompatibility are essential criteria. Nonetheless, it is a reality that the development and application of biomaterials have improved and have the potential to greatly enhance life quality. Additionally, tissue engineering can restore normal function to individuals who have organ failure or tissue damage, possibly removing the need for transplantation. Drug delivery systems can be used to precisely regulate the body's drug release, enhancing the safety and effectiveness of therapies employed in the creation of monitoring and diagnostic instruments. Currently, biomaterials are utilized in cosmetic and aesthetic medicine for procedures such as dermal fillers, facial implants, and skin rejuvenation. These procedures can help enhance the appearance and boost the self-confidence of individuals seeking aesthetic improvement. These materials also fabricate rehabilitation and assistive devices such as prosthetics, orthotics, and hearing aids. The clinically qualified bio implants that are now on the market (Co–Cr or Ti alloys) require a second surgery and are permanent, adding to the financial and physical strain on patients and decreasing the effectiveness of the entire healthcare system. Even though Co–Cr and Ti alloys are thought to be biocompatible, certain patients may still have negative reactions to them, like hypersensitivity or allergic reactions.

Future research directions are as follows:

1. **Design of Advanced Alloy**: Given the intrinsic constraints of individual metals in fulfilling a variety of therapeutic criteria, advanced alloy design should be included in the research horizon. The search for new alloys that include several key components could improve material qualities for particular medical procedures like cardiovascular and orthopaedic treatments.
2. **Advancement in Fabrication**: There should be more research done on the revolutionary potential of additive manufacturing methods, especially 3D printing. To enable greater customization, design adaptability, and the realization of novel material features, researchers should explore more complex uses of these approaches. These developments have the potential to completely transform applications in healthcare.
3. **Nanotechnology Revolutions**: Nanotechnology has the potential to produce new nanomaterials with special qualities. To expand surface area, improve medication delivery, and attain higher mechanical strength, research should concentrate on using nanoscale manipulation. These nanomaterials have the potential to significantly advance implant development, therapy, and diagnostics.
4. The careful creation of bioactive biomaterials that can actively interface with tissue and cellular components should be the main goal of future research projects. This calls for exact design techniques meant to increase the bioactivity of materials, which will greatly accelerate tissue regeneration and repair processes by eliciting certain biological reactions.
5. **Induction of AI and Machine Learning**: In biomaterials research, it is critical to strategically apply AI and machine learning methods. Large-scale database screening, predictive modelling of material properties, and the logical design of biomaterials with improved functions are all made possible by these sophisticated computational techniques.

6 Conclusion

- Due to its high strength, osteopromotive properties, biocompatibility, and biodegradability, magnesium (Mg) is a promising biomaterial for orthopaedic applications. Its swift degradation, however, can jeopardize mechanical integrity prior to the completion of tissue healing. Zinc (Zn) is also characterized by its high biocompatibility and its abundance in the human body. In vivo studies have shown that Zn exhibits good hemocompatibility, antiatherogenic properties, and is non-cytotoxic to endothelial cells. Nonetheless, the possible toxicity of certain Zn alloys restricts their use in biomedicine.
- Titanium is a biomaterial recognized for its biocompatibility, corrosion resistance, favourable strength-to-weight ratio, osseointegration potential, versatility, compatibility with imaging techniques, and long-term stability.

- In medical and dental applications, cobalt–chromium (Co–Cr) alloys are prized for their biocompatibility, mechanical properties, corrosion resistance, design versatility, radiopacity, and affordability.
- Materials such as molybdenum (Mo) exhibit distinctive characteristics, including robustness, resistance to corrosion, compatibility with biological systems, bioactivity, conductivity in both electrical and thermal contexts, radiopacity, and adaptability. However, when molybdenum is ingested or absorbed in excessive quantities, it can pose a risk of toxicity.
- Researchers are investigating methods to produce high-performance metals, with alloying proving to be an effective means of improving metal properties. Mo demonstrates potential as a future biomaterial; for clinical application, though, additional studies and tests are needed, as well as endorsement from regulators.
- To summarize, the field of biomaterials research is progressing due to interdisciplinary partnerships involving scientists, engineers, clinicians, and stakeholders. The goal of these endeavours is to create pioneering materials that can tackle intricate issues in sustainability, engineering, and healthcare.

References

1. Das P, Pathak DK, Sharma P, Pandey PM (2024) A review on the mechanical and biocorrosion behaviour of iron and zinc-based biodegradable materials fabricated using powder metallurgy routes. Corros Rev. https://doi.org/10.1515/CORRREV-2023-0169/ASSET/GRAPHIC/J_CORRREV-2023-0169_CV_004.JPG
2. Zheng YF, Gu XN, Witte F (2014) Biodegradable metals. Mater Sci Eng R Rep 77:1–34. https://doi.org/10.1016/j.mser.2014.01.001
3. Li H, Zheng Y, Qin L (2014) Progress of biodegradable metals. Prog Nat Sci Mater Int 24:414–422. https://doi.org/10.1016/j.pnsc.2014.08.014
4. Amukarimi S, Mozafari M (2022) Biodegradable magnesium biomaterials—road to the clinic. https://doi.org/10.3390/bioengineering9030107
5. Xing F, Li S, Yin D, Xie J, Rommens PM, Xiang Z, Liu M, Ritz U (2022). Recent progress in Mg-based alloys as a novel bioabsorbable biomaterials for orthopedic applications. https://doi.org/10.1016/j.jma.2022.02.013
6. Miao H, Huang H, Shi Y, Zhang H, Pei J, Yuan G (2017) Effects of solution treatment before extrusion on the microstructure, mechanical properties and corrosion of Mg-Zn-Gd alloy in vitro. Corros Sci 122. https://doi.org/10.1016/j.corsci.2017.01.001
7. Sharma P, Pandey PM (2019) Rapid manufacturing of biodegradable pure iron scaffold using amalgamation of three-dimensional printing and pressureless microwave sintering. Proc Inst Mech Eng C J Mech Eng Sci 233:1876–1895. https://doi.org/10.1177/0954406218778304
8. Pathak DK, Pandey PM (2020) An experimental investigation of the fabrication of biodegradable zinc–hydroxyapatite composite material using microwave sintering. Proc Inst Mech Eng C J Mech Eng Sci 234:2863–2880. https://doi.org/10.1177/0954406220910445
9. Pathak DK, Pandey PM (2022) Experimental investigation and empirical modeling for corrosion rate prediction of biodegradable zinc based composite considering the effect of constituent materials. Proc Inst Mech Eng C J Mech Eng Sci 236:4465–4476. https://doi.org/10.1177/09544062211055658
10. Mishra DK, Pandey PM (2021) Effects of morphological characteristics on the mechanical behavior of 3D printed ordered pore topological Fe scaffold. Mater Sci Eng A 804. https://doi.org/10.1016/j.msea.2021.140759

11. Singh J, Pandey PM, Kaur T, Singh N (2021) A comparative analysis of solvent cast 3D printed carbonyl iron powder reinforced polycaprolactone polymeric stents for intravascular applications. J Biomed Mater Res B Appl Biomater 109:1344–1359. https://doi.org/10.1002/JBM.B.34795
12. Kumar G, Preetam S, Pandey A, Birbilis N, Al-Saadi S, Pasbakhsh P, Zheludkevich M, Balan P (2025) Advances in magnesium-based bioresorbable cardiovascular stents: Surface engineering and clinical prospects. J Magnes Alloys 13:948–981. https://doi.org/10.1016/J.JMA.2025.01.025
13. Haude M, Erbel R, Erne P, Verheye S, Degen H, Böse D, Vermeersch P, Wijnbergen I, Weissman N, Prati F, Waksman R, Koolen J (2013) Safety and performance of the drug-eluting absorbable metal scaffold (DREAMS) in patients with de-novo coronary lesions: 12 month results of the prospective, multicentre, first-in-man BIOSOLVE-I trial. Lancet 381. https://doi.org/10.1016/S0140-6736(12)61765-6
14. Haude M, Erbel R, Erne P, Verheye S, Degen H, Böse D, Vermeersch P, Wijnbergen I, Weissman N, Prati F, Waksman R, Koolen J (2013) Safety and performance of the drug-eluting absorbable metal scaffold (DREAMS) in patients with de-novo coronary lesions: 12 month results of the prospective, multicentre, first-in-man BIOSOLVE-I trial. Lancet 381:836–844. https://doi.org/10.1016/S0140-6736(12)61765-6
15. Han HS, Loffredo S, Jun I, Edwards J, Kim YC, Seok HK, Witte F, Mantovani D, Glyn-Jones S (2019) Current status and outlook on the clinical translation of biodegradable metals. Mater Today 23:57–71. https://doi.org/10.1016/J.MATTOD.2018.05.018
16. Zhang D, Ni N, Su Y, Miao H, Tang Z, Ji Y, Wang Y, Gao H, Ju Y, Sun N, Sun H, Yuan G, Zhou H, Huang H, Gu P, Fan X (2020) Targeting local osteogenic and ancillary cells by mechanobiologically optimized magnesium scaffolds for orbital bone reconstruction in canines. ACS Appl Mater Interfaces 12:27889–27904. https://doi.org/10.1021/ACSAMI.0C00553/SUPPL_FILE/AM0C00553_SI_002.RAR
17. Kumar A, Pandey PM (2020) Development of Mg based biomaterial with improved mechanical and degradation properties using powder metallurgy. J Magnes Alloys 8:883–898. https://doi.org/10.1016/J.JMA.2020.02.011
18. Virtanen S (2011) Biodegradable Mg and Mg alloys: corrosion and biocompatibility. Mater Sci Eng, B 176:1600–1608. https://doi.org/10.1016/J.MSEB.2011.05.028
19. Akbarzadeh FZ, Sarraf M, Ghomi ER, Kumar VV, Salehi M, Ramakrishna S, Bae S (2024). A state-of-the-art review on recent advances in the fabrication and characteristics of magnesium-based alloys in biomedical applications. https://doi.org/10.1016/j.jma.2024.06.015
20. Agarwal S, Curtin J, Duffy B, Jaiswal S (2016) Biodegradable magnesium alloys for orthopaedic applications: a review on corrosion, biocompatibility and surface modifications. Mater Sci Eng, C 68:948–963. https://doi.org/10.1016/J.MSEC.2016.06.020
21. Tsakiris V, Tardei C, Clicinschi FM (2021) Biodegradable Mg alloys for orthopedic implants—a review. J Magnes Alloys 9:1884–1905. https://doi.org/10.1016/J.JMA.2021.06.024
22. Guo AXY, Cheng L, Zhan S, Zhang S, Xiong W, Wang Z, Wang G, Cao SC (2022) Biomedical applications of the powder-based 3D printed titanium alloys: a review. J Mater Sci Technol 125:252–264. https://doi.org/10.1016/J.JMST.2021.11.084
23. Venezuela J, Dargusch MS (2019) The influence of alloying and fabrication techniques on the mechanical properties, biodegradability and biocompatibility of zinc: a comprehensive review. Acta Biomater 87:1–40. https://doi.org/10.1016/J.ACTBIO.2019.01.035
24. Khan AR, Grewal NS, Zhou C, Yuan K, Zhang HJ, Jun Z (2023) Recent advances in biodegradable metals for implant applications: exploring in vivo and in vitro responses. Results Eng 20:101526. https://doi.org/10.1016/J.RINENG.2023.101526
25. Yan T, Tan L, Xiong D, Liu X, Zhang B, Yang K (2010) Fluoride treatment and in vitro corrosion behavior of an AZ31B magnesium alloy. Mater Sci Eng C 30:740–748. https://doi.org/10.1016/J.MSEC.2010.03.007
26. Zhi P, Liu L, Chang J, Liu C, Zhang Q, Zhou J, Liu Z, Fan Y (2022) Advances in the study of magnesium alloys and their use in bone implant material. Metals 12:1500. https://doi.org/10.3390/MET12091500

27. Mahmud S, Rahman M, Kamruzzaman M, Khatun H, Ali MO, Haque MM (2023) Recent developments in hydroxyapatite coating on magnesium alloys for clinical applications. Results Eng 17:101002. https://doi.org/10.1016/J.RINENG.2023.101002
28. Singh N, Batra U, Kumar K, Mahapatro A (2021) Investigating TiO_2–HA–PCL hybrid coating as an efficient corrosion resistant barrier of ZM21 Mg alloy. J Magnes Alloys 9:627–646. https://doi.org/10.1016/J.JMA.2020.08.003
29. Xu L, Liu X, Sun K, Fu R, Wang G (2022). Corrosion behavior in magnesium-based alloys for biomedical applications. https://doi.org/10.3390/ma15072613
30. Chi P, Yu W, Wu B, Gao M, Song K, Mao K, Li B, Liu X, Liu H, Zhang C, Liu J, Yu Y, Chen Q, Zhao R, Yu N, Mao K, Yang K, Tan L, Wang Y, Wang Z (2024) A novel Mg-Zn-Nd-Zr alloy lumbar interbody fusion cage: An in vitro and in vivo study. J Magnes Alloys. https://doi.org/10.1016/J.JMA.2024.06.017
31. Kirkland NT, Lespagnol J, Birbilis N, Staiger MP (2010) A survey of bio-corrosion rates of magnesium alloys. Corros Sci 52:287–291. https://doi.org/10.1016/J.CORSCI.2009.09.033
32. Witte F (2010) The history of biodegradable magnesium implants: a review. Acta Biomater 6:1680–1692. https://doi.org/10.1016/J.ACTBIO.2010.02.028
33. Witte F, Fischer J, Nellesen J, Vogt C, Vogt J, Donath T, Beckmann F (2010) In vivo corrosion and corrosion protection of magnesium alloy LAE442. Acta Biomater 6:1792–1799. https://doi.org/10.1016/J.ACTBIO.2009.10.012
34. Drelich AJ, Bowen PK, LaLonde L, Goldman J, Drelich JW (2016) Importance of oxide film in endovascular biodegradable zinc stents. Surf Innov 4:133–140. https://doi.org/10.1680/JSUIN.16.00014
35. Čapek J, Vojtěch D, Oborná A (2015) Microstructural and mechanical properties of biodegradable iron foam prepared by powder metallurgy. Mater Des 83:468–482. https://doi.org/10.1016/J.MATDES.2015.06.022
36. Heiden M, Walker E, Stanciu L (2015) Magnesium, iron and zinc alloys, the trifecta of bioresorbable orthopaedic and vascular implantation—a review. J Biotechnol Biomater 5:1–9 (2015). https://doi.org/10.4172/2155-952X.1000178
37. Di T, Xu Y, Liu D, Sun X (2022) Microstructure, mechanical performance and anti-bacterial activity of degradable Zn-Cu-Ag alloy. Metals 12:1444. https://doi.org/10.3390/MET12091444
38. Wątroba M, Bednarczyk W, Szewczyk PK, Kawałko J, Mech K, Grünewald A, Unalan I, Taccardi N, Boelter G, Banzhaf M, Hain C, Bała P, Boccaccini AR (2023) In vitro cytocompatibility and antibacterial studies on biodegradable Zn alloys supplemented by a critical assessment of direct contact cytotoxicity assay. J Biomed Mater Res B Appl Biomater 111:241–260. https://doi.org/10.1002/JBM.B.35147
39. Li, P., Schille, C., Schweizer, E., Kimmerle-Müller, E., Rupp, F., Han, X., Heiss, A., Richter, A., Legner, C., Klotz, U.E., Geis-Gerstorfer, J., Scheideler, L.: Evaluation of a Zn–2Ag–1.8Au–0.2V Alloy for Absorbable Biocompatible Materials. Materials 2020, Vol. 13, Page 56. 13, 56 (2019). https://doi.org/10.3390/MA13010056.
40. Hernández-Escobar D, Champagne S, Yilmazer H, Dikici B, Boehlert CJ, Hermawan H (2019) Current status and perspectives of zinc-based absorbable alloys for biomedical applications. Acta Biomater 97:1–22. https://doi.org/10.1016/J.ACTBIO.2019.07.034
41. Yang H, Qu X, Lin W, Wang C, Zhu D, Dai K, Zheng Y (2018) In vitro and in vivo studies on zinc-hydroxyapatite composites as novel biodegradable metal matrix composite for orthopedic applications. Acta Biomater 71:200–214. https://doi.org/10.1016/J.ACTBIO.2018.03.007
42. Coleman JE (1992) Zinc proteins: enzymes, storage proteins, transcription factors, and replication proteins. Annu Rev Biochem 61:897–946. https://doi.org/10.1146/ANNUREV.BI.61.070192.004341/CITE/REFWORKS
43. Bowen PK, Shearier ER, Zhao S, Guillory RJ, Zhao F, Goldman J, Drelich JW (2016) Biodegradable metals for cardiovascular stents: from clinical concerns to recent Zn-alloys. Adv Healthc Mater 5:1121–1140. https://doi.org/10.1002/ADHM.201501019
44. Jakab K, Norotte C, Marga F, Murphy K, Vunjak-Novakovic G, Forgacs G (2010) Tissue engineering by self-assembly and bio-printing of living cells. Biofabrication 2:022001. https://doi.org/10.1088/1758-5082/2/2/022001

45. Koo Y, Jang Y, Yun Y (2017) A study of long-term static load on degradation and mechanical integrity of Mg alloys-based biodegradable metals. Mater Sci Eng, B 219:45–54. https://doi.org/10.1016/J.MSEB.2017.02.009
46. Li T, He Y, Wu J, Zhou J, Tang S, Yang Y, Wang X (2018) Effects of scandium addition on the in vitro degradation behavior of biodegradable Mg–1.5Zn–0.6Zr alloy. J Mater Sci 53:14075–14086. https://doi.org/10.1007/S10853-018-2626-4/METRICS
47. Sikora-Jasinska M, Mostaed E, Mostaed A, Beanland R, Mantovani D, Vedani M (2017) Fabrication, mechanical properties and in vitro degradation behavior of newly developed ZnAg alloys for degradable implant applications. Mater Sci Eng C 77:1170–1181. https://doi.org/10.1016/J.MSEC.2017.04.023
48. Hermawan H (2018) Updates on the research and development of absorbable metals for biomedical applications. Prog Biomater 7:93–110. https://doi.org/10.1007/S40204-018-0091-4/FIGURES/1
49. Xiao X, Liu E, Shao J, Ge S (2021) Advances on biodegradable zinc-silver-based alloys for biomedical applications. J Appl Biomater Funct Mater 19. https://doi.org/10.1177/22808000211062407/ASSET/39817D27-1905-456C-94A0-5038B16B234A/ASSETS/IMAGES/LARGE/10.1177_22808000211062407-FIG5.JPG
50. Gao Y, Wang L, Li L, Gu X, Zhang K, Xia J, Fan Y (2019) Effect of stress on corrosion of high-purity magnesium in vitro and in vivo. Acta Biomater 83:477–486. https://doi.org/10.1016/J.ACTBIO.2018.11.019
51. Radwan-Praglowska J, Janus L, Szajna E, Galek T, Sierakowska A, Piatkowski M, Tupaj M, Radomski P, Michalec M, Bogdal D (2022) Biodegradable Mg-based implants obtained via anodic oxidation applicable in dentistry: preparation and characterization. J Market Res 20:1736–1754. https://doi.org/10.1016/J.JMRT.2022.07.064
52. Hagihara K, Fujii K, Matsugaki A, Nakano T (2013) Possibility of Mg- and Ca-based intermetallic compounds as new biodegradable implant materials. Mater Sci Eng, C 33:4101–4111. https://doi.org/10.1016/J.MSEC.2013.05.055
53. Xin Y, Huo K, Hu T, Tang G, Chu PK (2009) Corrosion products on biomedical magnesium alloy soaked in simulated body fluids. J Mater Res 24:2711–2719. https://doi.org/10.1557/JMR.2009.0323/METRICS
54. Zhang J, Hiromoto S, Yamazaki T, Huang H, Jia G, Li H, Yuan G (2017) Macrophage phagocytosis of biomedical Mg alloy degradation products prepared by electrochemical method. Mater Sci Eng, C 75:1178–1183. https://doi.org/10.1016/J.MSEC.2017.02.126
55. Kabir H, Munir K, Wen C, Li Y (2021) Recent research and progress of biodegradable zinc alloys and composites for biomedical applications: biomechanical and biocorrosion perspectives. Bioact Mater 6:836–879. https://doi.org/10.1016/J.BIOACTMAT.2020.09.013
56. Shi ZZ, Gao XX, Zhang HJ, Liu XF, Li HY, Zhou C, Yin YX, Wang LN (2020) Design biodegradable Zn alloys: second phases and their significant influences on alloy properties. Bioact Mater 5:210–218. https://doi.org/10.1016/J.BIOACTMAT.2020.02.010
57. Patel M, Sangral S, Murugesan J, Mutoh Y (2023) Effect of friction stir processing on plain fatigue and fretting fatigue behaviour of the AA6063 alloys. Tribol Int 186:108642. https://doi.org/10.1016/J.TRIBOINT.2023.108642
58. Wang W, Wu H, Sun Y, Yan J, Zhang L, Zhang S, Ni J, Song Y, Zhang X (2020) Local intragranular misorientation accelerates corrosion in biodegradable Mg. Acta Biomater 101:575–585. https://doi.org/10.1016/J.ACTBIO.2019.10.036
59. van Gaalen K, Quinn C, Benn F, McHugh PE, Kopp A, Vaughan TJ (2023) Linking the effect of localised pitting corrosion with mechanical integrity of a rare earth magnesium alloy for implant use. Bioact Mater 21:32–43. https://doi.org/10.1016/J.BIOACTMAT.2022.08.004
60. Marek R, Eichler J, Schwarze UY, Fischerauer S, Suljevic O, Berger L, Löffler JF, Uggowitzer PJ, Weinberg AM (2023) Long-term in vivo degradation of Mg–Zn–Ca elastic stable intramedullary nails and their influence on the physis of juvenile sheep. Biomaterials Advances. 150:213417. https://doi.org/10.1016/J.BIOADV.2023.213417
61. Tang H, Li Q, Li M, Gu X, Cheng C, Fan Y (2023) In vitro and in vivo evaluation of micro-alloyed magnesium for potential application in alveolar bone fixation screws. J Mater Sci Technol 144:62–69. https://doi.org/10.1016/J.JMST.2022.09.060

62. Ma D, Wang J, Zheng M, Zhang Y, Huang J, Li W, Ding Y, Zhang Y, Zhu S, Wang L, Wu X, Guan S (2023) Degradation behavior of ZE21C magnesium alloy suture anchors and their effect on ligament-bone junction repair. Bioact Mater 26:128–141. https://doi.org/10.1016/J.BIOACTMAT.2023.02.021
63. Yue R, Huang H, Ke G, Zhang H, Pei J, Xue G, Yuan G (2017) Microstructure, mechanical properties and in vitro degradation behavior of novel Zn-Cu-Fe alloys. Mater Charact 134:114–122. https://doi.org/10.1016/J.MATCHAR.2017.10.015
64. Tong X, Zhang D, Zhang X, Su Y, Shi Z, Wang K, Lin J, Li Y, Lin J, Wen C (2018) Microstructure, mechanical properties, biocompatibility, and in vitro corrosion and degradation behavior of a new Zn–5Ge alloy for biodegradable implant materials. Acta Biomater 82:197–204. https://doi.org/10.1016/J.ACTBIO.2018.10.015
65. Shi ZZ, Yu J, Liu XF, Zhang HJ, Zhang DW, Yin YX, Wang LN (2019) Effects of Ag, Cu or Ca addition on microstructure and comprehensive properties of biodegradable Zn-0.8Mn alloy. Mater Sci Eng C 99:969–978. https://doi.org/10.1016/J.MSEC.2019.02.044
66. Yang L, Guo P, Niu Z, Li F, Song Z, Xu C, Liu H, Sun W, Ren T (2019) Influence of Mg on the mechanical properties and degradation performance of as-extruded ZnMgCa alloys: In vitro and in vivo behavior. J Mech Behav Biomed Mater 95:220–231. https://doi.org/10.1016/J.JMBBM.2019.04.029
67. Li P, Zhang W, Dai J, Xepapadeas AB, Schweizer E, Alexander D, Scheideler L, Zhou C, Zhang H, Wan G, Geis-Gerstorfer J (2019) Investigation of zinc-copper alloys as potential materials for craniomaxillofacial osteosynthesis implants. Mater Sci Eng, C 103:109826. https://doi.org/10.1016/J.MSEC.2019.109826
68. Su Y, Wang K, Gao J, Yang Y, Qin YX, Zheng Y, Zhu D (2019) Enhanced cytocompatibility and antibacterial property of zinc phosphate coating on biodegradable zinc materials. Acta Biomater 98:174–185. https://doi.org/10.1016/J.ACTBIO.2019.03.055
69. Törne K, Larsson M, Norlin A, Weissenrieder J (2016) Degradation of zinc in saline solutions, plasma, and whole blood. J Biomed Mater Res B Appl Biomater 104:1141–1151. https://doi.org/10.1002/JBM.B.33458
70. Kraus T, Moszner F, Fischerauer S, Fiedler M, Martinelli E, Eichler J, Witte F, Willbold E, Schinhammer M, Meischel M, Uggowitzer PJ, Löffler JF, Weinberg A (2014) Biodegradable Fe-based alloys for use in osteosynthesis: outcome of an in vivo study after 52 weeks. Acta Biomater 10:3346–3353. https://doi.org/10.1016/J.ACTBIO.2014.04.007
71. Sam S, Joseph B, Thomas S (2023) Exploring the antimicrobial features of biomaterials for biomedical applications. Resul Eng 17:100979. https://doi.org/10.1016/J.RINENG.2023.100979
72. Murugan D, Rangasamy L (2022) The use of antimicrobial biomaterials as a savior from post-operative vascular graft-related infections: a review. Results Eng 16:100662. https://doi.org/10.1016/J.RINENG.2022.100662
73. Chen CZ, Shi XH, Zhang PC, Bai B, Leng YX, Huang N (2008) The microstructure and properties of commercial pure iron modified by plasma nitriding. Solid State Ion 179:971–974. https://doi.org/10.1016/J.SSI.2008.03.019
74. Kubásek J, Vojtěch D, Jablonská E, Pospíšilová I, Lipov J, Ruml T (2016) Structure, mechanical characteristics and in vitro degradation, cytotoxicity, genotoxicity and mutagenicity of novel biodegradable Zn–Mg alloys. Mater Sci Eng, C 58:24–35. https://doi.org/10.1016/J.MSEC.2015.08.015
75. Jablonská E, Vojtěch D, Fousová M, Kubásek J, Lipov J, Fojt J, Ruml T (2016) Influence of surface pre-treatment on the cytocompatibility of a novel biodegradable ZnMg alloy. Mater Sci Eng, C 68:198–204. https://doi.org/10.1016/J.MSEC.2016.05.114
76. He J, Li DW, He FL, Liu YY, Liu YL, Zhang CY, Ren F, Ye YJ, Deng XD, Yin DC (2020) A study of degradation behaviour and biocompatibility of Zn—Fe alloy prepared by electrodeposition. Mater Sci Eng, C 117:111295. https://doi.org/10.1016/J.MSEC.2020.111295
77. Zhou C, Li HF, Yin YX, Shi ZZ, Li T, Feng XY, Zhang JW, Song CX, Cui XS, Xu KL, Zhao YW, Hou WB, Lu ST, Liu G, Li MQ, Ma J ying, Toft E, Volinsky AA, Wan M, Yao X jun,

Wang C bin, Yao K, Xu S kun, Lu H, Chang SF, Ge JB, Wang LN, Zhang HJ (2019) Long-term in vivo study of biodegradable Zn-Cu stent: A 2-year implantation evaluation in porcine coronary artery. Acta Biomater 97:657–670. https://doi.org/10.1016/J.ACTBIO.2019.08.012
78. Yin YX, Zhou C, Shi YP, Shi ZZ, Lu TH, Hao Y, Liu CH, Wang X, Zhang HJ, Wang LN (2019) Hemocompatibility of biodegradable Zn-0.8 wt% (Cu, Mn, Li) alloys. Mater Sci Eng C 104:109896. https://doi.org/10.1016/J.MSEC.2019.109896
79. Chen Z, Jin M, He H, Dong J, Li J, Nie J, Wang Z, Xu J, Wu F (2023) Mesenchymal stem cells and macrophages and their interactions in tendon-bone healing. J Orthop Translat 39:63–73. https://doi.org/10.1016/J.JOT.2022.12.005
80. Ma J, Zhao N, Zhu D (2015) Endothelial cellular responses to biodegradable metal zinc. ACS Biomater Sci Eng 1:1174–1182. https://doi.org/10.1021/ACSBIOMATERIALS.5B00319/ASSET/IMAGES/MEDIUM/AB-2015-00319T_0009.GIF
81. Khan AR, Wang W, Altaf AR, Shaukat S, Zhang HJ, Rehman AU, Jun Z, Peng L (2023) Facial synthesis, stability, and interaction of Ti3C2Tx@PC composites for high-performance biocathode microbial electrosynthesis systems. ACS Omega 8:29949–29958. https://doi.org/10.1021/ACSOMEGA.2C08163/ASSET/IMAGES/LARGE/AO2C08163_0010.JPEG
82. Wan YZ, Xiong GY, Luo HL, He F, Huang Y, Wang YL (2008) Influence of zinc ion implantation on surface nanomechanical performance and corrosion resistance of biomedical magnesium–calcium alloys. Appl Surf Sci 254:5514–5516. https://doi.org/10.1016/J.APSUSC.2008.02.117
83. Keck CM, Müller RH (2013) Nanotoxicological classification system (NCS)—a guide for the risk-benefit assessment of nanoparticulate drug delivery systems. Eur J Pharm Biopharm 84:445–448. https://doi.org/10.1016/J.EJPB.2013.01.001
84. Khan A (2022) Introduction to special issue. Adv Geophys 63:1–7. https://doi.org/10.1016/BS.AGPH.2022.07.001
85. Huang T, Cheng Y, Zheng Y (2016) In vitro studies on silver implanted pure iron by metal vapor vacuum arc technique. Colloids Surf B Biointerfaces 142:20–29. https://doi.org/10.1016/J.COLSURFB.2016.01.065
86. Lin W, Qin L, Qi H, Zhang D, Zhang G, Gao R, Qiu H, Xia Y, Cao P, Wang X, Zheng W (2017) Long-term in vivo corrosion behavior, biocompatibility and bioresorption mechanism of a bioresorbable nitrided iron scaffold. Acta Biomater 54:454–468. https://doi.org/10.1016/J.ACTBIO.2017.03.020
87. Li X, Shi H, Pan K, Dai M, Wei W, Liu X (2022) Improved biocompatibility and antibacterial property of zinc alloy fabricated with γ-polyglutamic acid-g-dopamine/copper coatings for orthopedic implants. Prog Org Coat 173:107215. https://doi.org/10.1016/J.PORGCOAT.2022.107215
88. Sharma P, Kumar A, Kumar A (2025) An economically viable hybrid additive manufacturing process for the fabrication of hierarchically controlled porous biodegradable Fe-Mn scaffolds. Mater Lett 380:137701. https://doi.org/10.1016/J.MATLET.2024.137701
89. Hong D, Chou DT, Velikokhatnyi OI, Roy A, Lee B, Swink I, Issaev I, Kuhn HA, Kumta PN (2016) Binder-jetting 3D printing and alloy development of new biodegradable Fe-Mn-Ca/Mg alloys. Acta Biomater 45:375–386. https://doi.org/10.1016/J.ACTBIO.2016.08.032
90. Huang T, Cheng J, Zheng YF (2014) In vitro degradation and biocompatibility of Fe–Pd and Fe–Pt composites fabricated by spark plasma sintering. Mater Sci Eng, C 35:43–53. https://doi.org/10.1016/J.MSEC.2013.10.023
91. Wang X, Xu S, Zhou S, Xu W, Leary M, Choong P, Qian M, Brandt M, Xie YM (2016) Topological design and additive manufacturing of porous metals for bone scaffolds and orthopaedic implants: a review. Biomaterials 83:127–141. https://doi.org/10.1016/J.BIOMATERIALS.2016.01.012
92. Roman AM, Geantă V, Cimpoeşu R, Munteanu C, Lohan NM, Zegan G, Cernei ER, Ioniță I, Cimpoeşu N, Ioanid N (2022) In-vitro analysis of FeMn-Si smart biodegradable alloy. Materials 15:568. https://doi.org/10.3390/MA15020568
93. Schinhammer M, Hänzi AC, Löffler JF, Uggowitzer PJ (2010) Design strategy for biodegradable Fe-based alloys for medical applications. Acta Biomater 6:1705–1713. https://doi.org/10.1016/J.ACTBIO.2009.07.039

94. Ferretto I, Kim D, Lee WJ, Hosseini E, della Ventura NM, Sharma A, Sofras C, Capek J, Polatidis E, Leinenbach C (2023) Shape memory and mechanical properties of a Fe-Mn-Si-based shape memory alloy: effect of crystallographic texture generated during additive manufacturing. Mater Des 229:111928. https://doi.org/10.1016/J.MATDES.2023.111928
95. Huang T, Cheng J, Bian D, Zheng Y (2016) Fe–Au and Fe–Ag composites as candidates for biodegradable stent materials. J Biomed Mater Res B Appl Biomater 104:225–240. https://doi.org/10.1002/JBM.B.33389
96. Salama M, Vaz MF, Colaço R, Santos C, Carmezim M (2022) Biodegradable iron and porous iron: mechanical properties, degradation behaviour, manufacturing routes and biomedical applications. J Funct Biomater 13:72. https://doi.org/10.3390/JFB13020072
97. Sotoudehbagha P, Sheibani S, Khakbiz M, Ebrahimi-Barough S, Hermawan H (2018) Novel antibacterial biodegradable Fe-Mn-Ag alloys produced by mechanical alloying. Mater Sci Eng, C 88:88–94. https://doi.org/10.1016/J.MSEC.2018.03.005
98. Mandal S, Ummadi R, Bose M, Balla VK, Roy M (2019) Fe–Mn–Cu alloy as biodegradable material with enhanced antimicrobial properties. Mater Lett 237:323–327. https://doi.org/10.1016/J.MATLET.2018.11.117
99. Sharma P, Pandey PM (2019) Corrosion behaviour of the porous iron scaffold in simulated body fluid for biodegradable implant application. Mater Sci Eng, C 99:838–852. https://doi.org/10.1016/J.MSEC.2019.01.114
100. Mishra DK, Pandey PM (2022) Corrosion behavior and degradation mechanism of micro-extruded 3D printed ordered pore topological Fe scaffolds. J Biomed Mater Res B Appl Biomater 110:1439–1459. https://doi.org/10.1002/JBM.B.35011
101. Sharma P, Jain KG, Pandey PM, Mohanty S (2020) In vitro degradation behaviour, cytocompatibility and hemocompatibility of topologically ordered porous iron scaffold prepared using 3D printing and pressureless microwave sintering. Mater Sci Eng, C 106:110247. https://doi.org/10.1016/J.MSEC.2019.110247

MIX
Papier aus verantwortungsvollen Quellen
Paper from responsible sources
FSC® C105338

If you have any concerns about our products,
you can contact us on
ProductSafety@springernature.com

In case Publisher is established outside the EU,
the EU authorized representative is:
Springer Nature Customer Service Center GmbH
Europaplatz 3, 69115 Heidelberg, Germany

Printed by Libri Plureos GmbH
in Hamburg, Germany